AF870389

Festschrift in Honor of Rogério Cerqueira Leite

Festschrift in Honor of Rogério Cerqueira Leite

Editors

M. Balkanski

University of Paris

C. E. T. Gonçalves da Silva

LNLS and UNICAMP

J. M. Worlock

Bellcore

World Scientific

Singapore • New Jersey • London • Hong Kong

Published by

World Scientific Publishing Co Pte Ltd
5 Toh Tuck Link, Singapore 596224
USA office: 27 Warren Street, Suite 401-402, Hackensack, NJ 07601
UK office: 57 Shelton Street, Covent Garden, London WC2H 9HE

Library of Congress Cataloging-in-Publication data is available.

FESTSCHRIFT IN HONOR OF ROGÉRIO CERQUEIRA LEITE

Copyright © 1991 by World Scientific Publishing Co Pte Ltd

All rights reserved. This book, or parts thereof, may not be reproduced in any form or by any means, electronic or mechanical, including photocopying, recording or any information storage and retrieval system now known or to be invented, without written permission from the publisher.

For photocopying of material in this volume, please paye a copying fee through the Copyright Clearance Center, Inc., 222 Rosewood Drive, Danvers, MA 01923, USA. In this case permission to photocopy is not required from the publisher.

ISBN-13 9789810205720
ISBN-10 9810205724

PREFACE

In a recent interview for a Brazilian newspaper, Rogério told the reporter that one of his regrets in life is not to have pursued his research career. A detailed analysis of his other comments in this interview, specially those which refer to verifiable events and facts, would lead one to suspect that his regrets cannot be taken too seriously. Granted that a research career would probably have been intellectually more satisfying and considerably less stressful than the manifold activities - administrative, political, conspiratorial, entrepreneurial, journalistic - which mark Rogério's extremely active public life over the last twenty years. But, it is implausible that he could have kept himself away from the battle cries of his country and times. And, most certainly, little time has he had for regrets of any kind.

Born on July 14th, 1931, on the 142th anniversary of the Fall of the Bastille - the reference seems appropriate considering his revolutionary instincts - Rogério did his undergraduate studies at the prestigious Institute of Technology of the Air Force (in spite of its name, a school for civilian engineers) in São José dos Campos, then a sleepy provincial town some 60 miles from São Paulo, on the main road to Rio de Janeiro. He received his doctorate in Physics from the University of Paris, where his advisor was Pierre Aigrain. From 1962 until 1970, with a brief intermission in 1965, he was on the technical staff of Bell Labs. It was at this time that he made his most important contributions to Physics.

He took an active part in the glorious period following the invention of the laser and its pioneering application to light scattering studies by another Brazilian physicist, Sérgio Porto (Leite & Porto, J. Opt. Soc. Amer. **54**, 981 (1964)). He was one of the initiators of the study of resonant effects in light scattering (Leite & Porto, Phys. Rev. Lett. **17**, 538 (1966), Scott, Leite & Damen, Phys. Rev. **188**, 1285 (1969)). Probably, however, his most original contribution to Physics was in the study of hot excitations in semiconductors: phonons, electrons, and holes under conditions of high excitation (Shah, Leite & Scott, Solid State Commun. **8**, 1089 (1970)).

In the early seventies, Rogério returned to Brasil. As is the case with many men of universal learning, with a deep respect and understanding for the values of different cultures, he was always strongly attached to his own national roots. Some political opponents, when in a charitable mood, have even referred to him as a "ranting and raving nationalist". So, his return to Brasil was predictable. Almost as soon as he got back, together with Sérgio Porto and, later, José Ripper, he started to build up the

Institute of Physics at the State University of Campinas (UNICAMP). He created and was, during several years, simultaneously, Director of the Institute of Arts at UNICAMP. He served the University also as Dean of the Faculty.

Many years before it became fashionable, Rogério was worrying about and pushing for a stronger interaction between the University and industry. In the seventies, he created CODETEC, a non-profit company dedicated to the development of new technologies and to bridging the gap between research and commercial products, which has had an impact in such diverse areas as solar energy and pharmaceuticals.

As a member of the Editorial Board of Folha de São Paulo, the leading newspaper in Brasil, and a frequent writer in the same newspaper, Rogério has had a strong presence on the political scene in Brasil, not only on scientific and technological matters, but on social and economic issues as well. His independence, foresight, courage to express controversial opinions, and sharp style have produced some delightful and influential articles.

Anyone who followed Rogério's scientific career knows that he was a top rank physicist: creative, imaginative, and not very willing to stick to trodden paths. Physics lost a talented researcher when he directed his interests and energies elsewhere. He still is, more than a decade after quitting active research work, one of the most cited Brazilian scientists in the Science Citation Index. So, it is fitting that, for his 60th birthday, some of his friends from the Physics world decided to offer him this Festschrift to remind the younger generations that he is not only the political fighter they grew accustomed to, but, also, was an accomplished physicist, whose ideas have influenced the work of many of us.

Last, but not least, as everyone knows, behind every successful man there is a strong woman. All of Rogério's friends have come to appreciate the elegant, unflappable, patient woman, Ruth, who, as his wife, has shared his life and been the friendly and charming hostess of their home. In addition to the laurels of her own successful professional career, she is entitled to a substantial share of Rogério's accomplishments.

The Editors regret that, for lack of time and of information about all of Rogério's friends who would have liked to contribute to this volume, they were not contacted. They hope to set everything right for Rogério's 70th birthday.

The Editors

TABLE OF CONTENTS

COMMENTS ON THE DYNAMICS OF DISSIPATIVE PROCESSES

Roberto Luzzi
Departamento de Física do Estado Sólido e Ciência de Materiais
Instituto de Física "Gleb Wataghin"
Caixa Postal 6165 - 13081 - Campinas - SP - Brasil

To Rogério C.C. Leite
on his sixtieth birthday

I. PRELIMINARY WORDS

More than twenty years ago I had the pleasure of being an associate of Rogério C.C. Leite. At that time, on the one hand, Rogério was engaged in the development of the just-created UNICAMP and, mainly, its Institute of Physics, when I had the satisfaction of being a modest collaborator. On the other hand, on the scientific side, Rogério brought to the Institute ideas he pioneered with J. Shah and others at Bell Labs, related to the question of the physics of semiconductors under high levels of excitation. As a result, several of the laboratories at the emerging Institute (early seventies) became engaged in that type of research and, as a consequence, our theoretical group dedicated efforts to the study of such problems and has since maintained partial interest in them.

Semiconductors under high levels of excitation are a very interesting subject, not only for the connection of their behavior in such conditions with their functioning in devices but also, and not least important, for providing an excellent testing ground for theoretical ideas and methods in the field of nonequilibrium statistical thermodynamics. In this context, it is worth mentioning Ryogo Kubo's statement in the opening address of the Oji Seminar[1], pointing out that "Statistical mechanics of nonlinear nonequilibrium phenomena is just in its infancy [and] further progress can only be hoped for by close cooperation with experiment." This was in 1978 and much improvement has been attained since then.

Our group pursued this topic for several years, beginning with the concept of the hot carriers [2] and, going through the use of the seemingly powerful nonequilibrium statistical operator method [3], to obtain a description of the ultrafast time evolution of (i) relaxation processes [4], (ii) optical properties [5], and (iii) transport [6], in photoexcited semiconductors, to presently address the question of nonlinear phenomena in these far-from-equilibrium dissipative systems.

For the motivations mentioned above, I would like to contribute to Rogério's Festschrift with a few and simple comments on the actual relevance of the physics of dissipative processes for which, as stated, nonequilibrium highly excited semiconductors provide excellent examples.

II. ON THE QUESTION OF DISSIPATIVE PROCESSES

The last decades have provided, and the present one - the last of the century - will certainly provide, exciting new results in Physics. Some people speak of a New Physics which would be developing on three fronts: (1) the very small (largely, the age-old quest for the ultimate building blocks of matter), (2) the very large (astrophysics in its connections with cosmology and gravitation), and (3) the very complex (essentially the behavior of dynamical systems governed by nonlinear kinetic laws). [9] The question of the title of this section belongs to item (3).

It is certainly a truism to say that the physics of nonlinear phenomena is receiving a great deal of attention nowadays. It is difficult to define nonlinearity in an unambiguous way, aside, of course, from the character of the modelling equations. It has been noted that, in particular, Schroedinger's equation is linear, but has as a limit the classic Hamilton-Jacobi equation which is nonlinear, even for the free particle. The question is in many cases, and here the word complexity comes in, that nonlinearity is a way of description that deals with phenomena that admit qualitative changes (sometimes of a "catastrophic" character) when modifications on the constraints applied to the system are imposed (e.g. variations in the excitation imposed on the system). Chaos, today a high fashion in Physics after a long period of hybernation, is said to be considered a symptom of nonlinearity.

Nonlinear phenomena have been very important in technology, particularly in modern electrical engineering, and a typical case is the laser. Even in the case of the physics of condensed matter, nonlinear phenomena are not new. Suffice to mention the theory of the plastic flow of dislocations, or the case that harmonic vibrations of the lattice cannot explain thermal expansion, which requires the introduction of nonlinear forces, viz. the anharmonic interactions.

Nonlinearity is very ubiquitous and it is certainly in action in our daily lives and in all disciplines that deal with dynamical systems, e.g., in physics, chemistry, biology, engineering, ecology, economy, sociology, etc. But the renewed interest put forward by the physicists is of a basic character, i.e., in the direction to look for the fundamental principles that are behind nonlinear phenomena. Rolf Landauer has stated (Conference on Nonlinearity in Condensed Matter, Los Alamos National Laboratory, May 1986) that "What was missing until about a decade ago was not the sensitivity to nonlinear phenomena, but only an appreciation for the remarkable diversity of behavior available in nonlinear systems: What is new is the desire to celebrate nonlinearity."

Notwithstanding these comments, it must be said that linear physics, which received ample attention for a long time, is certainly not exhausted. During decades condensed matter physicists kept worrying with the implications of Quantum Mechanics - the major theory of our century - with a high degree of attention. Schroedinger, Heisenberg, Born, Dirac and many others taught us how to deal with the complicated dynamical play of arrays of atoms. Nowadays, besides existing theoretical applications, there exists an alive and growing interest in the nature of the quantum theory of measurement and its consequences.[10]

New and unexpected phenomena arise in nonlinear systems, as noted chaotic behavior in deterministic systems, and particular examples of self-organization in dissipative systems, illustrations often used are the convective cells in the Bènard effect, [11] chemical clocks, [12] and slime mold aggregation, [13] in Physics, Chemistry and Biology, respectively. These are just three examples of a myriad of additional candidates in which self-organized order may appear, which are many-body systems in far-from-equilibrium conditions, where highly nonlinear dissipative processes develop during their evolution, [14] being the source for the onset of such effects called, after Prigogine, dissipative structures. [15, 18]

It is not easy to give an 'a priori' rigorous definition of self-organization; we may say that it is the establishment in a dissipative nonequilibrium medium of some kind of ordered structure (either stationary or with cyclical time variations), determined by the inner properties of the medium itself, independent, or weakly dependent, on the characteristics of the source of nonequilibrium, the initial state of the medium and, often, on boundary conditions. It is thus of principle for self-organization to occur that the system lose any memory of the initial conditions and to have a direct link of the parameters of the structure with the properties of the dissipative medium. In brief, self-organization is a result of the development of instabilites in a disorganized (thermally chaotic) system with the stabilization of long-range coherent structures due to a balance between inner dissipative losses and inputs of energy, mass, etc., from an external source.

The emergence of order out of randomness and the formation of complex spatial structures in homogeneous nonequilibrium media came to the fore at the beginning of the second half of this century, mainly in connection with problems in chemical kinetics and biology. But self-organization is an ubiquitous and remarkably interesting subject that appears in a large class of systems, encompassing, among many others, hydrodynamic instabilites,[11] autocatalytic chemical and biological reactions,[12, 14, 17] cellular differentiation and morphogenesis,[19, 20, 21] neuronal networks,[22] population dynamics and evolution,[23, 24, 25] planet atmosphere,[26] functioning of semiconductor devices (like tunnel diodes, Gunn oscillators, lasers, etc.),[27] as well as nonequilibrium phase transitions - i.e. phase transitions modified by the presence of external pertubations, like the case of nonequilibrium superconductors. [28] All these different systems present, under appropriate conditions (i.e. basically when sufficiently far from equilibrium as noted) a marked coherent behavior extending on a macroscopic scale. Clearly, these ordered states can only occur in open systems since the second law of thermodynamics precludes low entropy structures.

We once again recall that formation of these kinds of Prigogine-dissipative structures is possible for open systems subjected to nonlinear kinetic laws and driven sufficiently far from equilibrium. Near equilibrium (in the so-called strictly linear regime or Onsager's regime), they are forbidden due to the stabilizing tendency of physico-chemical systems towards an attractor, consisting of a state of minimum production of entropy and regression of fluctuations under the prescribed boundary conditions (Cf. references 16 and 17).

It is worth noticing that the so-called Brussel's school has largely developed the thermodynamic theory of open systems, mainly for the study of autocatalytic chemical reactions (see e.g. references 14 through 18). The universal criterion for evolution and the criterion for (in)stability of systems arbitrarily away from equilibrium [16, 17] are capable of encompassing a large variety of situations in open systems. Several authors, mainly Ilya Prigogine and collaborators, have, on the basis of those results, advanced the idea that self-organization in open nonlinear systems may constitute a natural and compelling possibility of a mechanism for the emergence of life and for the puzzling problem of functional order in biological systems. [15, 25, 29, 30] Turing pioneered theoretical work on stationary spatial organization in a model chemical system, [19] a predecessor to Prigogine's Brusselator [17] and Noyes' Oregonator. [12] In these models there are critical points at which the normal spatial homogeneous state becomes unstable. The nonlinear chemical reaction evolves, beyond the critical point, in such a way as to promote a stationary state in which the concentrations of intermediate products vary sinusoidally in space. These authors suggest this phenomenon as a model for the physico-chemical bases of morphogenesis and they draw analogies with biological examples of spatial chemical organization and prebiological evolution. In connection with this, Prigogine and co-workers state that "if the system is to be able to evolve through successive instabilities, a mechanism must be developed whereby each new transition favors further evolution by increasing the nonlinearity and the distance from equilibrium." (Last of references 15). Thus, formalisms and approaches to the subject may represent very useful tools for theoretical biology and sources for ideas and new concepts in the area. For that reason, it is worthwhile pursuing this type of idea in the case of much simpler systems than the biological ones, as is the case of open semiconductors under high levels of excitation.

We have listed before a number of disciplines where self-organization may present itself. It is 'a priori' surprising, given such a variety of cases, the profound analogies of their overall behavior. This diversity of situations has been brought under a common phenomenological description by Generalized Irreversible Thermodynamics, which brings into evidence the common features and the common origin of these phenomena on the phenomenological and macroscopic level. The fundamental point has already been stressed: it is in the fact that we are dealing with open nonlinear systems, and the instability is what is called a branching point of solutions which is possible to appear only under the action of external constraints that lead the system to sufficiently far-from-equilibrium conditions (in general, well beyond the regime where, according to Prigogine's minimum entropy production theorem, the disordered thermally chaotic state is stable). In other words, the emergence of order may follow after a minimum level of dissipation has been obtained and a non-trivial connection between order, stability and dissipation appears, when feedback mechanisms could stabilize the coherent behavior of the system on the macroscopic scale.

In fluid dynamics, nonlinearity comes primarily from convection terms involving the form $\vec{u}.\nabla$ ($\vec{u}$ is the field of fluid velocity), and the instabilites appear when a certain control parameter - the Reynolds number - exceeds a critical value. In chemical reactions, nonlinearity is related to the concentration of reactants and the critical condition depends on the chemical affinities and, in homogeneous cases, also on the diffusion coefficients. In physical devices, nonlinearity is expected in the distribution functions of the elementary components (quasi-particles, plasmons, magnons, phonons, etc.), and the critical condition depends on the intensity of the applied fields. It is conjecturable that when (and if) self-organization follows, it is as a result that the pumped energy, instead of being totally used in thermal agitation, may be partially redirected to the production of collective motion and, therefore, to an increase of the degree of order because of the coherent character of the latter.

Two main questions can be associated to this kind of problem, namely

1. What is the microscopic origin of these transitions?

2. How are we to deal theoretically and be able to rigorously analyze these transitions?

The description of macroscopic dissipative processes in matter has been the subject of different approaches. We have already noted that phenomenological foundations are to be found in Generalized Irreversible Thermodynamics [16] and, of course, its improvements, like Extended Irreversible Thermodynamics[31] and Rational Thermodynamics[27]. Powerful stochastic theories are also available. Finally, in our view, the best approach for answering the questions above should be looked for in Nonequilibrium Statistical Mechanics which is not yet a fully developed discipline. It is our proposal that, in spite of as yet unresolved basic conceptual problems, a first good approach on the microscopic level for the evidencing and description of self-organization can be obtained for the case of many-body systems, governed by a Hamiltonian dynamics, resorting to mechanical-statistical formalisms, like the nonequilibrium statistical operator method (NSOM).[3]

We recall that the purpose of the Statistical Mechanics of systems away from equilibrium is to determine the thermodynamic properties and evolution in time of macroscopic observables of such systems in terms of the dynamical laws which govern the motion of their constitutive particles. The analysis of nonequilibrium systems presents far greater difficulties than those faced in the theory of equilibrium systems. This is due to the fact that a more detailed discussion is necessary to determine the temporal dependence of measurable properties and the calculation of time-dependent transport coefficients associated with the irreversible processes that take place in these systems. At present, several theoretical methods to study the macroscopic evolution of nonequilibrium systems exist. The usefulness of these methods can be ascertained through comparison of the results obtained by application of the theory with experimental data. However, the reason why these methods work well within their own dominions of application cannot be properly understood until a profound insight of the basic conceptual problems associated with the irreversible processes is obtained. Existing formalisms have inherent difficulties: typically those related to, mainly, introduction of coarse-grained procedures, the question of irreversibility and the definition of initial and boundary conditions.

Nonequilibrium statistical mechanisms has typically followed two directions:

1. The kinetics theory of diluted gases, where starting with a few hypotheses, although very controverted, one obtains a description of how simple systems approach equilibrium (the celebrated Boltzmann's H-theorem). An extension of these ideas to dense systems follows several paths like, for example, the construction of a generalized theory of kinetic equations [33] and the equations of the BBGKY hierarchy. [34]

2. A generalization of the theory of the Brownian motion. Here, the complicated dynamic equations - the generalized Newton-Langevin equations - that follow from the laws of Mechanics, used to describe the dynamics of the molecules (or particles, or quasi-particles) that constitute the system, are accompanied by statistical assumptions. The formalism of Mori's correlation function (or memory functions) and the master equation method [36], for example, belong to this approach.

The NSOM belongs in one sense or another to both of the above schemes. It provides a very promising technique which can be considered a far-reaching generalization of statistical methods based on the ideas of Boltzmann and Gibbs. When one resorts to a mechano-statistical approach like this one, as it is well known, the macroscopic state of the system is described in a contracted description in terms of some set of dynamical quantities, say, P1,P2,...,Pn, the statistical operator being a functional of them. The state of the system at the macroscopic level is characterized by a point in the thermodynamic phase space composed of the values at time t of the macroscopic variables Qj (t), j = 1, 2, ..., n, that are the average values of the Pj with the nonequilibrium statistical operator (NSO). The quantities Pj change in time with the microscopic evolution of the mechanical state of the system but an experiment does not follow this microscopic evolution: it follows the numerical values of Qj (t). The results of such an experiment are described by transport equations of the form

$$\frac{d}{dt} Q_j\ (t) = \phi_j \{Q_1\ (t), \ldots, Q_n\ (t); t\}\,, \tag{1}$$

where the ϕ_j are functionals of the macrovariables $Q_j(t)$ which, in general, are expected to be nonlinear, non-local and with memory effects, i.e., depending on the past history of variables Q_j up to the time t when the measurement is performed.

The above considerations immediately give rise to several questions that need to be addressed:

1) How to choose the basic variables? At present, there seems to be no wholly satisfactory theory to generate this information, allowing one to make a unique decision. Some authors suggest that this basic set of variables must include all approximate integrals of motion, or quasi-invariant variables that change very slowly on a molecular time scale. Others argue that one should include enough variables to make the functionals ϕ "almost Markoffian", i.e., almost instantaneous in time. In the NSOM, the criterion for such choice consists of first separating the total Hamiltonian H into two parts, H_o+H_1, so that H_o contains the kinetic energies and the strong interactions that produce very rapid relaxation effects, leading to the rapid decay of a large family of correlations in times much smaller than the characteristic time of observation (typically the time resolution of the detecting device in the experiment). On the other hand, H_1 contains the interactions that produce long relaxation processes. Quantities P_j are chosen such as to be those conserved under H_o, satisfying the so-called Zubarev-Peletminskii symmetry condition, namely

$$[P_j, H_o] = \sum_k \Omega_{jk} P_k\,, \tag{2}$$

in an appropriate quantum representation and where Ω_{jk} are c-numbers. In the last of references 3, it is shown how this criterion is closely linked to the kind of choice of the basic variables done in phenomenological extended irreversible thermodynamics.

2) How are the functionals ϕ_j in Eq. (1) obtained? In other words, what is the form of the nonlinear transport equations for macrovariables Q_j? There are several approaches available for the different methods; they are classified by Zwanzig.[37] In the NSOM, the answer is straightforward: once the NSO $\rho(t)$ is given, then

$$\frac{d}{dt} Q_j(t) = \mathrm{Tr}\left\{\frac{1}{i\hbar}[P_j,H]\,\rho(t)\right\} \tag{3}$$

3) The question of initial conditions. Transport equations, Eq. (1) or (3), are of first order in the time derivative and therefore require an initial condition for a unique solution to be obtained. In other words, it is necessary to provide an initial condition $\rho(t_o)$ at a given time t_o, which should show a large degree of reproducibility. Many times this is done using initial conditions $Q_j(t_o)$ that appear to be reasonable and well-suited to theoretical analysis. The ideal should be to have experimental access to the values $Q_j(t_o)$ but this is seldom feasible. Thus, in any particular problem, the sensitivity of the results to the details of the chosen initial state needs to be carefully considered.

Finally, it should be noted that for isolated systems, the NSO must satisfy the Liouville equation

$$\frac{\delta}{\delta t}\rho(t) + i\mathcal{L}\rho = 0\,, \tag{4}$$

where $\mathcal{L}$ is the Liouville operator of the system (in the quantum case $i\mathcal{L}\rho = (1/i\hbar)[\rho,H]$). This equation is said to be $t\mathcal{L}$ -invariant, i.e., it remains unaltered by the transformation $(t,i\mathcal{L}) \rightarrow (-t,(i\mathcal{L})^{\dagger})$. This poses another fundamental question:

4) How to obtain irreversible behavior in the evolution of the macroscopic state of the system? This is sometimes referred to as the time arrow problem, [38,39] associated to the statement of the second law of thermodynamics. To overcome this problem, Prigogine proposed the principle of dynamic condition for dissapativity. [38,40] This principle is

introduced resorting to a non-unitary theory of transformations that leads to a breaking of the time-reversal symmetry of the Liouville equation which is cast in the form

$$\frac{\delta}{\delta t}\tilde{\rho}(t) + i\wedge\tilde{\rho}(t) = 0\,, \tag{5}$$

where $\wedge$ is a modified Liouvillian composed of an odd and an even part under time-reversal and $\tilde{\rho}$ is a transformed NSO in which an irreversible character for the macroscopic evolution of the macroscopic state of the system has been built. The NSOM incorporates irreversibility from the outset, using a non-mechanical ad hoc hypothesis that produces modified Liouville equations like Eq. (5).[3]

So far we have stated the four main difficulties associated with the construction of a mechano-statistical formulation for nonequilibrium processes, but we have not discussed the fundamental question of how to build the NSO. Several approaches to this question have been developed: NSO's have been derived either on the basis of heuristic approaches [41,44] or time-dependent projection operator techniques. [45,47] As shown, [3] these approaches can be brought together under the unifying method of reasoning put forward in Jaynes' Predictive Statistical Physics. [48] In such a construction, first, according to Jaynes, "we must enunciate a rather basic principle, which might be dismissed as an obvious triviality were it not for the fact that it is not recognized in any of the literature known to this writer: if any macrophenomenon is found to be reproducible, then it follows that all microscopic details that were not under the experimenter's control must be irrelevant for understanding and predicting it." Further, "the difficulty of prediction from microstates lies [...] is our own lack of the information needed to apply them. We never know the microstate; only a few aspects of the macrostate. Nevertheless, the aforementioned principle of reproducibility convinces us that this should be enough; the relevant information is there, if only we can see how to recognize it and use it." This connects us with our comments, questions 1 to 3 stated previously, concerning the question of contraction of information and the choice of macrovariables Q_j, a question that makes contact with Bogoliubov's assertion that a hierarchy of relaxation times exists, which leads to successive contractions of the description as the system evolves in time. [49,50]

The next technical problem is, as put by Jaynes, "how shall we use probability theory to help us do plausible reasoning in situations where, because of incomplete information, we cannot use deductive reasoning?" In other words, the main question associated with this approach is how to obtain the probability assignment compatible with the available information and avoiding unwarranted assumptions. This is answered by Jaynes who formulated the criterion that: the least biased probability assignment $\{p_j\}$ for a set of mutually exclusive events $\{x_j\}$ is that which maximizes the statistical "entropy"

$$S = -\sum_j p_j \ell n p_j \ , \tag{6}$$

subjected to the constraints imposed by available information. [5] This is based on Shannon's ideas, who first demonstrated that for an exhaustive set of mutually exclusive propositions there exists a unique function measuring the uncertainty of the probability assignment. [52]

This is precisely a variational principle that, as noted, provides a unifying theoretical framework for the NSOM. [3] In short, one looks for an extreme of the functional (the so-called Gibbs statistical entropy)

$$I(t) = -\mathrm{Tr}\{\rho(t)\ell n\ \rho(t)\} \ , \tag{7}$$

for any time in the interval between t_o, the time at which the initial condition is given, and t, the time a measurement is performed. This extreme is subjected to the constraints

$$Q_j(t') = \mathrm{Tr}\{P_j\ \rho(t')\} = \mathrm{Tr}\{P_j\,(t'-t)\rho(t)\} \ , \tag{8}$$

for $t_o \leq t' \leq t$, and the normalization of the NSO at all times. Of course, Eq. (8) implies that the choice of variable Q_j has been performed, as described previously. Eqs. (8) introduce a dynamical character into the information but it ought to be noted that the information-gathering interval can (and should) be reduced to information recorded at a unique time, say t_0, when the initial values $Q_j(t_0)$ are provided, in some way as discussed, to be used to solve uniquely the equations of evolution, Eqs. (3). Details of the process and how to retrieve the different approaches to the NSOM are given in reference 3. We simply comment that an appropriate choice of the Lagrange multipliers that the variational procedure introduces is done in order (1) to include Prigogine's dynamical condition for dissipativity, fixing an initial condition from which follows the irreversible evolution of the macroscopic state of the nonequilibrium many-body system, and (2) to introduce a set of functions $F_j(t)$, $j = 1,2,...,n$, so that they will have the role of intensive variables, thermodynamically conjugated to the extensive variable $Q_j(t)$, in such a way as to produce a close connection with existing phenomenological irreversible thermodynamics (see last of references 3).

Once the NSO is obtained, we may construct the equations of evolution for the basic variables, Eqs. (3), in order to follow the evolution of the macroscopic state of the system. These equations involve a collision operator of a very complicated structure. We cannot go into detail here but suffice to say that the separation of the Hamiltonian in the form H_0+H_1 we mentioned before, the use of the symmetry condition of Eq. (2), and the fact that H_1 contains weak interactions, producing slow relaxation rates, allows that such collision operators can be written in the form of an infinite series of collision integrals of ever-increasing order in the interaction strengths contained in H_1. For a long time, calculations were restricted to the lowest order (Markoffian) approximation called the quasi-linear theory of relaxation.[53] A procedure to go beyond this approximation is proposed in the second of references 3 (also reference 8). Because of the form of the expansion of the collision operator, the NSOM nonlinear transport theory can be considered [37] a far-reaching generalization of the Hilbert-Chapman-Enskog method in the kinetic theory of gases, and also a far-reaching generalization of the Mori-Langevin equations method, able to deal with systems in quite arbitrary far-from-equilibrium conditions. Thus, the NSOM, as was stated previously, is capable of dealing theoretically at the microscopic level with Prigogine's dissipative structures in complex systems, thus providing mechano-statistical foundations for the existing phenomenological theories, once the system admits a Hamiltonian-like dynamics. This is certainly a quite interesting and engaging theoretical subject to pursue.

Returning to the question of the physics of highly-excited semiconductors, we comment in conclusion of this Section that the nonequilibrium plasma (electrons, holes and the ionic background) in these systems is a good candidate to show nonlinear phenomena. In fact, it seems that nonlinearities involving the coupling of Fourier components of different wavelengths of the carrier charge density and, as a result of collective plus dissipative effects, may lead to the formation of a spatial structuration of the carrier density in doped semiconductors and to a metallic-non-metallic nonequilibrium phase transition in the photoinjected carrier system in intrinsic semiconductors.[7,54,55] Also, free carrier absorption in doped materials is a source of energy to the LO phonons; at sufficiently high intensities of the laser light, a nonequilibrium Bose-Einstein-like condensation of excitations may occur in the mode of the lowest energy[8, 56] As a consequence, the system of polar lattice vibrations becomes unstable and tends to form a Prigogine-like dissipative structural organization (a kind of antiferroelectric state).[57] This seems to be a manifestation of an effect H. Frölich conjectured as possible in biological systems.[58] Then we may say that the physics of highly excited semiconductors is an interesting and attractive field in the realm of the physics of complex systems and thus worth pursuing.

III. CONCLUDING REMARKS

The last Section contains a series of comments on the aspects of the physics of dissipative systems. Attention is called to the fact of its connection with the emerging theory of complex systems and the attached nonlinearity of the governing kinetic laws which, in particular, play a relevant role in the behavior of open systems in stationary states in sufficiently far-from-equilibrium conditions, which may lead to the formation of self-organizing, macroscopically-ordered dissipative structures (after Prigogine's terminology). The thermodynamics of irreversible processes far from equilibrium have led to the discovery that the fluxes passing through certain physico-chemical systems and shifting them away from equilibrium can give rise to phenomena of spontaneous self-organization, symmetry breaking and a growing tendency towards complexity and diversity, with all the underlying possibilities of such behavior in, besides physico-chemical, all dynamic systems, biological and social as well. Dissipation, contrary to what was thought earlier, is not a source of decay but, it must be emphasized, has quite a constructive role, perhaps including the emergence of life, natural evolution and the outstanding functioning of living systems.

For that reason, it was stated that "...our vision of Nature is undergoing a radical change toward the multiple, the temporal and the complex. Curiously, the unexpected complexity that has been discovered in Nature has not led to a slowdown in the progress of science but, on the contrary, to the emergence of new conceptual structures that now appear as essential to our understanding of the physical world - the world that includes us [...] Indeed, today we are beginning to go beyond [...] the world of quantity into the world of qualities and thus of becoming [...] We believe it is precisely this transition to a new description that makes this moment in the history of science so exciting. Perhaps it is not an exaggeration to say that it is a period [...] in which a new view of Nature is being born." [59] (It is worth mentioning that the epistemological approach of Prigogine and the Brussel's school to the "new science" has been much discussed and also criticized.[60]

The aspect of coherence in dissipative structures is amazing. The system behaves as a whole: it is structured as though each component were "informed" about the overall state of the system. The general mechanism capable of producing a symmetry break (a new structure) is connected with transport and/or chemical reactions. A concept of complex organization is required to connect the various levels of description and account for the relationship between the whole and the behavior of the parts.[18,38,59] It is interesting to note that structures in complex systems arising at bifurcation points seem to have been predicted more than a century ago by J.C. Maxwell, who pointed out that "Every existence above a certain rank has its singular points. At these points, influences whose physical magnitude is too small to be taken account of by a finite being may produce results of the greatest importance." [61] Also, in connection with the synergetic [27] behavior of these dissipative structures, it is interesting to mention Volkenstein's assertion that "Darwin developed the principles of synergetics in the case of living nature a long time before its general formulation [...] For the first time in the history of natural sciences, the mechanism of the formation of an ordered, directed process - the biological evolution - in a system with randomly varying properties was established" (Second of references 29).

As discussed in Section II, the so-called thermodynamic branch of solutions (the one that develops continually from equilibrium with increasing values of the external constraint) is stable, because of Prigogine's theorem of minimum entropy production, eventually until the system is shifted to a certain distance from equilibrium. At a critical point, a branching out of solutions occurs and the thermodynamic branch becomes unstable against the formation of an ordered (in some sense) structure. At the branching point, a particular fluctuation (that associated with the degree of order to follow) that regresses in the thermodynamic branch while stable, increases, leading to the ordering of the system on the macroscopic scale. Hence, chance and necessity have a delicate interplay in self-organization: near a bifurcation, fluctuations (a random element, viz. chance) would play an important role, while between bifurcations the deterministic aspects (necessity) would be dominant. From a point of view of biological systems, this establishes a sharp difference between Monod's and Prigogine's arguments. For Monod, there is no theory of organization, with the living beings merely carrying out a program that has already been written and on whose origins it is not possible to formulate hypotheses that are subject to scientific theorization. In Prigogine's perspective, the living system does not appear as a rigidly conservative mechanism but rather as a system that itself contains the potentiality and the stimuli for evolving and, indeed, in a certain sense cannot help evolving to subsist (Second of references 60). Perhaps it can be said that Prigogine's approach conciliates the warring parties in the reductionist vs. vitalist controversy in biological sciences, providing a generalized and broad scheme.

Following in Section II, we propose that for systems governed by a Hamiltonian dynamics, a statistical method, namely the nonequilibrium statistical operator method, can provide foundations at the microscopic levels for the characterization and rigorous analysis of dissipative structures. The NSOM is based on a scheme put forward by Jaynes, which is considered a powerful approach, being based on a Bayesian approach to probability instead of the usual and restricted frequential approach. It is a predictive statistics that should be built only on the basis of the access the observer has to information relevant for the description of the system. As pointed out by Jaynes, "How shall we best think about Nature and most efficiently predict her behavior, given only our incomplete knowledge" (of the microscopic details of the system). "...we need to see it, not as an example of the N-body equations of motion, but as an example of the logic of scientific inference, which by-passes all the details by going directly from our macroscopic information to the best macroscopic predictions that can be made from that information... Predictive Statistical Mechanics is not a physical theory, but a method of reasoning that accomplishes this by finding, not the particular that the equations of motion say in any particular case, but the general things that they say, in "almost all" cases consistent with our information; for those are the reproducible things." [48] As discussed and briefly described in Section II, the NSOM is based on Jaynes' scheme and the mathematical apparatus it requires is the conjunction of Hamiltonian dynamics with the construction of the NSO in terms of Jaynes' principle of maximization of the statistical entropy.

The NSOM, as noted, seems to be an extremely powerful formalism to deal with a Hamiltonian dynamical system arbitrarily away from equilibrium. Besides providing the expected instantaneous average values of mechanical quantities, it allows for the derivation of a nonlinear quantum transport theory, i.e., the construction of the equations of evolution for the basic variables that permits one to describe the evolution of the macroscopic state of the system. As shown elsewhere (first of references 3), it also allows for the derivation of a response function theory for a system far away from equilibrium and an accompanying nonequilibrium thermodynamic Green function technique. Also, it has been shown that the NSOM can provide appropriate mechano-statistical foundations for existing phenomenological irreversible thermodynamics (last of references 3).

The mathematical handling of the nonlinear quantum transport equations that follow from the NSOM is extremely difficult. Approaches to deal with them have been proposed, like the quasi-linear theory of relaxation, which is the lowest approximation and makes the equations

Markoffian in character. In the second of references 3, a mechanism to go beyond LTR is proposed, at the price of rewriting the very complicated collision operators of the NSOM transport theory in the form of an infinite series of partial collision integrals, instantaneous in time of ever-increasing level of complexity with increasing order in the series. Resorting to this NSOM nonlinear quantum transport theory, we have begun to test our proposal that the NSOM can provide microscopic foundations for the theory of self-organization. As previously commented, we choose the case of highly-excited semiconductors because, on the one hand, of the scientific and technological interest that accompany them and, on the other hand, because they are well known (Hamiltonian and all intrinsic parameters given) and studied. In preliminary results - already mentioned in Section II - we have predicted a transition, between stationary uniform states, to an organized (excitonic-like) non-metallic phase of the photoinjected carrier system in an intrinsic inverted band polar semiconductor,[7,54] a morphological transition (steady state charge density waves of carriers) in n-doped semiconductors,[7,55] and a nonequilibrium Bose-Einstein-like condensation in the polar optical vibration system in photoexcited, doped semiconductors.[8,56] In all this preliminary work, we put into evidence the instability of the stationary, uniform (thermally chaotic) thermodynamic branch; additional work is necessary in order to describe the new dissipative structure, and the careful analysis in the neighborhood of the critical point where the deterministic approach needs to be complemented with a stochastic study of fluctuations.[63]

Nonlinearities, and the eventually accompanying dissipative structures, in the plasma in nonequilibrium semiconductors seem quite worth exploring to their limit, if only to see what their limits may be and particularly to provide a much simpler and better-known picture than the complex biological systems, enabling us, we think, to extend and clarify ideas and methods in this engaging field of the influence of dissipative effects on complex systems.

As a final word, I would like to thank the editors of this book for their kind invitation to participate in this Festschrift in honor of Rogério Cerqueira Leite; I congratulate Rogério on his sixtieth birthday and certainly all of us expect him to continue to contribute to the support of science and technology in this Country.

REFERENCES

1. Kubo, R., Prog. Theor. Phys. Suppl. **64**, 1 (1978)
2. E.g. Vasconcellos, A.R., Ph.D. thesis (UNICAMP, 1976); Ferrari, C.A., Ph.D. thesis (UNICAMP, 1977); also the review Luzzi, R. and Miranda, L.C., "Optical Properties of Semiconductors Under Intense Laser Radiation", in *Physics Reports Reprint Book Series*, Vol. 3, pp. 423-453 (North Holland, Amsterdam, 1978)
3. Luzzi, R. and Vasconcellos, A.R., Fortschritte der Physik/Progress of Physics, **38**, 887 (1990). In this article, we present a review of the nonequilibrium statistical operator method (NSOM) and a unifying theory for the different approaches to it, based on Jaynes' Predictive Statistical Physics, is described. A practical method to deal with the nonlinear quantum transport theory derived from the NSOM is described in Lauck, L., Vasconcellos, A.R. and Luzzi, R., Physica **168A**, 789 (1990). The connection of the NSOM and the phenomenological irreversible thermodynamics is discussed in Vasconcellos, A.R., Luzzi, R., and Garcia-Colin, L.S., Phys. Rev. A (in press).
4. E.g. Algarte, C.A., Ph.D. thesis (UNICAMP, 1983); also the review Luzzi, R. and Vasconcellos, A.R., "Relaxation Processes in Plasma in Nonequilibrium Semiconductors", in *Semiconductors Probed by Ultrafast Laser Spectography*, Vol. 1, pp. 135-189; R. Alfano, ed. (Academic, New York, 1984).
5. E.g. Sampaio, A.J., Ph.D. thesis (UNICAMP, 1983); also the review Luzzi, R., "Nonequilibrium Plasma in Highly Photoexcited Semiconductors", in *Recent Developments in Nonequilibrium Thermodynamics*, J. Casas-Vazquez and D. Jou, eds. (Springer, Berlin-Heidelberg, 1986).
6. E.g. Freire, V.N., Ph.D. thesis (UNICAMP, 1988)
7. E.g. Tome, T., Ph.D. thesis (UNICAMP, 1988); Esperidião, A.S., Ph.D. thesis (UNICAMP, 1991)
8. E.g. Lauck, L., Ph.D. thesis (UNICAMP, 1991)

9. Davies, P., "The New Physics: a synthesis", in *The New Physics* ; P. Davies, ed. (Cambridge Univ. Press, Cambridge, 1987)
10. E.g. Greenberger, D.M., editor: *New Techniques and Ideas in Quantum Measurement Theory* (Annals of the New York Academy of Sciences, Vol. 480, New York, 1986)
11. E.g. in Normand, C., Pomeau, Y., and Velarde, M.G., Rev. Mod. Phys. **49**, 581 (1977)
12. E.g. Field, R.J., American Scientist **73**, 142 (1985); Epstein, I.R., "Pendulums in the Test Tube", in *1986 Yearbook of Science and the Future* (Encyclopaedia Britannica, Chicago, 1985)
13. E.g. in Babloyantz, A., *Molecules, Dynamics and Life,* p. 276ff (Wiley, New York, 1986); Robison, G.A., "Cyclic AMP", in *1974 Yearbook of Science and the Future* (Encyclopaedia Britannica, Chicago, 1973)
14. Nicolis, G., Rep. Prog. Phys. **49**, 873 (1986)
15. Prigogine, I., Nature **246**, 67 (1973); Int. J. Quantum Chem. Symp. **9**, 443 (1975); Prigogine, I., Nicolis, G., and Babloyantz, A., Physics Today 25, 28 and 38 (1972)
16. Glansdorff, P., and Prigogine, I., *Thermodynamics Theory of Structure, Stability and Fluctuations* (Wiley, New York, 1971)
17. Nicolis, G. and Prigogine, I., *Self-organization in Nonequilibrium Systems* (Wiley, New York, 1977)
18. Nicolis, G. and Prigogine, I., *Exploring Complexity* (Freeman, San Francisco, 1989)
19. Turing, A.M., Phil. Trans. Roy. Soc. **B237**, 37 (1952)
20. Thom. R., *Structural Stability and Morphogenesis* (Benjamin, Readings, MA, 1975)
21. Meinhardt, H., "The Random Character of Bifurcations and Embryonic Development", in *Bifurcation Theory and Applications* ; O. Gurel and O.E. Rössler, eds. (Annals of the New York Academy of Sciences, Vol. 316, New York, 1979)
22. E.g. Grossberg, S., "Developmental, Cognitive, and Motivational Theory", in *Brain and Information* ; R. Karrer, J. Cohen, and P. Tueting, eds. (Annals of the New York Academy of Sciences, Vol. 425, New York, 1984)

23. Haken, H., editor: *Synergetics: Cooperative Phenomena in Multicomponent Systems* (Teubner, Stuttgart, 1973)
24. Allen, P.M. and McGlade, J.M., Found. Phys. **17**, 723 (1987)
25. Prigogine, I., "Structure, Dissipation and Life", in *From Theoretical Physics to Biology*, M. Marois, ed. (North Holland, Amsterdam, 1969)
26. E.g. Ingersol, A.P., Scientific American, pp. 90-108 (December, 1981); "The Weather of Other Planets", in *1984 Yearbook of Science and the Future* (Encyclopaedia Britannica, Chicago, 1983)
27. E.g. Haken, H., *Synergetics* (Springer, Berlin-Heidelberg, 1978)
28. E.g. Chang, J.J. and Scalapino, D.J., in *Superconductors: Machines and Devices*, S. Foner and B. Schwartz, eds. (Plenum, New York, 1974)
29. Volkenstein, M.V., "Physics and Biology", in Usp. Fiz. Nauk **109**, 499-515 (1973) [Soviet Phys. Usp. **16**, 207 (1974)]; "The Evolutionary Triad", in *Self-Organization by Nonlinear Irreversible Processes;* W. Ebelling and H. Ulbricht, eds. (Springer, Berlin-Heidelberg, 1986)
30. Haken, H., "Synergetics: From Physics to Biology", in *Physics and Development Series* (International Centre for Theoretical Physics, Trieste, 1986)
31. E.g. Jou, D., Casas-Vazquez, J. and Lebon, G., Rep. Prog. Phys. **51**, 1105 (1988); Garcia-Colin, L.S., Rev. Mex. Fis. (Mexico) **34**, 344 (1988)
32. Truesdell, C., *Rational Thermodynamics* (McGraw Hill, New York, 1985)
33. E.g. Dorffmann, J.R., "Advances and Challenges in the Kinetic Theory of Gases", in Proc. XIV Int. Conf. on Thermodynamics and Statistical Mechanics; J. Stephenson, ed. (North Holland, Amsterdam, 1981): Physica **106A**, 77 (1981)
34. E.g. Bogoliubov, N.N., *Lectures on Quantum Statistics*, Vol 1 (Gordon and Breach, New York, 1967); see also reference 36
35. Mori, H., Prog. Theor. Phys. **33**, 423 (1965); also, Forster, D., *Hydrodynamic Fluctuations, Broken Symmetry and Correlation Functions* (Benjamin, Reading, MA, 1975)
36. Kreuzer, H.J., *Nonequilibrium Thermodynamics and its Statistical Foundations* (Clarendon, Oxford, 1981)
37. Zwanzig, R., "Where do we go from here", in *Perspectives in Statistical Physics* ; H.J. Revechè, ed. (North Holland, Amsterdam, 1981); Kinam (Mexico) **3**, 5 (1981)

38. Prigogine, I., *From Being to Becoming* (Freeman, San Francisco, 1980)
39. E.g. Penrose, O., Rep. Prog. Phys. **42**, 1237 (1979); Coveney, P.V., La Recherche **20**, 190 (1989); Hawking, S., "The Arrow of Time", in *1990 Yearbook of Science and the Future*, pp. 44-53 (Encyclopaedia Britannica, Chicago, 1989); Feynman, R., *The Character of Physical Law* (MIT Press, Cambridge, MA, 1965); Lestienne, R., Scientia **113**, 313 (1980)
40. Prigogine, I., Acta Phys. Austriaca, *Suppl. X*, 401 (1973); also the first two references 15
41. Zubarev, D.N., *Neravnovesnaia Statisticheskaia Termodinamika* (Izd. Nauka, Moskow, 1971) [English Transl.: *Nonequilibrium Statistical Thermodynamics* (Plenum, New York, 1974)]
42. McLennan, J.A., Phys. Fluids **4**, 1319 (1961)
43. Peletminskii, S.V. and Yatsenko, A.A., Zh. Eksp. Teor. Fiz. **53**, 1327 (1967) [Soviet Phys. - JETP **26**, 773 (1968)]
44. Mori. H., Oppenheim, I., and Ross. J., "Wigner Function and Transport Theory", in *Studies in Statistical Mechanics*, Vol 1; J. de Boer and G.E. Uhlenbeck, eds. (North Holland, Amsterdam, 1962)
45. Robertson, B., Phys. Rev. **144**, 151 (1966); "Applications of Maximum Entropy to Nonequilibrium Statistical Mechanics", in *The Maximum Entropy Formalism*, R.D. Levine and M. Tribus, eds. (MIT Press, Cambridge, MA, 1978)
46. Kawasaki, K. and Gunton, J.D., Phys. Rev. **A8**, 2048 (1972)
47. Grabert, H., *Projection Operator Techniques in Nonequilibrium Statistical Mechanics* (Springer, Berlin-Heidelberg, 1982)
48. Jaynes, E.T., "Macroscopic Prediction", in *Complex Systems: Operational Approaches* ; H. Haken, ed. (Springer, Berlin-Heidelberg, 1985); "Predictive Statistical Mechanics", in *Frontiers of Nonequilibrium Statistical Physics* ; G.T. Moore and M.O. Scully, eds. (Plenum, New York, 1986)
49. Bogoliubov, N.N., in *Studies in Statistical Mechanics*, Vol. 1, p. 5ff ; I. de Boer and G.E. Uhlenbeck, eds. (North Holland, Amsterdam, 1962)
50. Uhlenbeck, G.E., in *Lectures in Statistical Mechanics*, pp. 183-203; M. Kac, ed. (American Mathematical Society, Providence, RI, 1963); an example taken from a many-spin system is given in Buishvili, L.L. and

Zviadadse, M.D., Physica **59**, 697 (1972); and in semiconductor physics in Vasconcellos, A.R., Algarte, A.C. and Luzzi, R., Physica **166A**, 517 (1990)

51. Jaynes, E.T., Phys. Rev. **106**, 620 (1957); ibid. **108**, 171 (1957); "Where Do We Stand on Maximum Entropy", in *The Maximum Entropy Formalism* ; pp. 15-118; R.D. Levine and M. Tribus, eds. (MIT Press, Cambridge, MA, 1978)
52. Shannon, C.E., Bell System Tech. Journ. **27**, 379 (1948); also, Beck, A.H.W., *Statistical Mechanics, Fluctuations and Noise*, Ch. 10 (Arnold, London, 1976)
53. In reference 41, see Ch. IV, Section 25.1; also, Kalashnikov, V.P., Teor. Mat. Fiz. **35**, 127 (1978) [Theor. Math. Phys. (USSR) **35**, 362 (1978)]
54. Esperidião, A.S., Vasconcellos, A.R. and Luzzi. R., Solid State Commun. **73**, 673 (1990)
55. Esperidião, A.S., Vasconcellos, A.R. and Luzzi. R., Solid State Commun. **74**, 889 (1990)
56. Lauck, L., Vasconcellos, A.R. and Luzzi, R., Solid State Commun. **76**, 11 (1990); J. Theor. Biol. (submitted)
57. Vasconcellos, A.R., private communication
58. Fröhlich, H., Int. J. Quantum Chem. Symp. **2**, 641 (1968); Nature **228**, 1093 (1970); Adv. Electronics and Electron Phys. **53**, 85 (1980)
59. Prigogine, I. and Stengers, I., *La Nouvelle Alliance : Metamorphose de la Science* (Gallimard, Paris, 1979) [English Transl.: *Order Out of Chaos : Man's New Dialogue with Nature* (Bantam, Toronto, 1984)]
60. E.g. Toffler, A., "Science and Change", foreward to the English edition of reference 59; Bocchi, G., Scientia **117**, 21 (1982); Paty, M., Scientia **117**, 53 (1982); Pagels, H.R., Physics Today, January 1985, p. 97; Engel, P., "Against the Current of Chaos", in The Sciences, September/October 1984, p. 50; See also Festschrift on the occasion of I. Prigogines's seventieth birthday in Found. Phys. Vol. 17, Nºs 5, 6, and 7 (1987)
61. Maxwell, J.C., op. cit. in *The Life of J.C. Maxwell* ; L. Campbell and W. Garnett, eds. (McMillan, London, 1882)

62. Monod, J., *O Acaso e a Necessidade* (Vozes, Petrópolis, 1976) [Translation to Portuguese of Le Hasard et la Necessité (Editions du Seuil, Paris, 1970)]

63. Nicolis, G. and Turner, J.W., "Effects of Fluctuations on Bifurcation Phenomena", in *Bifurcation Theory and Applications* ; O. Gurgel and O.E. Rössler, eds. (Annals of the New York Academy of Sciences, Vol. 316, New York, 1979)

THE PHYSICS OF SLOW PROCESSES*

Hans J. Queisser

Max-Planck-Institut für Festkörperforschung,

D-7000 Stuttgart 80, Germany

Almost all excited complex systems return to their equilibria in a continually decelerated rate. Various generalizations have been proposed to repair the simple exponential time law in order to fit experimental data. A logarithmic dependence on time is shown to represent a description with much more physics insight than most of the empirical formulae. Universal dielectric response, described by a logarithmic hierarchy, serves as an example.

Equilibrium is a fundamental concept of physics, essential not only for classical mechanics and thermodynamics. Yet its definition remains circuitous and vague. Feynman states that thermal equilibrium is reached if *"all the fast things have happened and all the slow things not."*[1] The supply of relatively slow processes in nature is rapidly enlarged presently by our experimental ability to probe extremely fast processes, such as by picosecond spectroscopy of well-defined solid state specimens, especially of semiconducting materials.[2] Leite and his colleagues, for example, pioneered work on high excess densities of excitons in compound semiconductors.[3]

Semiconductors are beautiful systems to establish wide excursions from equilibrium. The long lifetime of minority carriers, first found in germanium, surprised and delighted the early researchers[4] hunting for transistor realization. The *p-n* junction, paradigm of semiconductor electronics,[5] represents a frozen-in nonequilibrium of donor and acceptor ions, firmly fixed in the lattice at the operating temperature,

*Dedicated in friendship to Rogério Leite on the occasion of his relaxing into the sexagenarian state.

while subject to extremely high spatial gradients of dopant concentrations.[2]

Nonequilibrium prevails everywhere and supports our lives, it *"is the source of order, it brings order out of chaos."*[6] Hence it behooves us physicists to consider especially the slow processes, which by their very nature support the nonequilibria. Kohlrausch[7] was one of the first to describe such processes, which bothered him in his electrical measurements, because the electrometer fibers with their very sluggish viscoelastic return to their new deflection prevented him from acquiring accurate data within reasonable times of observation. He suggested that the simple time law $I = I_o exp(-t/\tau)$, which obviously could no longer be used, should be replaced by

$$I = I_o exp(-t/\tau)^{\beta}, \tag{1}$$

where the parameter β has no definitive or intuitive meaning, but is just a fit parameter, $0 < \beta < 1$, to make the decay of a quantity I consecutively slower as time t progresses. Today, we call this "Kohlrausch formula" also the "time-extended exponential."

Another view on the omnipresent slow phenomena reveals the $1/f$ - noise in the frequency domain.[8] Noise power increases quite universally as frequency f is lowered. No unique theoretical foundation has as yet been established, which can really prove that we here face a general property of nature. A trend, however, is obvious: after an arbitrary cut-off for the fast processes, there is a bottomless arsenal of slow processes, which only asymptotically return a disequilibrated system to equilibrium. Only for very dilute systems of independent relaxors can we safely apply the pure exponential law with one and only one common relaxation time τ; all complex systems with many-body interactions and interlocking mechanisms of relaxation cannot be appropriately described in this fashion. Complex systems may actually be defined by this feature of decelerated relaxation.

The Kohlrausch formula has been awakened recently from a long dormancy. Much theoretical effort is currently expended to deduce it or, at least, to render it plausible. There seem to be two different approaches. The first one is purely pragmatic and empirical. One may assume a statistical distribution of relaxation times in complex, interactive systems – especially in condensed matter physics – to replace the unique common relaxation time. Such an assumption is justifiable for many instances; a fit to experiments then yields a mean relaxation time and a statistical spread.[9] Very often, however, the microscopic details remain unresolved.

This approach asumes a multitude of parallel relaxation paths; the emphasis is on different local conditions, rather than on mutual interactions.

The second approach assumes a hierarchical ordering of the relaxation.[10] First, the fast processes must and can relax and attain therefore a different configuration. This different configuration then acts on a lower hierarchy of processes; these become unlocked by the earlier processes and can now relax themselves. In turn, these relaxors then release the next level of hierarchy; a continually decreasing rate of relaxation is thus established. This model takes internal constraints as the most significant conditions. The external stimulus, such as a polarizing field or a mechanical stress, merely serves to unlock the initial hierarchy levels, which then determine the fate of the successive generations more strongly than does the external force. This approach is thus serial in nature. Palmer and coworkers[10] published a seminal paper on this approach and gave a survey of mathematical tools, especially recursion formulae, linking one hierarchy level to the succeeding one.

A lucid experimental example of such hierarchies was demonstrated by the layered structure of epitaxial gallium arsenide specimens.

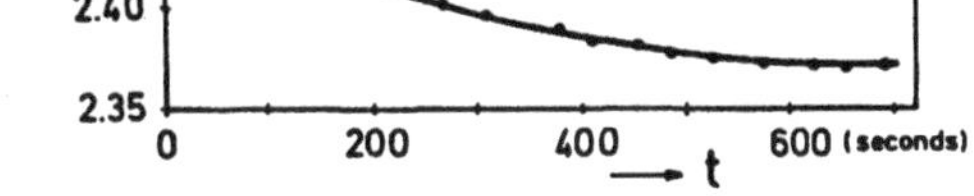

Fig. 1 Nonexponential decay of photoconductivity, see Ref. 12. Fit is with Kohlrausch formula, Eq. (1), but a more physical interpretation results from a model with a spatio-temporal hierarchy in the recombination of electrons (see inset: minus signs in left n-layer), with trapped holes depicted as black dots in right hand substrate.

The frozen-in nonequilibrium here is the photoconductivity. At low temperatures, photoconductivity persists after the light stimulus has terminated.[11] The return to equilibrium can be made imperceptibly slow. At higher temperatures, relaxation quickens. Figure 1 shows relaxation data, which may be force-fitted with Kohlrausch's law, Eq. (1).[12] However, we could show that a better and physically justified fit could be made with a relation that can be derived from considerations of a spatio-temporal hierarchy. The persistent photoconductivity here results from a spatial separation of the light-generated electron-hole pairs. The electrons remain in an n-type layer to generate the excess current, while the holes migrate to the p-type substrate – just as in a solar cell – where they are localized at traps. This spatial separation of the excess pairs transforms into a temporal hierarchy of the relaxation by recombining electrons and holes: closer pairs recombine first, more distant pairs follow, successively slower in rate. The amount of persistent photoconductivity σ decayed through this recombination sequence depends logarithmically on time:

$$\sigma(t=0) - \sigma(t) = C\ ln\ (t/\tau_o), \tag{2}$$

where the constant C and the initial recombination time τ_o can be measured or derived from the spatial distribution of the localizing traps.[13]

Such logarithmic time laws are frequently found for slow processes.[14,15] I have recently suggested to use this functional dependence in order to explain the so-called "universal dielectric response" of condensed matter. Jonscher has collected the data of frequency-dependence of the dielectric susceptibility for wide diversities of materials.[16] He found that the convenient explanation with a Debye model – using again just one common dielectric relaxation time – fails completely. This failure does not come as a surprise, because Debye's model was made for dilute solutions and cannot be expected to hold for solids with their strong many-body interactions. The Debye model predicts for the ratio of imaginary to real part of the susceptibility χ

$$\chi''_D/\chi'_D = \omega\tau, \tag{3}$$

implying a proportionality on frequency ω. The experiments, however, display a completely different picture. The ratio of imaginary to real part of χ is frequency independent and can be described with one parameter n:

$$\chi''/\chi' = cot(n\pi/2), \tag{4}$$

in accordance with the Kramers-Kronig constraint for the complex susceptibility. I propose to explain this relation with a recursion formula in a logarithmic hierarchy for the dimensionless relaxation rate $\rho = R/\omega$ given by:[14,15]

$$\rho_{m+1} = ln \rho_m \tag{5}$$

for the relaxation rates ρ in the successive levels m of the hierarchy. The initial rates still follow the Debye assumption of Eq. (3). As soon as a rate ρ_v falls below 1, however, its logarithmization yields a complex value. The physics behind this mathematical formality is clear: a phase shift must arise between external stimulus and internal response, which is governed more strongly by the hierarchical sequence than by the external events. Hence the proper description arises in a complex rate-plane ρ', ρ'', as shown in Fig. 2.[14,15]

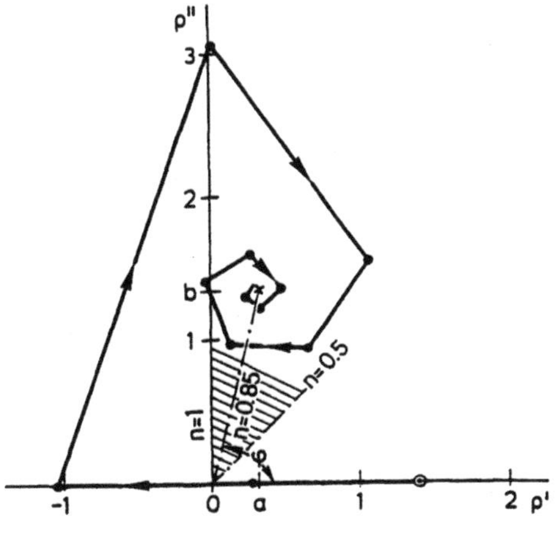

Fig. 2 Path of one relaxation hierarchy in the complex rate plane, see Refs. 14,15. The spiral of the successive hierarchies converges to an attractor, which corresponds to the universality of response, suggested by Eq. (4). Shaded cone embraces the experimentally found values of the parameter n in the universal behavior of Eq. (4).

All hierarchies in this complex plane for the continually decelerated relaxation rates spiral towards an attractor point having the coordinates a and b (see Fig. 2). It is the point where $a + ib = ln(a + ib)$. The existence of the attractor makes plausible the experimentally observed universality of Eq. (4) and also shows that the parameter n falls into the range as experimentally observed, indicated by the shaded cone in Fig. 2. The model[14,15] is, however, idealistic in assuming a strict sequential hierarchy; real systems will show a mixture of parallel and serial relaxations. Yet, even in this more realistic view, the logarithmic law of Eq. (5) lets us understand the observed tendency of Eq. (4) with a narrowly confined range of parameter n.

Slow processes and their evolution in time have hitherto been mostly ignored, simply by setting some more or less arbitrary limit in the convention to define appropriate thermal or electrochemical equilibria. Complex systems suffer under this inadequate simplification; Prigogine assigns specific times to specific physical processes;[6] aficionados of music[17] are long aware of it, if only by acknowledging the tenet: *"die Zeit, die ist ein sonderbar Ding."*[18]

References

[1] Feynman, R.P., *Statistical Mechanics*, D. Pines, ed. (Benjamin, Reading, 1972), p. 1.

[2] Rühle, W., Proc. Int. Conf. Physics of Semiconductors, Thessaloniki, 1990 (to be published).

[3] Freitas, J.A., Mattos, J.C.V., Lluesma, E.G., and Leite, R.C.C., Proc. Phys. Semicond., M.H. Pilkuhn, ed. (Teubner, Stuttgart, 1974), p. 467.

[4] Haynes, J.R. and Shockley, W., Phys. Rev. 81, 835 (1951).

[5] Sze, S.M., *Physics of Semiconductor Devices*, 2nd ed. (Wiley, New York, 1981), ch. 2.

[6] Prigogine, I. and Stengers, I., *Order out of Chaos* (Bantam, New York, 1984), p. 286.

[7] Kohlrausch, R., Ann. Phys. (Leipzig) 72, 393 (1847).

[8] Chapter III of: *Noise in Physical Systems*, D. Wolf, ed. (Springer, Heidelberg, 1978).

[9] Ross Macdonald, J., J. Appl. Phys. 62, R51 (1987).

[10] Palmer, R.G., Stein, D.L., Abrahams, E., Anderson, P.W., Phys. Rev. Lett. 53, 958 ('84).

[11] Queisser, H.J., Proc. Int. Conf. Semic., J. Chadi, ed. (Springer, New York, 1985), p. 1303.

[12] Queisser, H.J., Phys. Rev. Lett., 54, 234 (1985).

[13] Queisser, H.J. and Theodorou, D.E., Phys. Rev. B33, 4027 (1986).

[14] Queisser, H.J., Proc. 20th Int. Conf. Physics of Semicond., Thessaloniki, 1990 (to be publ.).

[15] Queisser, H.J., in press.

[16] Jonscher, A.K., Nature 267, 673 (1977).

[17] I gratefully acknowledge conversations on musical themes with R.C.C. Leite.

[18] For a consideration of the characteristic rate differences between youth and advanced age, listen to this aria of the Feldmarschallin to Octavian in: von Hofmannsthal, H. and Strauß, R. *"Der Rosenkavalier,"* Act 1.

OPTICAL SPECTROSCOPY OF NONEQUILIBRIUM CARRIERS IN SEMICONDUCTORS

Jagdeep Shah
AT&T Bell Laboratories
Holmdel, N. J. 07733, U. S. A.

The ability of optical techniques to create and probe non-equilibrium carrier distributions has been used to investigate a variety of phenomena in semiconductor physics. We trace the history of the development of this field and review some of the highlights.

1. INTRODUCTION

It was more than 20 years ago that the first experiment [1] demonstrating the use of photoexcitation for *creating as well as detecting* non-equilibrium carrier distributions in semiconductors was performed. This field has continued to grow and has led to many interesting results. In this Festschrift in honor of Rogerio's 60th birthday, it is appropriate to review how this field has developed. This article is not meant to be comprehensive. The reader is referred to other reviews for more extensive references to literature [2, 3].

2. EARLY STUDIES OF CARRIER RELAXATION

The first experiment demonstrating the heating of the carriers by photoexcitation was performed on GaAs using excitation by a cw Argon ion laser [1]. Luminescence spectra were recorded over several decades of intensity with particular emphasis on the high energy tail. Typical spectra for n and p-GaAs show that the interband luminescence intensity decreases exponentially with the photon energy. It was concluded from this behavior that the carrier distribution function (DF) is Maxwellian with an effective temperature T_c, and that T_c increases with an increase in excitation power. Similar experiments have now been reproduced in many different laboratories and the use of luminescence technique to investigate carrier DF is now well established.

These data showed that (1) photoexcitation creates hot carriers, (2) the DF for these hot carriers is Maxwellian in the steady state, and (3) optical phonon scattering is the dominant energy loss mechanism for these hot carriers in this temperature range. In subsequent years, these measurements were extended to higher excitation powers, to ternary and quaternary semiconductors, and to the investigation of the dependence of hot carrier temperature on the excitation photon energy. Acoustic phonon energy loss rates were investigated by Ulbrich and coworkers [4] by investigating the electron to acceptor luminescence.

While these were significant achievements, the use of this technique to obtain absolute values of carrier-phonon scattering rates was frustrated by several difficulties: (1) the fraction of excess photon energy given to the carrier system can not be determined with certainty, (2) one must consider both electrons and holes, and the fact that holes couple to the lattice by polar as well as deformation potential interaction, and (3) the density of carriers may vary with the excitation power in a nonlinear manner, introducing another variable in the problem.

3. TIME RESOLVED OPTICAL SPECTROSCOPY

A quantitative determination of energy loss rates to the lattice became possible with the use of ultrafast lasers that were developed in the late 1960s. These lasers also provided an ideal means for investigating the *dynamics* of carrier relaxation. We discuss some highlights of these measurements in this section.

3.1 Cooling of Thermalized Hot Carriers

Time resolved studies of the cooling of thermalized hot carrier distributions (times > 1 ps) have provided considerable information about quasi-2D and bulk semiconductors.

3.1.1 Experimental Results on Bulk Semiconductors

Photoexcited carriers thermalize into a hot plasma in ≈ 1 ps after photoexcitation by an ultrafast laser pulse. The first time resolved studies of hot carriers in semiconductors investigated the cooling of a hot, thermalized carrier system as a function of time following excitation by a short laser pulse [5, 6]. The rate of change of the average carrier energy (<E>) is given by the rate of loss of energy to the lattice per carrier, $P(T_c)$. In the simplest case of Maxwellian distributions for electrons and holes, and $T_e = T_h = T_c$, $\langle E \rangle = 3k\,T_c$, where k is the Boltzmann constant. Therefore, the time evolution of T_c can be calculated starting from any initial condition. Comparison of experimental and theoretical cooling curves would then give information about the carrier-phonon scattering rates without a knowledge of the fraction of excess energy given to the carrier system. Such cooling curves were obtained for bulk GaAs by excite-and-probe as well as luminescence spectroscopy. It was deduced from such studies that at low excitation density the scattering rate was close (within a factor of two) to that expected theoretically, but that there was *a reduction in the scattering rate at high excitation density* .

3.1.2 Experimental Results on Quasi-2D Semiconductors

The first such study in a quasi-2D system was reported by Shank et al [7] who measured the carrier cooling curves in undoped GaAs/AlGaAs multiple quantum well sample using the excite and probe technique. They observed a reduction in the cooling rates at high density just as for bulk GaAs and found that, for comparable excitation density, the cooling rates are the same for the bulk and the quasi-2D system.

Ryan and coworkers [8] were the first to use time resolved luminescence spectroscopy to investigate relaxation of hot thermalized carriers in quasi-2D systems. They carried out a detailed investigation of the dependence of the cooling curves on excitation intensity excitation photon energy, quantum well width and magnetic field. These measurements and subsequent measurements by many different groups [9] have confirmed the reduction in the cooling rate, but shown that it is a function of excitation density. It should be emphasized that similar reductions in cooling rates have also been observed in bulk GaAs at high excitation densities. *Therefore, the reduction in the cooling rates at high density is not a phenomenon unique to quasi-2D system.*

3.1.3 Experiments in Modulation-Doped Quantum Wells

The use of modulation doped quantum wells provided an ideal system for measuring energy loss rates by two techniques. The first is a steady state technique that uses electric field to heat carriers in quantum wells and luminescence spectroscopy to measure T_c . Since all the power supplied by the field is given to the carriers, the power input to the carrier system is known precisely and therefore the rate of energy loss to the lattice at a given T_c can be determined directly [10]. The second makes use of time resolved spectroscopy in modulation doped samples so that only one type of carriers contributes to energy loss rates and the analysis becomes more precise [11].

Both of these studies provided experimental evidence that the reduction in energy loss rates was caused by hot phonon effects rather that screening. This is discussed further in the next section.

3.1.4 Screening and Hot Phonon Effects

Cooling of hot carriers occurs primarily through inelastic collisions between carriers and phonons. Carrier-phonon interactions in bulk semiconductors have been the subject of investigation for over three decades and expressions for these scattering rates are given in books on hot carriers as well as in review chapters [12]. Most of the experiments were performed in the range where optical phonons dominated the energy loss process. For electrons, only polar optical phonon interaction is important in III-V semiconductors. Since the reduction in cooling rates occurred at high excitation densities, it was proposed that the reduction is a result of the *screening* of this polar interaction. Such conclusions were supported by simple estimates of the screening effect considering Debye or Thomas-Fermi screening, and by early calculations.

Another model for the reduction in the cooling rates was proposed by Kocevar and co-workers [13] and Collet et al [14] for the case of bulk semiconductors. They proposed that a large number of optical phonons are emitted by the hot plasma in the initial stages of cooling. These phonons decay by anharmonic interaction into two large wavevector acoustic phonons. If this decay time is relatively long, and if the generation rate for these optical phonons is large, then a ***non-equilibrium*** population of phonons, much larger than the population at T_L , can build up. One must then consider not only the emission of phonons by the hot plasma, but also the reabsorption of the non-equilibrium phonons by the plasma in calculating the net cooling rate for the plasma. It is clear that the effect of hot phonons is to reduce the net cooling rate for the hot plasma.

There is ample experimental evidence that photoexcitation leads to a generation of substantial hot phonon population in III-V semiconductors. Light scattering from optical phonons under steady state photoexcitation conditions [15] found that the ratio of anti-Stokes to Stokes light scattering from LO phonons was a function of excitation density. The phonon occupation number deduced from such ratio shows a linear dependence on the excitation intensity. A number of time resolved light scattering experiments have since been reported and have directly measured the decay time of these hot phonons [16].

The generally accepted view at the present is that these hot phonons are the primary reason why energy loss rate is reduced in bulk as well as quasi-2D semiconductors and that the effects of screening and dimensionality are small.

3.2 Non-Thermal Distributions

Monte-Carlo simulations [17] predict non-thermal distributions at higher photon energies where no signal was observed in these luminescence experiments. More sensitive techniques with better time resolutions are needed to obtain such non-thermal DF. While luminescence spectroscopy provides the best means for accurately determining the DF, it does not match the time resolution and the sensitivity of excite-and-probe transmission measurements. Non-thermal DFs near the fundamental bandgap energy have been observed by several groups using these techniques [18, 19]. Particularly interesting among these are recent measurements with 6 fs pulses [20] and the application of the concept of using modulation-doped quantum wells, discussed above in connection with carrier-phonon interactions, to investigating carrier-carrier scattering [21].

3.3 Coherent Processes

Immediately following excitation with an ultrashort pulse, the photo-created carriers retain some coherence among themselves as well as with the local electromagnetic field. The first relaxation process that occurs is the loss of this coherence, perhaps as a result of a momentum scattering process. The availability of pulses as short as 6 fs have made it possible to investigate this coherent regime in semiconductors for the first time. The reader is referred to the Proceedings of a recent conference [22] for the latest developments in the field.

4. SUMMARY

Considerable insight into hot carrier phenomena has been gained in the last 20 years by using optical techniques for creating and monitoring non-equilibrium carrier distributions in semiconductors. The early work was based on the use of steady-state techniques; advances in ultrafast lasers have allowed us to build upon this foundation. Picosecond and femtosecond optical techniques provide one of the most powerful means for investigating the dynamics of hot carriers in semiconductors. Investigations of carrier-carrier scattering, intervalley scattering and non-thermal populations using ultrafast optical techniques will continue to bring new results and understanding to this field.

REFERENCES

[1] Jagdeep Shah, and R. C. C. Leite, "Radiative recombination from photoexcited hot carriers in GaAs", Phys. Rev. Lett. **22**, 1304 (1969).

[2] Jagdeep Shah, and R. F. Leheny, "Hot carriers in semiconductors probed by picosecond techniques", in "Semiconductors probed by ultrafast laser spectroscopy", edited by R. R. Alfano, Academic Press, New York; p. 45 (1984).

[3] Jagdeep Shah, "Hot Carriers in quais-2D polar semiconductors", IEEE J. Quantum Electronics **QE-22**, 1728 (1986).

[4] R. Ulbrich, "Low density photoexcitation phenomena in semiconductors: aspects of theory and experiment", Solid State Electron. **21**, 51 (1978).

[5] C. V. Shank, R. L. Fork, R. F. Leheny and Jagdeep Shah, "Picosecond relaxation ofhot carriers in GaAs", Phys. Rev. Lett. **42**, 112 (1979).

[6] R. F. Leheny, Jagdeep Shah, R. L. fork and C. V. Shank, "Dynamics of hot carrier cooling in photoexcited GaAs", Solid State Commun. **31**, 809 (1979).

[7] C. V. Shank, R. L. Fork, R. Yen, Jagdeep Shah, B. I. Greene, and C. Weisbuch, "Picosecond dynamics of hot carrier relaxation in highly excited multi-quantum well structures", Solid State Commun. **47**, 981 (1983).

[8] J. F. Ryan, R. A. Taylor, A. J. Turberfield, A. Maciel, J. M. Worlock, A. C. Gossard and W. Wiegmann, Phys. Rev. Lett. **53**, 1841 (1985).

[9] K. Leo, Ruhle, W.W.; Ploog, K. , "Hot-carrier energy-loss rates in GaAs/Al/sub x/Ga/sub 1-x/As quantum wells ", Phys. Rev. B **38**, 1947 (1988).

[10] Jagdeep Shah, A. Pinczuk, A. C. Gossard, and W. Wiegmann, "Energy loss rates for hot electrons and holes in GaAs quantum wells", Phys. Rev. Lett. **54**, 2045 (1985).

[11] Kathleen Kash, Jagdeep Shah, D. Block, A. C. Gossard and W. Wiegmann, "Picosecond luminescence measurements of hot carrier relaxation in III-V semiconductors using sum-frequency generation", Physica **134B**, 189 (1985).

[12] E. M. Conwell, "High Field Transport in Semiconductors", in "Solid State Physics, Supplement 9", edited by F. Seitz, D. Turnbull and H. Ehrenreich, Academic Press, New York; p. (1967).

[13] W. Potz, and P. Kocevar, "Electronic power transfer in pulsed laser excitation of polar semiconductors", Phys. Rev. **B 28**, 7040 (1983).

[14] J. Collet, A. Cornet, M. Pugnet and T. Amand, "Cooling of high density electron-hole plasma", Solid State Commun. **42**, 883 (1982).

[15] Jagdeep Shah, R. C. C. Leite and J. F. Scott, "Photoexcited hot LO phonon in GaAs", Solid State Commun. **8**, 1089 (1970).

[16] D. von der Linde, J. Kuhl and H. Klingenburg, "Raman scattering of nonequilibrium LO phonons with picosecond resolution", Phys. Rev. Lett. **44**, 1505 (1980).

[17] P. Lugli, and S. M. Goodnick, "Monte carlo probe of ultrafast phenomena in polar semiconductors", in "Proc. of SPIE: Ultrafast laser probe phenomena in bulk and microstructure semiconductors II", edited by R. R. Alfano, SPIE; p. 10 (1988).

[18] J. L. Oudar, A. Migus, D. Hulin, G. Grillon, J. Etchepare and A. Antonetti, "Femtosecond orientational relaxation of photoexcited carriers in GaAs", Phys. Rev. Lett. **53**, 384 (1984).

[19] W. H. Knox, C. Hirliman, D. A. B. Miller, J. Shah, D. S. Chemla and C. V. Shank, "Femtosecond excitation of nonthermal carrier populations in GaAs quantum wells", Phys. Rev. Lett. **56**, 1191 (1986).

[20] P. C. Becker, H. L Fragnito, C. H. Brito-Cruz, R. L. Fork, J. E. Cunningham, J. E. Henry and C. V. Shank, "Femtosecond photon echoes from band to band transitions in GaAs", Phys. Rev. Lett. **61**, 1647 (1988).

[21] W. H. Knox, D. S. Chemla, G. Livescu, J. E. Cunningham and J. E. Henry, "Femtosecond carrier thermalization in dense Fermi seas", Phys. Rev. Lett. **61**, 1290 (1988).

[22] Proceedings of the VIIth International Conference on Ultrafast Phenomena, edited by C. B. Harris, E. P. Ippen, G. A. Mourou and A. H. Zewail, Springer-Verlag; (1990).

GENERATION OF NONEQUILIBRIUM OPTICAL PHONONS IN BULK AND MICROSTRUCTURE FABRICATED FROM GaAs

Peter Y. Yu and Dai-sik Kim
Department of Physics, University of California, Berkeley, CA 94720
and
Material Science Division, Lawrence Berkeley Laboratory
1 Cyclotron Road, Berkeley, CA 94720

ABSTRACT

Back in the 1970's Leite and coworkers demonstrated that non-thermal equilibrium distributions of optical phonons (or hot phonons) can be generated optically and studied easily by Raman scattering. Since then this technique has become an important method for studying the properties of phonons and of electron-phonon interactions. Some of these results in bulk and quantum wells of GaAs are reviewed.

1. INTRODUCTION

In 1973 a paper appeared in Solid State Communications entitled: "Photoexcited Hot LO Phonons in GaAs". The authors were J. Shah, R.C.C. Leite and J.F. Scott.[1] This paper demonstrated for the first time that a non-thermal equilibrium distribution of longitudinal optical (LO) phonons can be generated in GaAs by focussing a cw Ar ion laser onto a bulk GaAs sample. The occupation numbers of both the transverse optical (TO) and LO phonons were measured from the ratio of the Anti-Stokes (AS) to Stokes (S) Raman intensities. As the laser power density on the sample was increased, the TO phonon occupation number was found to increase only slightly while the LO phonon occupation number increased by a factor of 3! The authors explained this difference in behavior between the LO and TO phonons by the strong interaction between electrons and LO phonons via the Fröhlich interaction. The processes via which the laser excites the non-thermal LO phonons (or abbreviated as hot phonons by these three authors) are as follows. The incident photons excite electrons from the valence band into the conduction band. Since the photon energy is much larger than the band gap of GaAs the photo-excited electrons have kinetic energies much larger than k_bT (where k_b is the Boltzmann's constant and T is the lattice temperature). These energetic electrons (also known as hot electrons) relax by emission of LO phonons. The time to emit one LO phonons is about 0.2 picosecond (ps).[2] If the LO phonon lifetime were much longer than 0.2 ps, a non-thermal population of LO phonon can be excited even by a cw laser. Knowing the phonon generation rate, the LO phonon lifetime can be determined from the hot phonon population. The LO phonon lifetime of GaAs at room temperature was determined by Shah, Leite and Scott[1] to be 5 ps.

In subsequent papers, Leite and coworkers[3,4] demonstrated that non-thermal distribution of hot phonons could be excited optically in many other semiconductors. In general, the more ionic semiconductors have stronger Fröhlich interaction and hence larger hot phonon populations. This pioneering work of Leite and coworkers has demonstrated a new method for studying the properties of phonons and of interaction between electrons and phonons. The beauty of this technique is its simplicity. The same laser used to excite the hot phonons is used to determine the hot phonon population by Raman scattering. The technique can be easily extended to become a time-resolved experiment by using two short pulses. For example, phonon lifetimes can be determined directly by exciting a non-thermal population with a pump pulse and probing the decay of the hot phonon population with a suitably delayed probe pulse.[5] The fact that Raman scattering is an instantaneous process implies that the time resolution of this technique is limited only by the length of the laser pulses. In this article we will discuss some of the applications of photo-excited hot LO phonons in GaAs to study electron-phonon interaction. We will also discuss the limitations of this technique in studying electron-phonon interaction in semiconductor quantum wells and propose methods to overcome these limitations.

2. NON-EQUILIBRIUM PHONON SPECTROSCOPY

In 1983 Collins and Yu[6] proposed to use the efficiency of photons in exciting hot phonons as a means to probe electronic band structure and intervalley scattering in semiconductors. Their idea was simply that the efficiency of a photon in producing hot phonons depended on the photon energy and the band structure. The amount by which the photon energy exceeds the band gap of the semiconductor determines the kinetic energy of the hot electrons. The number of hot phonons which will be emitted by a hot electron is given by its kinetic energy divided by the phonon energy E_{LO}. For hot electrons with a given kinetic energy, the distribution in momentum space of hot phonons emitted by these hot electrons depends on the band structure. Typically this phonon distribution has a minimum and a maximum value determined by the band dispersion.[7] On the other hand, the range of wave vectors of hot phonons probed by Raman scattering in a particular scattering geometry is well defined. Thus the variation of the hot phonon population with the excitation photon energy can be used to probe the band dispersion. In semiconductors like GaAs, besides the conduction minimum at the Brillouin zone center (Γ) there are higher conduction band minima at the L and X points of the Brillouin zone. When a hot electron has kinetic energy higher than these conduction minima, it can relax into these higher minima by *inter*valley scattering (IVS). If these IVS rates are faster than the *intra*valley electron-LO phonon scattering rate, the electrons are removed from the Γ valley to the zone-edge valleys. Since the zone-center hot electrons are responsible for generating hot LO phonons we expect a drop in the hot phonon generation efficiency when the photon energy becomes larger than the onset energies for IVS. Thus the location of the higher conduction minima and the IVS rates can be deduced from such decreases in the hot LO phonon generation efficiency in GaAs. Since this technique involves measuring the hot phonon generation efficiency as a function of photon energy, Collins and Yu[6] has labeled this technique: hot phonon spectroscopy.

To apply hot phonon spectroscopy to GaAs, Collins and Yu[6] used tunable picosecond laser pulses from a synchronously pumped dye laser to excite a piece of bulk GaAs. The laser pulse length was 4 ps and comparable to the hot LO phonon lifetime while the separation between the laser pulses was about 12 ns. Compared to the continuous wave Ar ion laser used by Shah et al. higher population of hot LO phonons could be generated without heating the sample. The hot LO phonon occupation number measured by Collins and Yu[7] at a constant electron density of 4.8×10^{18} cm^{-3} are shown in Fig. 1 as the solid circles. The experimental results are compared with two different

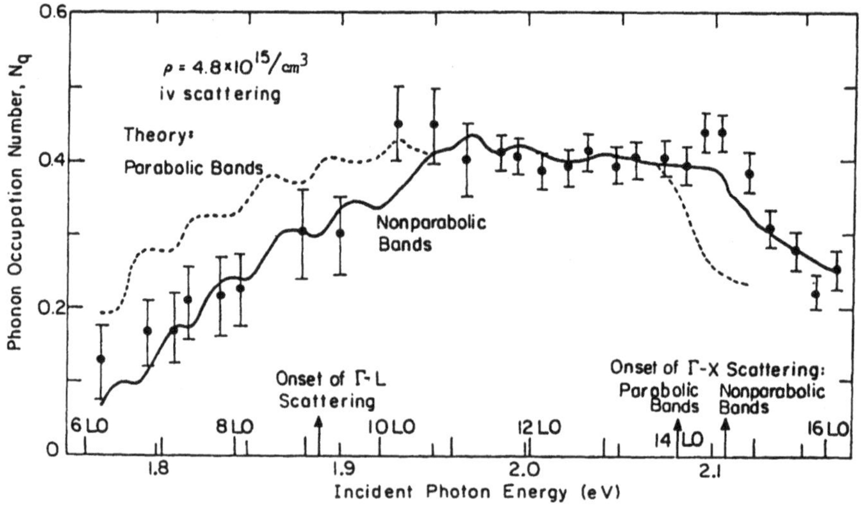

Fig. 1 Hot LO phonon occupation number excited in bulk GaAs excited by picosecond laser pulses. The solid circles are experimental points. The solid and dashed curves are calculated from two different band models. Reproduced from Ref. [7].

model calculations. In one model the conduction band of GaAs was assumed to be parabolic (dashed curve in Fig. 1). In the second model (shown as the solid curve in Fig. 1) the conduction band was assumed to be nonparabolic with dispersion given by:

$$\hbar^2k^2/2m=E+\alpha E^2+\beta E^3, \quad (1)$$

where k and E were the electron wave vector and energy respectively. Obviously the second model agrees much better with the experimental results. In particular the energy for the onset of IVS from Γ to X is quite sensitive to the conduction band nonparabolicity. From the magnitude of the decrease in the LO phonon population due to IVS to the X conduction minima, Collins and Yu[7] deduced the Γ-X IVS deformation potential ($D_{\Gamma\text{-}X}$) to be 10^9 eV/cm. Collins and Yu did not observe any significant decrease in the hot phonon population at the onset of Γ-L IVS. This led them to conclude that the Γ-L IVS deformation potential ($D_{\Gamma\text{-}L}$) is less than 1.5×10^{18} eV/cm. More recently, $D_{\Gamma\text{-}L}$ has been determined by several independent methods[8-10] and the measured values are all $>4\times10^8$ eV/cm. This discrepancy between the value of $D_{\Gamma\text{-}L}$ estimated by Collins and Yu[7] and more recent results can be explained by the fact that there are two possible interpretations of the absence of a drop in LO phonon population at the onset of Γ-L IVS. One explanation is that the deformation potential is small so IVS does not remove a significant number of hot electrons from the Γ valley and hence the LO phonon generation rate is not changed. The other possibility is that $D_{\Gamma\text{-}L}$ is large so that many electrons are removed in less than 1 ps from the Γ valley by IVS. However, the electrons in the L valley have no where to relax other than to return to the Γ valley. If these electrons return to the Γ valley within the duration of the laser pulse of 4 ps, the number of hot LO phonons emitted by the Γ electrons is unchanged by IVS to the L valley. Recently it has been shown that this return of hot electrons from the L valley to the Γ valley occurs in about 7 ps[8] and therefore explains the absence of a decrease in the hot LO phonon generation rate at the onset of the Γ-L IVS. The situation is different for IVS to the X valleys. In this case hot electrons scattered into the X valleys do not return to the Γ valley directly. Instead they are scattered into the L valley. The X-L scattering rate is larger than the X-Γ scattering rate because the L valley density-of-states is higher. As a result, a decrease in the hot phonon generation rate is observed at the onset of Γ-X IVS. The relaxation paths for electrons in the X and L conduction valleys are shown schematically in Fig. 2.

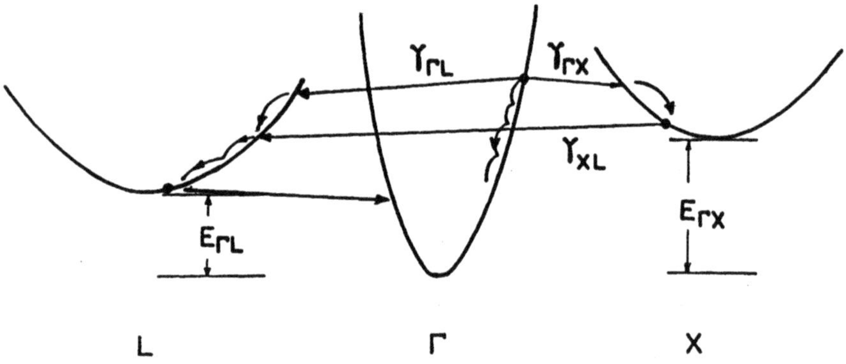

Fig. 2 Interband and Intraband scattering of hot electrons by phonons.

3. THE 'HOT PHONON EFFECT' AND PHONON TEMPERATURE OVERSHOOT

It was assumed that a photoexcited hot electron gas in GaAs cools predominantly by emission of LO phonons provided the electron temperature is high enough to emit LO phonons. Measurement of hot electron cooling curves in GaAs using picosecond laser pulses were first reported in 1979.[11,12] However, the experimental cooling curves agree well with theory only for low excitation densities. For photo-excited electron density higher than 10^{17} cm^{-3} it was found that the cooling rate was significantly reduced.[13] Two explanations have been proposed to explain this density dependence of hot electron cooling rate. One explanation is based on screening of the Fröhlich interaction by the high density of electron-hole pairs.[13] The second explanation is based on the generation of a high non-thermal population of LO phonons (this is now known as the 'hot phonon effect').[14] Much theoretical and experimental work has since been devoted to understand the effects of screening and of hot phonons on the cooling of hot electrons in GaAs at high electron densities.[15] It is now generally believed that screening is not important at densities below 10^{19} cm^{-3}. However, the evidence in support of the hot phonon explanation has mostly been indirect.

We have applied Raman scattering to investigate the hot phonon effect in reducing the hot electron cooling rate in bulk GaAs under high excitation conditions. The idea is an extension of the experiment of Shah, Leite and Scott.[1] Using a single beam of picosecond or subpicosecond laser pulses, one can excite hot electrons and hot phonons. The photo-excited hot electron will scatter light inelastically via single-particle excitations (SPE).[16] The electron temperature can be determined directly from the ratio of AS to S Raman intensities of the SPE spectra:

$$I_{AS}/I_S=\exp(\hbar\omega/k_bT_e) \tag{2}$$

where I_{AS} and I_S are , respectively, the AS and S Raman intensities of the SPE spectra at frequency ω, and T_e is the electron temperature. In discussing the 'hot phonon effect', it is convenient to define a 'phonon temperature' T_{LO} in terms of the phonon population N_{LO} via the equation:

$$N_{LO}=[\exp(E_{LO}/k_bT_{LO})-1]^{-1} \tag{3}$$

The advantage of the Raman technique is that both the electron and phonon temperatures can be determined in the same experiment under identical conditions. If the 'hot phonon effect' explanation were correct, we would expect that at low excitation intensities, no hot LO phonon would be observed. As the excitation density is increased we expect that the phonon temperature will increase while the hot electron temperature will also be increased due to the reduced cooling rate. At sufficiently high excitation level, the electrons and phonons should reach thermal equilibrium and have the same temperature.

In the experiment performed by Huang and Yu[17] on GaAs using laser pulses of 4 ps duration, it was found that at photoexcited electron densities of 10^{18} cm^{-3} the electron and phonon temperatures were both equal to about 800 K. This suggested that large amount of hot phonons were indeed generated by the relaxation of the dense, hot electron gas. Furthermore, the hot electrons and hot phonons reached thermal equilibrium within 4 ps, the duration of the laser pulses. In a more recent experiment, Kim and Yu[18] used subpicosecond laser pulses generated by a colliding-pulse modelocked (CPM) dye laser to excite and probe hot electrons and hot phonons in GaAs. In this case they found that the LO population in GaAs increased with photoexcited electron density as expected. However, at electron densities above 10^{18} cm^{-3} they found that the phonon temperature became <u>higher</u> than the electron temperature as shown in Fig. 3.

Both the low electron temperature and the "overshoot" of the electron temperature by the phonon temperature observed with subpicosecond laser pulses cannot be explained by a simple model in which hot electrons cool by emission of LO phonons only. The results calculated from this simple model are shown as the broken curves in Fig. 3. The reason for the failure of this model in explaining the experimental results is that, within the subpicosecond duration of the laser pulses, the hot electrons have time to emit one or two LO phonons only. Thus LO phonon emission cannot explain the rapid cooling of the hot electrons from the 2500 K temperature at excitation to the observed temperature of about 600 K in fraction of a ps. On the other hand, the low electron temperature can be

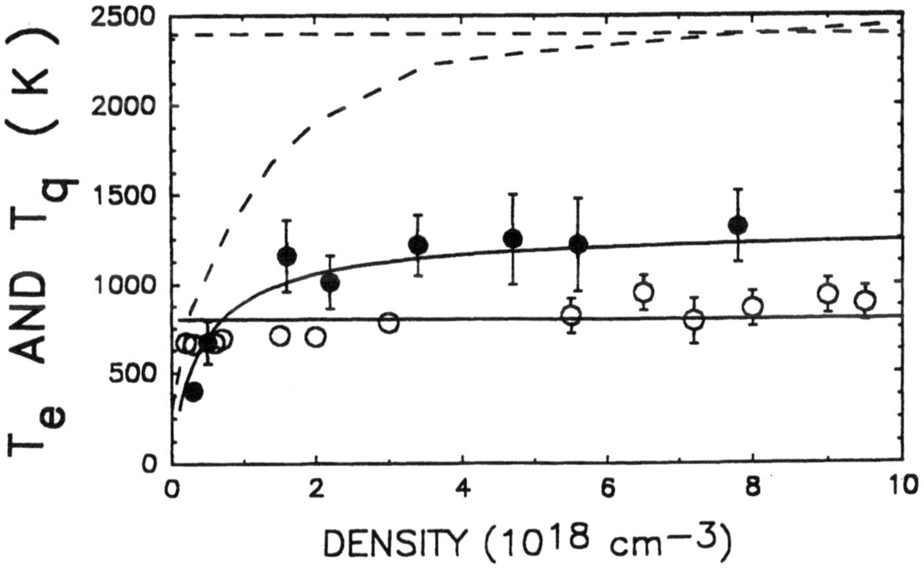

Fig. 3 Hot electrons (open circles) and hot phonon temperatures (closed circles) in GaAs excited by subpicosecond laser pulses of 0.6 ps duration plotted as a function of the photoexcited electron densities. The broken curves are the electron and phonon temperatures calculated with a conduction band model neglecting IVS. The solid curves are calculated with a model including IVS. The deformation potential $D_{\Gamma\text{-}L}$ was adjusted to 7×10^8 eV/cm in order to fit the data points. Reproduced from Ref. [18].

explained by IVS which removes the high energy electrons in about 0.1 ps. By including IVS in the model the experimental results could be explained quantitatively as shown by the solid curves in Fig. 3. Thus our result shows that: *for high electron densities and for time scales shorter than a ps hot electrons in GaAs cool predominantly by scattering into the higher conduction band valleys at X and L.* The high LO phonon temperature is explained by the small phonon density-of-states near zone-center. Thus the so-called 'hot-phonon effect' is effective in reducing the hot electron cooling rate only for times longer than a picosecond and involves the larger wave vector LO phonons. The value of the deformation potential $D_{\Gamma\text{-}L}$ in GaAs we deduced from fitting the experimental points in Fig. 3 was 7×10^8 eV/cm. This value of $D_{\Gamma\text{-}L}$ is in good agreement with the values determined by other techniques.[8-10]

4. NONEQUILIBRIUM PHONON GENERATION IN QUANTUM WELLS

There has been much interest in the properties of quasi-two-dimensional microstructures fabricated from GaAs and AlAs. These structures are usually referred to as quantum wells (QW) or multiple quantum wells (MQW) because electrons and holes are confined inside the GaAs layers (whose band gap is smaller than that of AlAs). What is less well known is that the optical phonons are also confined in these QW.[19,20] The phonons are confined because the optical phonon frequencies in GaAs and in AlAs do not overlap. As a result optical phonons cannot propagate from one layer to the next. The confinement of electrons and phonons in QW has significant influence on the electron-phonon interaction and hence on hot electron relaxation in QW.

Figure 4 shows schematically a symmetric MWQ where both GaAs and AlAs layer thicknesses are given by L. The z-axis is taken to be perpendicular to the layers. Let q denote the GaAs LO phonon wave vector in the QW. Due to confinement of the LO phonon inside the GaAs layer, the q_z component is

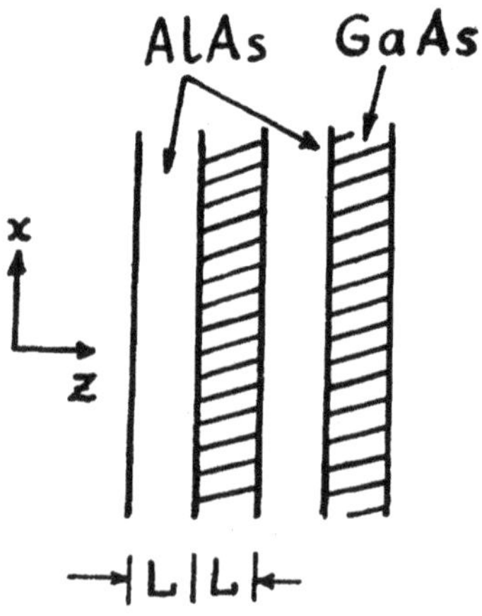

Fig. 4 Schematic diagram of a symmetric GaAs/AlAs MQW.

quantized into: $q_z = m\pi/L$ where m is an integer. The Fröhlich interaction matrix element $|F|^2$ is proportional to $1/q^2$ so the immediate effect of quantization of the LO phonon is to make the Fröhlich interaction dependent on the QW thicknesses:[21]

$$|F|^2 \approx [q_x^2 + q_y^2 + (m\pi/L)^2]^{-1} \qquad (4)$$

As we have mentioned earlier, the LO phonons emitted during hot electron relaxation have a well defined range of wave vectors determined by the electron dispersion. In case of QW, only the components of the wave vector <u>parallel</u> to the layers (q_x and q_y) are conserved in electron-phonon scattering. For GaAs, q_x and q_y typically have values between 3×10^5 to 2×10^6 cm^{-1}. Thus for QW layer thicknesses of 10 nm or less, the q_z component becomes much larger than the component parallel to the layers. As a result $|F|^2$ decreases with L as $(m\pi/L)^{-2}$. In addition to its dependence on L^2, $|F|^2$ also depends on the symmetries of the confined electron and phonon envelope wave functions. For example the wave function of the electron subbands are either symmetric or anti-symmetric with respect to reflection about the center of the layer. Similarly the confined LO phonon modes produce electric potentials which have definite parity under this reflection. These symmetries determine which quantized phonon modes can mediate an *inter*subband scattering and which modes can mediate an *intra*subband scattering.

We have studied the dependence of $|F|^2$ on quantum well thicknesses by measuring the efficiency of hot phonon generation by photoexcited hot electrons in a series of GaAs/AlAs MQW with varying thicknesses.[21] Our results, shown in Fig. 5, confirmed the prediction that, as L decreases, the photo-excited hot electrons become less efficient in emitting hot LO phonons. We have also found an additional effect which decreased the amount of hot phonons observable by Raman scattering in QW. Due to confinement, wave vector conservation in Raman scattering holds only for the component of wave vectors parallel to the layers. In a typical backscattering geometry where the incident and scattered radiations are both propagating almost perpendicular to the layers, the component of wave

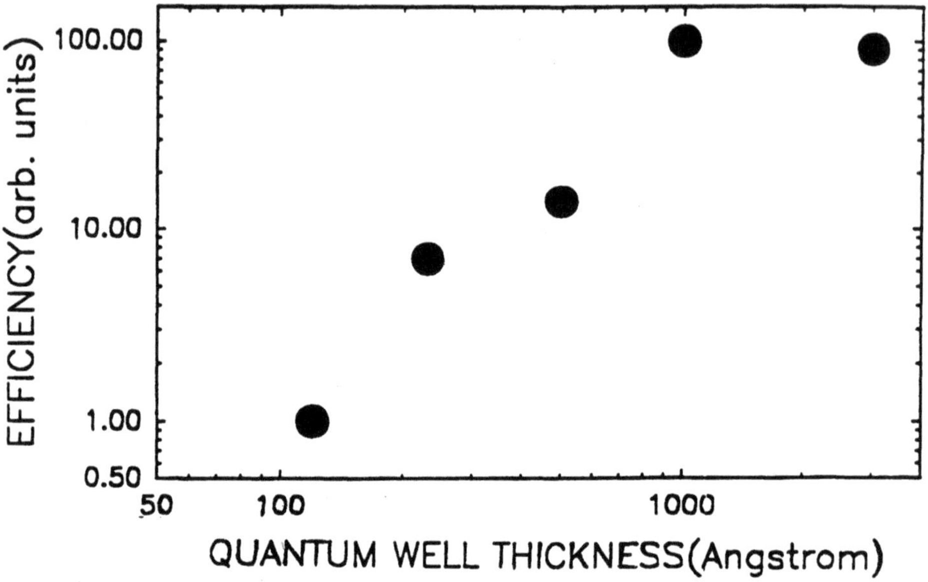

Fig.5 Efficiency of hot electrons excited by subpicosecond laser pulses in generating hot phonons in GaAs/AlAs MQW as a function of well thickness. The electron densities have been kept below 10^{18} cm^{-3} in the measurements to avoid saturation effects.

vector in the xy-plane of the Raman phonon is very small (around $0.5x10^5$ cm^{-1} in our experiment). On the other hand the hot phonons emitted by inter- and intra-subband hot electron relaxation have typically wave vectors larger than $3x10^5$ cm^{-1} in the xy-plane. These considerations imply that hot phonons are more difficult to observe in QW and hence for studying electron-phonon interaction in QW.

To overcome these limitations we suggest the following solutions. The distribution of phonon wave vectors generated by *inter*subband relaxation of hot electrons depends very much on the separation (Δ) between the two subbands involved. This situation is clear from examination of Fig. 6. When the two subband separation is close to the phonon energy $\hbar\omega$, the phonons emitted during

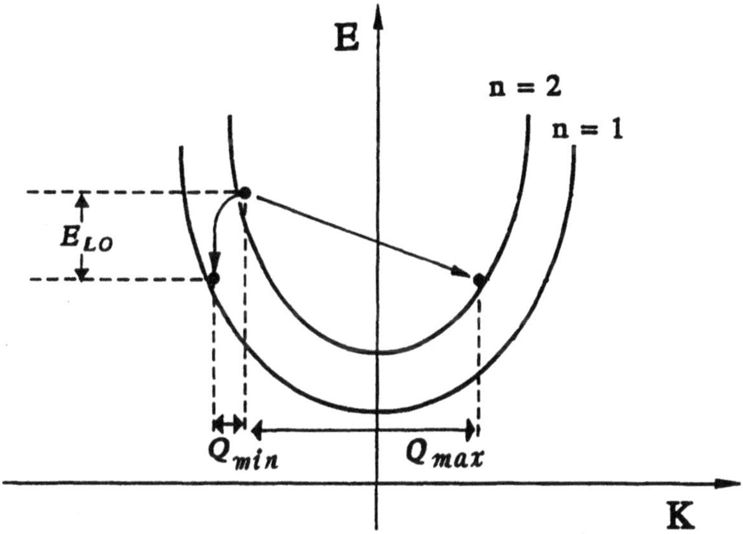

Fig. 6 A schematic diagram showing how energy and momentum conservation in the xy-plane during intersubband electron-phonon scattering determines the minimum and maximum wave vectors of phonon emitted.

intersubband scattering have very small wave vectors. Thus one expects a 'resonance' in the hot phonon generation efficiency when the subband separation is 'tuned' to be equal to the phonon energy. One way to achieve this resonance is to change the QW layer thickness. To estimate the QW layer thickness for this resonance to occur, we have calculated the electronic subband energies in GaAs/AlAs QW using an effective-mass model. The electrons are assumed to have effective mass of 0.67 times the free electron mass while the potential well has a depth of 1.1 eV. Within this simple model, the separation between the two lowest subbands becomes equal to the LO phonon energy (36 meV) when the well thickness L is around 20 nm. To estimate the width of this resonance, we assume that hot electrons obey Boltzmann statistics and are in quasi-equilibrium with a temperature of 1500 K. The phonon observable in the experimental Raman scattering has a wave vector of $0.5x10^5$ cm^{-1} in the xy-plane. The calculated full-width-at-half-maxima of this 'resonance" is only about 1.4 nm.[22] In order to see this resonance the QW thickness has to be controlled to within one monolayer. This is achievable nowadays with growth techniques such as Molecular Beam Epitaxy.

Another way to overcome the limitation of Raman scattering in observing hot phonon in QW is to grow the MQW on a grating fabricated on the substrate. If the grating period is D, wave vector conservation in the xy-plane during electron-phonon scattering will be relaxed by $2\pi/D$. By varying D it should be possible to determine the distribution of hot phonons emitted by hot electrons in QW.

5. CONCLUSIONS

After Leite and coworkers have demonstrated that non-thermal distribution of optical phonons can be generated and detected readily in bulk semiconductors, much progress has been made in applying these hot phonons to study electron-phonon interactions and hot electron relaxation in semiconductors such as GaAs. While it is desirable to extend the technique to study hot phonons in QW, there are difficulties which need to be overcome. We have proposed two ways to overcome these difficulties and we plan to test these ideas experimentally. If these ideas turn out to be feasible, we expect that hot phonons will play an important role in the study of electron-phonon interactions in QW also.

ACKNOWLEDGEMENTS

PYY acknowledges a Miller Research Professorship from the Miller Institute for Research in Science. This research is supported by the Director of Energy Research, Office of Basic Energy Sciences, Materials Science Division, of the U.S. Department of Energy under Contract No. DE-AC03-76SF00098.

REFERENCES

[1] Shah, J., Leite, R.C.C. and Scott, J.F., Solid State Commun. 8, 1089 (1970).

[2] Conwell, E.M., and Vassell, M.O., IEEE Trans. on Elect. Devices, ED-13, 22 (1966).

[3] Mattos, J.C.V., and Leite, R.C.C., Solid State Commun. 12, 465 (1973).

[4] Lluesma, E.G., Mendes, G.,Arguello, C.A., and Leite, R.C.C., Solid State Commun. 14, 1195 (1974).

[5] Von der Linde, D., Kuhl, J. and Klingenberg, H., Phys. Rev. Lett. 44, 1505 (1980).

[6] Collins, C. and Yu, P.Y., Phys. Rev. B27, 2602 (1983).

[7] Collins, C. and Yu, P.Y., Phys. Rev. B30, 4501 (1984).

[8] Shah, J., Pinczuk, A, Gossard, A.C., and Wiegmann, W., Phys. Rev. Lett. 54, 2045 (1985).

[9] Ulbrich, R.G., Kash, J.A., and Tsang, J.C., Phys. Rev. Lett. 62, 949 (1989).

[10] Fasol, G., Hackenberg, W., Hughes, H.P., Ploog, K., Bauser, E., and Kano, H., Phys. Rev. B41, 1461 (1990).

[11] Shank, C.V., Fork, R.L., Leheny, R.F., and Shah J., Phys. Rev. Lett. 42, 112 (1979).

[12] Von der Linde, D., and Lambrich, R., Phys. Rev. Lett., 42, 1090 (1979).

[13] Leheny, R.F., Shah, J., Fork, R.L., Shank, C.V., and Migus, A., Solid State Commun. 31, 809 (1979).

[14] Van Driel, H.M., Phys. Rev. B19, 5928 (1979).

[15] Shah, J., Pinczuk, A., Gossard, A.C., and Wiegmann, W., Phys. Rev. Lett. 54, 2045 (1985); Pötz, W., and Kocevar, P., Phys. Rev. B28, 7040 (1983); Ruhle, W.W., Leo, K., and Bauser, E., Phys. Rev. B40, 1756 (1989).

[16] Klein, M.V., in Light Scattering in Solids ,Topics in Applied Physics, Vol. 8, ed. by M. Cardona (Springer-Verlag,New York, 1975) p.148.

[17] Huang, Y.H. and Yu, P.Y., Solid State Commun. 63, 109 (1987).

[18] Kim, D.S. and Yu, P.Y., Phys. Rev. Lett. 64, 946 (1990); Phys. Rev. B43 (in press).

[19] Klein, M.V., IEEE Quantum Electron. QE-22, 1760 (1990).

[20] Sood, A.K., Menendez, J., Cardona, C., and Ploog, K., Phys. Rev. Lett. 54, 2111 (1985).

[21] Kim, D.S. and Yu, P.Y., in **Ultrafast Laser Probe Phenomena in Bulk and Microstructure Semiconductors III**, ed. by Robert R. Alfano (SPIE Proceeding Series, Vol. 1282, 1990) p.39.

[22] Details of these calculations will be published elsewhere.

LIGHT SCATTERING FROM PHONONS IN ULTRATHIN Si/Ge HETEROSTRUCTURES

David J. Lockwood

Institute for Microstructural Sciences
National Research Council
Ottawa, K1A 0R6 Canada

ABSTRACT

The use of Raman scattering spectroscopy for characterizing the physical properties of ultrathin semiconductor heterostructures is explored for similar and dissimilar $(Si_mGe_n)_p$ atomic-layer superlattices and Ge_n epilayers buried in Si.

1. INTRODUCTION

Modern crystal growth techniques such as molecular beam epitaxy (MBE) are now so sophisticated that layered crystal structures may be grown one atomic layer at a time using a wide variety of semiconductor materials. Just as in the case of the exploratory bulk-semiconductor investigations of 25 years ago when Rogério Leite and his co-workers were studying resonance enhancement effects,[1] laser Raman spectroscopy has proved to be a versatile technique for investigating the dynamical, electronic and structural properties of these artificial semiconductor crystals.[2] Thick layer heterostructures have been widely studied by Raman spectroscopy in the past 10 years.[3] In the case of superlattices, for example, the acoustic phonon spectrum yields information on the superlattice period, the individual layer thickness, the alloy composition, and the interface sharpness, while the optic phonon spectrum provides information on the local structure and the strain within the layers.[4,5] The theoretical models used to interpret these thick layer acoustic-phonon[6,7] and optic-phonon spectra[8] are generally in good agreement with experiment, which is why Raman spectroscopy is now a valuable semiconductor characterization tool.[2]

More recently, ultrathin or atomic layer superlattices (ALSs) have attracted considerable attention, because they have been shown to exhibit novel electronic/optical properties of potential use in devices. These ALSs are of the form $(A_mB_n)_p$ comprising p repeats of m and n monolayers of materials A and B, respectively, grown epitaxially on an appropriate substrate/buffer layer. For small m and n, new crystallographic unit cells are formed artificially along the growth direction whose physical properties are quite different from the bulk materials A and B and depend strongly on m and n. Thus the vibrational properties of ALSs along the growth direction can be quite different from the thick layer case. Raman scattering has already been used to probe the lattice dynamics and structure of a wide variety of ALSs[2)] and here I review Raman results from our recent investigations[9,10)] of Si/Ge artificial structures comprising both ALSs and buried epilayers.

2. EXPERIMENTAL DETAILS

The Raman spectra of the samples in a He gas atmosphere were recorded at room temperature using the quasi-backscattering geometry described earlier.[11)] In this geometry first-order scattering from longitudinal phonons dominates the Raman spectrum. The spectra were excited with 300 mW of 457.9 nm argon laser light, while the Raman scattered light was dispersed with a Spex 14018 double monochromator set usually at a resolution of 3 cm^{-1} and detected with a cooled RCA 31034A photomultiplier. The incident light was polarized in the plane of the incident and scattered light, whereas the scattered light was not analyzed before being directed through a polarization scrambler into the spectrometer. The focussed laser beam illuminated a slit-shaped area of approximately 0.1×1 mm^2 on the sample surface.

The various Si/Ge heterostructures were grown by J.-M. Baribeau in a VG Semicon V80 MBE system on 100 mm (100) Si Czochralski wafers. Details of the substrate preparation and growth methods are given elsewhere.[9,12)] The structural parameters of the ALSs and Ge_n epilayers investigated in this study, as determined by x-ray diffraction and other techniques,[9,10)] are given in Tables 1, 2 and 3.

3. ATOMIC LAYER SUPERLATTICES

The usefulness of Raman spectroscopy for characterizing the epitaxial quality of ALSs is demonstrated here from a consideration of two cases: comparisons of Raman spectra obtained as a result of (a) different growths of the same nominal ALS and (b) deliberate variations in the composition of the ALS.

Table 1. Structural parameters of two $(Si_mGe_n)_8$ ALSs grown on a 300 nm Ge buffer on Si. The target structure was m = 8, n = 12.[10]

Sample	m	n	Period (nm)
861	6.9	19.6	3.68
868	7.2	17.3	3.39

Table 2. Structural parameters of three $(Si_mGe_n)_{100}$ AlSs grown on a 200 nm $Si_{0.4}Ge_{0.6}$ buffer on Si.[10]

Sample	Nominal		Measured	
	m	n	m	n
879	6	4	6.2	4.5
880	7	3	7.5	4.4
881	8	2	5.4	2.8

Table 3. Structural parameters of three buried Ge_n epilayers grown on a 150 nm Si buffer on Si and capped with ~ 33 nm of Si.[9]

n		Thickness	
Nominal	Measured (±5%)	Nominal (nm)	Measured (±0.05 nm)
4	3.8	0.58	0.55
6	5.2	0.87	0.77
12	12.1	1.70	1.70

3.1. Similar ALSs

The Raman spectra of the two similar samples 861 and 868 (see Table 1) are given in Fig. 1. The peaks near 300 and 500 cm^{-1} in both samples arise from optical vibrational modes confined largely to the Ge and Si layers, respectively.[13] The peak near 400 cm^{-1} is a lower frequency Si-layer mode whose frequency and Raman intensity are sensitive to blurred (or disordered) Si-Ge interfaces.[13] A prominent peak due to a folded acoustic mode is visible near 50 cm^{-1}. The frequencies of these optical and acoustic modes for samples 861 and 868 are given in Table 4. The intensities of the folded acoustic modes are a sensitive indicator of interface quality.[13,14] The higher intensities and narrower linewidths for the folded modes in sample 868 compared with sample 861 indicate a superior interface quality. The lower peak frequency of the unresolved doublet near 50 cm^{-1} in sample 861 versus sample 868 (see Fig. 1 and Table 4) is quantitatively consistent with the larger ALS period found by x-ray diffraction (see Table 1). Details of the acoustic modes in sample 868 are shown in Fig. 2.

According to the elastic continuum theory of Rytov[6] for layered media, the folded acoustic phonon dispersion is given approximately by[7]

$$\omega = v \; |(2\pi m/d) \pm q| \;, \tag{1}$$

where d is the ALS period, m = 0, 1, 2 . . . is the following index, and q the ALS wave vector. The ALS acoustic velocity is given in Rytov's theory by

$$v = d\left[\frac{d_{Si}^2}{v_{Si}^2} + \frac{d_{Ge}^2}{v_{Ge}^2} + \left(r + \frac{1}{r}\right)\frac{d_{Si}\, d_{Ge}}{v_{Si}\, v_{Ge}}\right]^{-1/2} \tag{2}$$

Table 4. Frequencies (in cm^{-1}) of peaks in the Raman spectra of samples 861 and 868 (see Fig. 1).

Sample	Folded mode	Ge layer	Si layer	Si layer
861	49.6	301.4	396.2	490.1
868	52.7	302.4	397.6	489.3

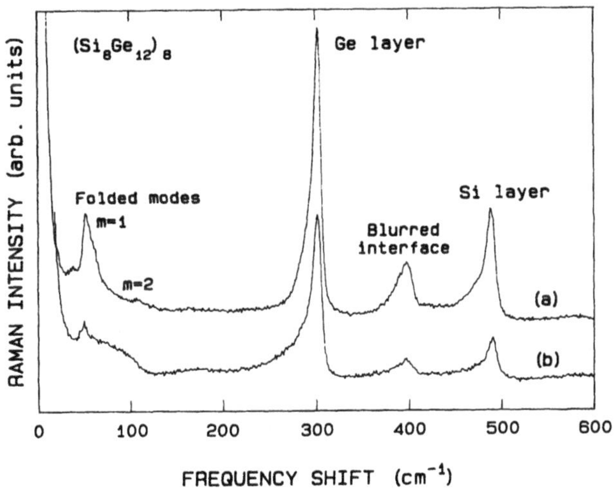

Fig. 1. Raman spectra of samples (a) 868 and (b) 861.

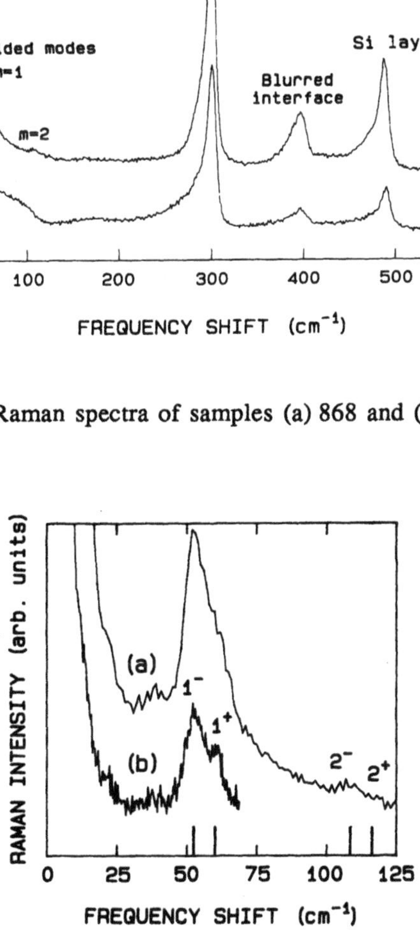

Fig. 2. Raman spectrum of the folded acoustic modes in sample 868 recorded at spectral resolutions of (a) 3 cm^{-1} and (b) 1.5 cm^{-1}. The solid vertical lines mark the m = 1 and 2 doublet frequencies obtained from theory.

Table 5. Experimental and calculated frequencies (in cm^{-1}) for the folded acoustic modes in sample 868.

Folding index m	Frequency Experiment	Theory
1−	52.7 ± 0.4	52.4
1+	60.5 ± 0.8	60.0
2−	108 ± 1	108.6
2+	116 ± 1	116.2

with

$$r = \rho_{Ge}\, v_{Ge} / \rho_{Si}\, v_{Si}\,, \tag{3}$$

where the ALS period $d = d_{Si} + d_{Ge}$, d_{Si} and d_{Ge} are the Si and Ge layer thicknesses, v is the layer sound velocity and ρ the layer density. For the pseudo-backscattering geometry used in these experiments, the component of the scattering wave vector perpendicular to the layers is given by[11)]

$$q \equiv \frac{4\pi n}{\lambda}\left(1-\frac{1}{4n^2}\right), \tag{4}$$

where n is the refractive index of the material at wavelength λ. Using $v_{Si} = 8440$ m/s, $v_{Ge} = 4900$ m/s, $\rho_{Si} = 2.33$ g/cm^3, $\rho_{Ge} = 5.36$ g/cm^3, $d = 3.39$ nm (as given in Table 1) and the Si value of 4.595 for n at $\lambda = 457.9$ nm, a best fit to the experimental frequencies for the $m = 1$ and 2 doublets in sample 868 was obtained with $d_{Si} = 1.2$ nm. As Table 5 shows, the agreement obtained between theory and experiment with just one adjustable parameter is remarkably good, and the value of 1.2 nm for d_{Si} is reasonably close to the Si layer thickness determined by x-ray diffraction (see Table 1).

Besides verifying the ALS parameters, the Raman technique has shown that sample 868 is of superior quality to sample 861 and this agrees with conclusions drawn from x-ray diffraction measurements.[10)] Although it is not possible to obtain quantitative information about the differences between the two ALSs from Raman spectroscopy, sufficient information is obtained to decide which of the two growths of the same nominal structure is superior.

3.2. Different ALSs

The Raman spectra of three quite different $(Si_mGe_n)_{100}$ samples (see Table 2) are shown in Fig. 3. As for the other ALSs considered in Section 3.1, three optical phonon peaks are observed in each sample at frequencies near 300 cm^{-1} (Ge layers), 400 cm^{-1} (lower-frequency Si mode/blurred interface sensitive), and 500 cm^{-1} (Si layers), respectively. Broad and generally weak peaks due to folded acoustic modes are visible at frequencies below 250 cm^{-1} in all three samples. Sample 879 (Si_6Ge_4) exhibits a relatively sharper and much more intense line near 154 cm^{-1} compared with the 95 and 160 cm^{-1} lines in sample 880 (Si_7Ge_3) and the 105 and 205 cm^{-1} lines in sample 881 (Si_8Ge_2). This indicates that the interfaces of sample 879 are of better quality than those of the other samples. However, the large widths of the folded acoustic modes and the optic modes indicate poor sample quality overall and suggest the existence of variations in the layer thicknesses along the growth direction. The frequency of the Si-layer optical mode shifts from 495 cm^{-1} in sample 879 to 501 cm^{-1} in the other samples. On the other hand, the Ge-layer vibration shifts down in frequency from 295 cm^{-1} in sample 879 to 290 cm^{-1} in the others. This greater (lower) strain in the Si (Ge) layers of sample

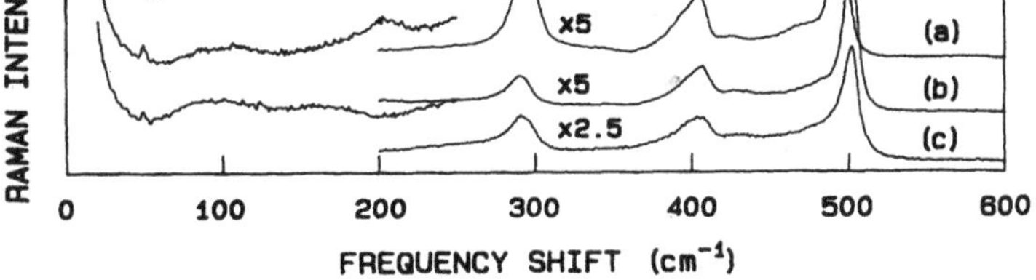

Fig. 3. Raman spectra of samples (a) 879, (b) 881 and (c) 880.

879 compared with samples 880 and 881 is consistent with the closer lattice match of the sample 879 ALS structure to the $Si_{0.4}Ge_{0.6}$ buffer layer.

For these three samples, x-ray diffraction[10] and Raman scattering have both shown that the actual ALS structures are far from the ideal ones. The samples have relatively poor interfaces, their crystallinity is suspect, and there are significant fluctuations in their layer thickness/periodicities along the growth direction. In addition, the Raman results have shown that of the three different growths sample 879 is clearly the best.

4. BURIED Ge_n EPILAYERS

Earlier experimental and theoretical work on thick and ultrathin layer superlattices of Si-Ge[9,15] has shown that current lattice dynamical models for the acoustic and optic phonon frequencies in these structures are in good accord with experiment. For thick layer superlattices, the layer thicknesses and intralayer strain may be deduced from the Raman spectrum with the aid of these models. The same information may also be obtained from theoretical analyses of the ALS spectra, but here the interpretation is made more complicated by the now significant effects of interface blurring on phonon frequencies and intensities.[13] The analysis of the Raman spectrum is dependent to some extent on the model used to represent the physical form of the interface in a particular sample. Nevertheless, the experience gained in the earlier studies, particularly of phonons in structures with abrupt interfaces where strain and confinement effects may be clearly delineated,[15] provides a useful starting point for analyzing the Raman spectra of Ge_n epilayers buried in Si.

The optical-phonon Raman spectra of the three Ge_n samples (see Table 3) with nominal n values of 4, 6 and 12, respectively, are shown in Fig. 4. The dominant peak at 520 cm^{-1} is largely due to the Si buffer layer, as the 457.9 nm exciting light is strongly absorbed in Si[9] and does not reach the Si substrate. A small contribution to the signal near 520 cm^{-1} also comes from the relatively thin Si capping layer. The optical phonon peak from the capping layer may be at a slightly different frequency from the buffer/substrate line if the capping layer is strained. However, it is difficult to extract any information about the capping layer from its contribution to the Raman spectrum, because of the dominant contribution from the underlying bulk Si. The weaker peak near 300 cm^{-1} is a mixture of contributions from the pure Si second-order spectrum (see Fig. 4) and Ge-Ge vibrations within the Ge_n layer.[16,17] Finally, the even weaker peaks

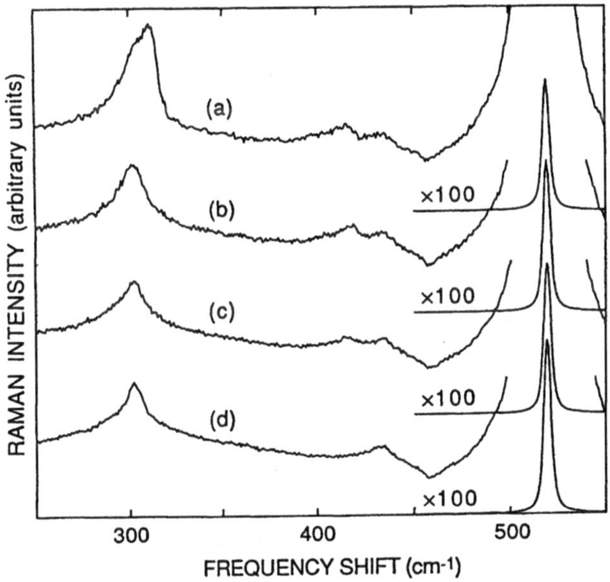

Fig. 4. Raman spectra of the optic phonons of the (a) n = 12, (b) n = 6 and (c) n = 4 Ge_n samples, and (d) the Si substrate. The intense peak at 520 cm^{-1} is due to the Si buffer and capping layers.

near 400 and 425 cm^{-1} are respectively due to Ge-Si vibrations associated with Si next to (or within) the Ge_n layers[18,19] and a second-order Si feature (see Fig. 4).

The spectral features arising from the Ge_n layer simply add on to other features due to the Si cap/buffer layers and may be revealed by subtracting the Si spectrum. In the present case the spectrum from the Si substrate material shown in Fig. 4 was scaled to the 520 cm^{-1} peak in the Ge_n -layer spectra of Fig. 4 and subtracted. The results of the subtractions are given in Fig. 5. The subtraction of the strong 520 cm^{-1} line is not perfect, as evidenced by the large oscillation near 520 cm^{-1}, which could be due to the fact that the Si cap/buffer-layer spectrum is possibly not identical to the Si substrate spectrum. Nevertheless, the subtractions have eliminated the weaker second-order Si features and have clearly revealed the intrinsic optical phonons in the Ge_n layers. The peak frequencies of these phonons are listed in Table 6.

It has been shown from theoretical studies of Si/Ge ALSs that a Ge-Si Raman peak near 400 cm^{-1} should not be observable in the backscattering geometry for systems with perfect interfaces.[19,20] Such peaks are, in fact, commonly seen in the Raman spectra of

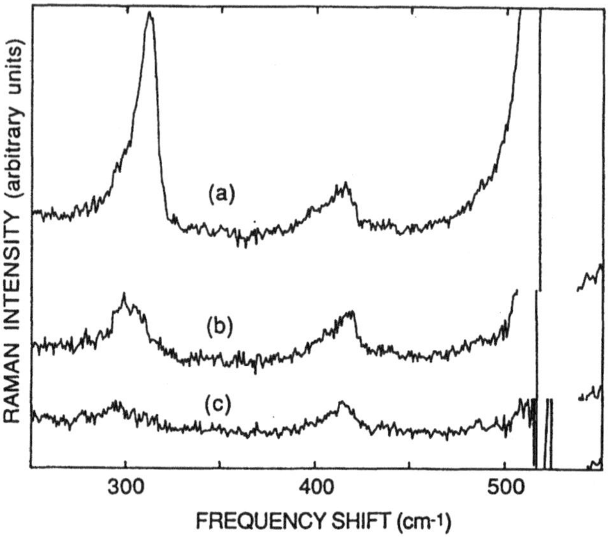

Fig. 5. Raman spectra of the Ge_n layers for (a) n = 12, (b) n = 6 and (c) n = 4 samples revealed by the subtraction process described in the text.

Table 6. Peak frequencies (in cm^{-1}) of longitudinal optical phonons in the Raman spectra of Ge_n layers buried in Si.

n	Ge-Ge line	Ge-Si line
4	294.4	414.7
6	298.3	417.0
12	311.4	414.5
bulk Ge	300.6	—

ALSs, because the as-grown structures do not have perfectly abrupt Si-Ge interfaces.[13)] Thus the Ge-Si line observed near 415 cm^{-1} in our samples is an indicator of imperfect Si-Ge interfaces. The line is relatively weak and of similar frequency and intensity in each sample suggesting that the degree of interface disorder is small and that it occurs to a similar extent in each sample. The peak is asymmetric in shape having a shoulder on the low frequency side in each case (see Fig. 5). The double-humped disposition of the interface peak is characteristic of interface roughness.[21)]

The Ge-Ge peak on the other hand shows considerable variations in the frequency and intensity with n. For n = 12, the peak frequency lies above the bulk Ge value of 301 cm^{-1}, but decreases below 301 cm^{-1} for n = 6 and 4. The confinement effect of sandwiching the Ge epilayer within the Si layers lowers the Ge-Ge phonon frequency from the bulk value, while both strain and interface roughness act to increase the Ge-Ge phonon frequency.[13,16)] In the thicker n = 12 layer, where the Ge-Ge line frequency exceeds the bulk value by ~4%, the effect of confinement is less significant and it is clear that lattice strain and/or interface roughness are dominating and are responsible for the overall upward shift in frequency. A transmission electron microscope analysis has shown that the n = 12 layer is not homogeneous and that the Si cap is heavily dislocated.[9)] The Raman sampling area (~0.1×1 mm^2) is large enough that the observed spectrum contains features due to all of the different configurations within the Ge layer, which range from epitaxial strained layers near the Si buffer/Ge layer interface to relaxed three-dimensional islands near the Si cap. It is thus not surprising that the line at 311 cm^{-1} in Fig. 5 has an asymmetric lineshape. This line appears to comprise largely two components, one of which could be due to the islands (the shoulder near 300 cm^{-1}) and the other (the main peak) due to partially relaxed epitaxial layers. Such an interpretation is consistent with the Ge-on-Si growth model of Eaglesham and Cerullo[22)] and with the Macdonald et al[23)] x-ray diffraction analysis of the Ge strain relaxation. If this interpretation is correct, then the 311 cm^{-1} line is the composite Raman signal of epitaxial Ge layers strained to varying degrees, but always less than that expected from the Si-Ge lattice mismatch. In this case it is probable that interface roughness is the major factor in producing the upward shift of the Ge-Ge line.

The n = 4 and 6 cases are more problematic, because all three factors (strain, confinement and interface roughness) must be considered. Recent calculations[16)] for a perfect bulk-Si/Ge_n/5-nm-Si-cap sandwich without strain indicate a Ge-Ge peak frequency of 284 and 295 cm^{-1} for the n = 4 and n = 8 cases, respectively, with a strain imposed by a 10% increase in the Ge layer force constants raising the respective frequencies to 297 and 307 cm^{-1}. The experimental results given in Table 6 lie within this range.

Assuming that the Raman intensity of the Ge-Ge line is proportional to the Ge layer thickness, the n = 12 line is approximately twice as strong as would be expected from the n = 4 and 6 results. This implies that not all of the Ge_n layers are pure Ge. If we assume that the observed intensity variation is due to interface blurring only, then the Raman results show that the interface roughness on the Ge side of each Si-Ge interface due to Si inclusions is not more than two monolayers in size.

More detailed physical information about the buried Ge_n layer system could be obtained from a theoretical modelling of the Raman spectra, as has been done for Si/Ge ALSs.[13] Such an approach necessarily entails the development of an appropriate model for the Si-Ge interfaces, which is not an easy task at present.

5. CONCLUSIONS

This study of ultrathin layer Si/Ge heterostructures has shown that Raman scattering spectroscopy can provide useful information about the physical properties of such structures. However, in order to fully characterize such thin heterostructures it is necessary to apply complementary techniques that are sensitive to different physical features. In that regard, x-ray diffraction and Raman spectroscopy are closely related techniques in the kind of information they provide.[10] X-ray diffraction provides direct information on layer thickness and strain (for thicker layers). Raman spectroscopy of optic phonons can also provide information on intralayer strain, although in a less direct manner. For thicker ALSs, analysis of the folded acoustic phonon features in the Raman spectra yields directly the average superlattice period and layer thicknesses, and provides qualitative information on interface quality. Both techniques allow a rapid and non-destructive determination of the heterostructure epitaxial quality from different points of view. Nevertheless, additional analytical techniques including high-resolution transmission electron microscopy, Rutherford backscattering, secondary ion mass spectrometry and extended x-ray absorption fine structure analysis are usually needed to fully determine such properties as the thin film morphology, the extent of interdiffusion at Si/Ge interfaces and the crystalline quality.

Subsequent to the pioneering work of Rogério Leite, the renewed interest shown recently in the use of laser Raman spectroscopy for investigating the properties of semiconductor structures and devices[2] serves to demonstrate 25 years later the continuing importance of the technique in semiconductor research. The challenge of the future is to adapt the Raman scattering technique to probe ever smaller structures at the nanometer length scale.

6. ACKNOWLEDGEMENTS

The invaluable contributions to the work reviewed here of my colleagues Geoff Aers, Jean-Marc Baribeau, John McCaffrey, Chandre Dharma-wardana, Stephen Rolfe and Tom Jackman at the Institute for Microstructural Sciences, NRCC are gratefully acknowledged. I particularly thank Henri J. Labbé for his outstanding technical assistance with the Raman measurements.

7. REFERENCES

1. Leite, R.C.C. and Porto, S.P.S., Phys. Rev. Lett. 17, 10 (1966); Leite, R.C.C., Scott, J.F. and Damen, T.C., Phys. Rev. Lett. 22, 780 (1969).
2. Lockwood, D.J. and Young, J.F., eds., "Light Scattering in Semiconductor Structures and Superlattices", Plenum, New York (1991).
3. See, for example, Cardona, M. and Güntherodt, G., eds., "Light Scattering in Solids", Vol. V, Springer, Heidelberg (1989).
4. Jusserand, B. and Cardona, M., Topics in Applied Physics 66, 49 (1989).
5. Sapriel, J. and Djafari Rouhani, B., Surface Science Reports 10, 189 (1989).
6. Rytov, S.M., Acoust. Zh. 2, 71 (1956).
7. Colvard, C., Gant, T.A., Klein, M.V., Merlin, R., Fisher, R., Morkoç, H. and Gossard, A.C., Phys. Rev. B31, 2080 (1985).
8. See, for example, Cardona, M., Superlattices and Microstructures 5, 27 (1989).
9. Baribeau, J.-M., Lockwood, D.J., Jackman, T.E., Aebi, P., Tyliszczak, T. and Hitchcock, A.P., Can. J. Phys. 69, xxx (1991).
10. Lockwood, D.J. and Baribeau, J.-M., in Ref. 2 above.
11. Lockwood, D.J., Dharma-wardana, M.W.C., Baribeau, J.-M. and Houghton, D.C., Phys. Rev. B35, , 2243 (1987).
12. Baribeau, J.-M., Jackman, T.E., Maigné, P., Houghton, D.C. and Denhoff, M.W., J. Vac. Sci. Technol. A5, 1898 (1987).
13. Dharma-wardana, M.W.C., Aers, G.C., Lockwood, D.J. and Baribeau, J.-M., Phys. Rev. B41, 5319 (1990).
14. Brugger, H., Friess, E., Abstreiter, G., Kasper, E. and Kibbel, H., Semicond. Sci. Technol. 3, 1166 (1988).

15. See, for example, several papers on this topic in Fasol, G., Fasolino, A. and Lugli, P., eds., "Spectroscopy of Semiconductor Microstructures", Plenum, New York (1989).
16. Dharma-wardana, M.W.C., Aers, G.C., Lockwood, D.J. and Baribeau, J.-M., in Ref. 2 above.
17. Tsang, J.C., in Ref. 15 above, p. 175.
18. Menéndez, J., Pinczuk, A., Bevk, J. and Mannaerts, J.P., J. Vac. Sci. Technol. B6, 1306 (1988).
19. Wilke, S., Solid State Commun. 73, 399 (1990).
20. Molinari, E. and Fasolino, A., Appl. Phys. Lett. 54, 1220 (1989).
21. White, J., Fasol, G., Ghanbari, R., Gibbings, C.J. and Tuppen, C.G., Thin Solid Films 183, 71 (1989).
22. Eaglesham, D.J. and Cerullo, M., Phys. Rev. Lett. 64, 1943 (1990).
23. Macdonald, J.E., Williams, A.A., van Silfhout, R., van der Veen, J.F., Finney, M.S., Johnson, A.D. and Norris, C., Proc. NATO ARW on "Kinetics of Ordering and Growth at Surfaces", to be published.

TEMPERATURE DEPENDENCE OF THE RAMAN SPECTRA OF DIAMOND AND DIAMOND-LIKE THIN FILMS ON SILICON

D. Yan[(a)], T.-I. Park[(a)], Fred H. Pollak[(a)], A.R. Srivatsa[(b)], T.A. Skotheim[(b)], S.C. Kuo[(c)], A. Badzian[(d)], T. Badzian[(d)], and R. Messier[(d)]

(a) Department of Physics, Brooklyn College of CUNY, Brooklyn, NY 11210 USA

(b) Moltech Corporation, Chemistry Building, SUNY at Stony Brook, Stony Brook, NY 11794 USA

(c) Weber Research Institute, Polytechnic University, Farmingdale, NY 11735 USA

(d) Material Research Laboratory, Pennsylvania State University, State College, PA 16802 USA

ABSTRACT

We have measured the first order Raman spectra of diamond and diamond-like thin films for temperatures between 80K and 770K. The samples were prepared by Chemical Vapor Deposition (CVD) on silicon substrates. The diamond film exhibited a sharp peak at ~ 1332cm^{-1}, which is the first order Raman mode of crystalline material. A small amount of amorphous graphite was also observed. The temperature dependence of the 1332cm^{-1} Raman mode agreed very well with existing results on natural diamond. The diamond film was found to be stable up to the highest temperature. The Raman spectra of the diamond-like sample had an asymmetric broad band at about 1570cm^{-1}, which originated from the sp^2 bonding (graphite-like) components within the sample. We did not detect any signals from either crystalline or amorphous diamond in the CVD diamond-like film. Above about 700K, the diamond-like film became unstable and evaporated completely from its silicon substrate.

INTRODUCTION

The unique combination of several important properties make diamond an attractive candidate for a wide variety of applications. Its extreme hardness, high thermal conductivity, wide band gap (5.5eV) and chemical inertness are useful attributes for diverse applications such as wear resistant coatings, heat sinks for integrated circuits, radiation hardened, high power, high temperature semiconductor and protective coatings for infrared optical materials[1-3]. The recent discovery of low pressure Chemical Vapor Deposition (CVD) synthesis of diamond[4-5] has spurred considerable research activity in this technologically important material. However, to achieve high growth rate, it is necessary to heat the substrate to temperatures in excess of about 800°C. This limits the range of substrates compatible with CVD diamond synthesis. Amorphous diamond-like carbon (or a-C:H) films have many of the attributes of diamond films, such as hardness, thermal conductivity, chemical inertness and optical transparency[6-7]. These films, with extremely smooth surface morphology, can be synthesized at temperatures close to room temperature making them compatible with a wide variety of substrate materials. Consequently, there is considerable interest in the preparation and properties of diamond-like carbon materials.

Raman scattering has proven itself to be a powerful tool in characterizing carbon materials, including effects that disrupt the translational symmetry. For example, the first order Raman spectra for crystalline diamond and graphite exhibit sharp peaks at 1332cm^{-1} and 1581cm^{-1}, respectively[8-9]. In glassy carbon and pyrolytic graphite, which consist of amorphous sp^{2} structures, two broad bands are observed at 1360 and 1590cm^{-1} [10]. An asymmetric broad band is observed at about 1530cm^{-1} in diamond-like amorphous carbon films[11-12]. While there have been a number of room temperature Raman studies of thin film diamond and diamond-like materials, relatively little work has been done as a function of temperature, particularly to elevated temperatures.

In this paper, we report a study of the temperature dependence of the first order Raman spectra of CVD diamond and diamond-like thin films (deposited on silicon substrates) for temperatures ranging from 80K to 770K. The CVD diamond film had a sharp peak in the position of Raman frequency of natural diamond and was stable throughout the temperature range. The temperature shift of this diamond Raman line closely matched that of natural diamond but its line width was still considerably larger. A small amount of amorphous graphite was also present in the CVD diamond film. The diamond-like sample had an asymmetric broad band at ~ $1570cm^{-1}$, which originated from the graphitic regions of the sample, but with no evidence of diamond in either crystalline or amorphous form. We also observed that the diamond-like film was not stable at temperatures above about 700K and evaporated from its silicon substrate.

EXPERIMENTAL DETAILS

The diamond films used in this study were synthesized by microwave plasma deposition of a gas mixture containing 0.5% methane in hydrogen at $975^{o}C$ and 80 Torr. Depositions were carried out on silicon substrates. The diamond-like films were prepared by electron cyclotron resonance (ECR) microwave plasma CVD. A pure methane plasma at a pressure of 10^{-4} Torr was employed and the substrate was held at a negative bias of about 150V during the deposition.

Raman spectra were measured in the backscattering geometry, with a SPEX 1403 double monochromator, a Products For Research GaAs photo-multiplier tube and associated photon-counting circuit. The instrumental resolution was about $2cm^{-1}$. The excitation sources were the 514.5nm and 488.0nm lines of an Ar-ion laser. The laser power on the sample was kept below 0.1W. A cylindrical lens was used to focus the laser beam into a vertical line of approximately the size of 200μm by 3mm to avoid heating. No change in line shape with respect to laser intensity was found. No polarization measurements were attempted. The samples were mounted on the molybdenum heater block of a high temperature furnace, near the window of a vacuum chamber. The temperature of the sample was monitored by a type J (iron-constantan) thermocouple and was controllable to within about 5°C. The whole system was kept at a pressure below 10^{-6} Torr by a turbo-molecular pump. The low temperature measurements were performed in a crystat with liquid nitrogen as the coolant.

EXPERIMENTAL RESULTS

The Raman spectra of a CVD diamond film on silicon are shown in Fig. 1 for temperatures between 80K and 770K. The line at ~ $522cm^{-1}$ is from the silicon substrate. The sharp peak at ~ $1332cm^{-1}$ corresponds to the first order ($q \approx 0$) Raman mode of crystalline diamond. Its line width was ~ $9cm^{-1}$ at room temperature and ~ $11cm^{-1}$ at 770K. To produce such a peak at the Raman frequency of single crystal diamond, it is necessary that the crystal domain size within the sample be larger than the phonon decay length (~ 1μm)[13]. Therefore, the ~ $1332cm^{-1}$ peak a originated from regions of crystalline diamond in the sample that were at least 1μm in size and is indicative of the high quality of the film. Our measured width of the $1332cm^{-1}$ peak is in close agreement with that of Ref. 11. However, it is still a factor of 4 larger than that of natural diamond[14-15].

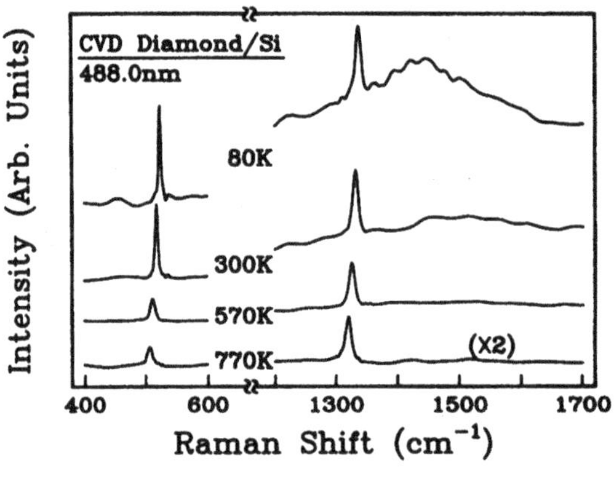

Figure 1. Raman spectra of a CVD diamond film on silicon for temperatures between 80K and 770K. The ~ $1332cm^{-1}$ peak is the first order Raman line of crystalline diamond while the peak at ~ $522cm^{-1}$ is from the silicon substrate.

A broad band structure at about $1500cm^{-1}$ was also visible in Fig. 1 for temperatures below 300K. It has been shown that, under similar circumstances, this broad band may originate from the amorphous graphite (sp^2 bonding) regions of the sample[13,16]. Because of its high optical absorption, the Raman sensitivity of the graphite-like structures are much enhanced in relation to diamond. The actual amount of amorphous graphite material in the sample is thus much less than what the relative intensities would indicate[13]. The broad band disappeared completely above 600K. We wish to point out that the diamond film was stable even at the highest temperature (770K), and the change in the broad band structure with temperature also was reversible. No evaporation process was detected.

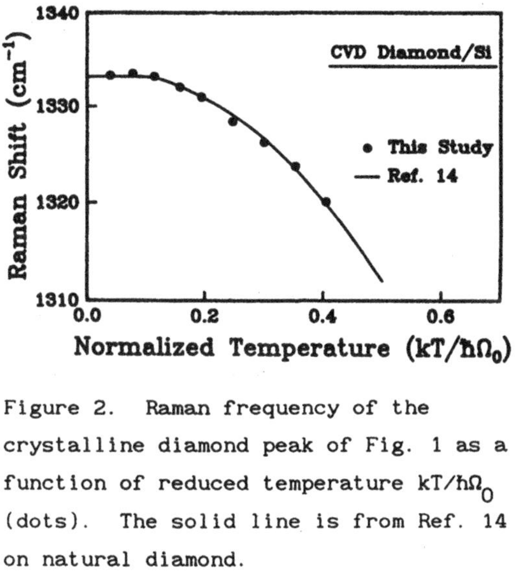

Figure 2. Raman frequency of the crystalline diamond peak of Fig. 1 as a function of reduced temperature $kT/\hbar\Omega_0$ (dots). The solid line is from Ref. 14 on natural diamond.

Figure 2 is the plot of the Raman frequency of the CVD diamond on silicon as a function of the normalized temperature $kT/\hbar\Omega_0$, where Ω_0 is the extrapolated Raman frequency for T=0[15]. Our data is in very good agreement with an earlier measurement on natural diamond by Anastassakis *et al*[14] throughout this temperature range, another indication that the CVD diamond film closely resembles many of the properties of bulk crystalline diamond.

The Raman spectra of a CVD diamond-like film on silicon as a function of temperature are shown in Fig. 3. No sharp peaks were detected in the spectra. Also there was no signal from the substrate

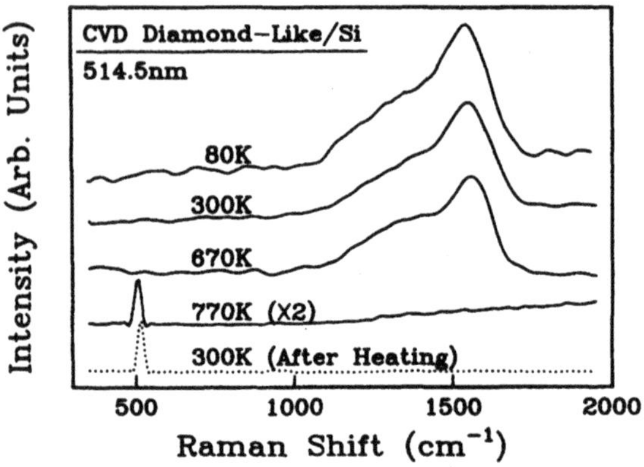

Figure 3. Raman spectra of CVD diamond-like film on silicon for temperatures between 80K and 770K. Note that the film evaporated at 770K. The dashed line was taken at 300K after the high temperature measurements.

silicon, an indication that the diamond-like sample has higher optical absorption in the visible wavelength region than the diamond film of Fig. 1. There was an asymmetric broad band structure peaked at ~ 1570cm^{-1}, which was identified as originating from the graphite-like sp^2 bonding regions of the sample[10-11]. Nothing in the spectra could be attributable to signals from either crystalline or amorphous diamond. However, other studies of diamond-like films by electron energy loss spectroscopy have shown features attributable to diamond structures[13]. Because the Raman spectra of these diamond-like films are overwhelmingly dominated by the highly absorptive graphitic components, diamond regions of crystalline or amorphous form may very well exist in the sample but not be detected by Raman scattering.

The asymmetric broad band structure changed very little with temperature, except for a small blue shift of the peak position as the temperature increased. By 770K, it disappeared completely and the Raman line of the substrate silicon at ~ 522cm^{-1} became visible. This was an irreversible process. An examination of the sample surface by eye after high temperature measurements revealed that the diamond-like film was completely evaporated. Evidently the diamond-like film was not stable at high temperatures. This is in sharp contrast to the diamond film which was stable up to 770K, and maintained its sharp and clean spectra throughout the temperature range.

In conclusion, Raman spectra were measured as a function of temperature on CVD prepared diamond and diamond-like thin films on silicon substrates. The diamond thin film had a sharp peak at $1332cm^{-1}$, the first order Raman frequency of single crystal diamond. The temperature dependence of this Raman line was in very good agreement with that of natural diamond. A small amount of graphitic structure was also detected. The diamond film was found to be stable at the highest temperature. The diamond-like film displayed an asymmetric broad band at about $1570cm^{-1}$ from its graphitic components. Raman measurements were not able to detect any diamond components although other measurements have shown that they can exist. It was also discovered that the diamond-like film was not stable for temperatures above about 700K.

A.R. Srivatsa and T.A. Skotheim acknowledge the support of SDIO under SBIR contract number N00014-89-C-0212, managed by ONR.

REFERENCES

1. Bachmann, P.K., and Messier, R., Chem. Eng. News, 67, 24 (1989).
2. Kamo, M., Sato, Y., Matsumoto, S., and Setaka, N., J. Cryst. Growth, 62, 642 (1983).
3. Kurihara, K., Sasaki, K., Kawarada, M., and Koshino, N., Appl. Phys. Lett. 52, 437 (1988).
4. Spitsyn, B.V., and Derjaugin, B.V., USSR Inv. Certif. No. 339134, Oct. 7, 1956 (in Russian); Official Bulletin of Inventions of USSR, 17, 323 (1980).
5. Spitsyn, B.V., Bouilov, L.L., and Derjaugin, B.V., J. Cryst. Growth, 52, 219 (1981).
6. Franks, J., J. Vac. Sci. Tech. A7, 229 (1989).
7. Zou, J.W., Reichelt, K., Schmidt, K., and Dischler, B., J. Appl. Phys. 65 3914 (1989).
8. Solin, S.A., and Ramdas, A.K., Phys. Rev. B1, 1687 (1970).

9. Dresselhaus, M.S., and Dresselhaus, G., "Light Scattering in Solids, III", ed. by Cardona, M., and Guntherodt, G., Springer, Berlin, 1982, p.3.
10. Nemanich, R.J., and Solin, S.A., Phys. Rev. B20, 392 (1979).
11. Yoshikawa, M., Katagiri, G., Ishida, H., Ishitani, A., Ono, M., and Matsumura, K., Appl. Phys. Lett. 55, 2608 (1989).
12. Yoshikawa, M., Katagiri, G., Ishida, H., Ishitani, A., and Akamatsu, T., J. Appl. Phys. 64, 6464 (1988).
13. Nemanich, R.J., Glass, J.T., Lucovsky, G., and Shroder, R.E., J. Vac. Sci. Technol. A6, 1783 (1988).
14. Anastassakis, E., Hwang, H.C., and Perry, C.H., Phys. Rev. B4, 2493 (1971).
15. Menéndez, J., and Cardona, M., Phys. Rev. B29, 2051 (1984).
16. Shroder, R.E., Nemanich, R.J., and Glass, J.T., Phys. Rev. B41, 3738 (1990).

RAMAN SPECTROSCOPIC STUDY OF PrS

E. Anastassakis and Y. S. Raptis
National Technical University
Physics Department, GR 157 73 Zografou, Greece.

The Raman spectrum of the rare-earth metallic compound PrS was measured. The observed bands are attributed to defect-activated one- and two-phonon Raman scattering from acoustical and optical branches.

Praseodymium sulfide is a rare-earth metallic compound which, like other rare-earth monochalcogenides, crystallizes in the rocksalt structure (space group O_h^5). Essentially, it is an ionically-bound Van-Vleck paramagnet; its metallic conductivity arises from the fact that the trivalent Pr^{3+} ions contribute one electron each to the (5d,6s)-like conduction band[1]. Weak ferromagnetic ordering has been reported around 10K (Ref. 2), but not reproduced in subsequent works[3,4]. Like other monochalcogenides with even number (2) of 4f-electrons in their cations, PrS should normally exhibit no such ordering. Its crystal-field energy level diagram has been examined by inelastic neutron scattering techniques and accounted for, quantitatively, by a nearest neighbour point-charge model with an effective charge of -2 (Ref. 1). These results are also consistent with high-pressure susceptibility measurements[3,5]. PrS has been regarded as a potential candidate for hyperfine-enhanced nuclear adiabatic cooling down to 0.3 mK (Ref. 4).

We are not aware of any lattice dynamical studies in PrS, from either spectroscopic or neutron scattering sources. Pr-based compounds, isomorphic to YBaCuO superconductors, like $PrBa_2Cu_3O_{7-\delta}$, have recently attracted interest concerning their phonon spectra[6]. Spectroscopic investigations have been reported in related rare-earth compounds, e.g., phonons and two-magnon excitations in Eu_2CuO_4 and Gd_2CuO_4 through Raman scattering[7], phonons in Nd_2CuO_4 through Raman and IR spectroscopy[8], and phonons in their isomorphic Pr_2CuO_4 through IR spectroscopy[9]; no data on two-magnon Raman scattering in single crystals of either Pr_2CuO_4 or Nd_2CuO_4 appear to exist.

Reflection spectra from PrS have been reported in the past, covering the region 100-50,000 cm^{-1} (Ref. 10) and 400-50,000 cm^{-1} (Ref. 11). The dispersion of optical constants and also the presence of a reflection minimum in the visible region of the spectrum were determined from such measurements[11]. As far as Raman spectroscopy is concerned, PrS has not been investigated to our knowledge. On the contrary, there have been several reports concerning Gd-based monochalcogenides[12–14] and the metallic compound TiN (Ref. 15). Finally, PrS exhibits a lattice constant of 5.735 Å (Ref.1) and an isothermal compressibility of (0.99 ±0.12) $Mbar^{-1}$ (Ref. 16). Epitaxial growth of PrS on NaCl substrates was reported more recently[17].

Because of its rocksalt structure, PrS is expected to show no first-order Raman activity. Its metallic character prevents the use of infrared spectroscopy for the study of phonons. Unlike natural Gd, which at 1.8 Å exhibits a thermal-neutron capture cross section of 49,000 barn and is therefore a strong absorber of thermal neutrons, natural Pr exhibits at 1.8 Å a cross section of only 11.5 barn[18], which would pose no serious problem in neutron scattering experiments. No such results have been reported, however. Advanced lattice dynamical models could, in principle, be employed for the calculation of phonon dispersion curves. Unfortunately, lack of vital input information, such as elastic constants, would make the results rather ambiguous.

Raman spectroscopy appears, therefore, to be the only possibility at present, to obtain first-hand information about phonons in PrS. The symmetry limitations imposed by the rocksalt structure of PrS are overcome by the relaxation of the $q \approx 0$ restriction of normal Raman scattering. This is a consequence of the reduced translational symmetry which, in turn, is due to the presence of lattice imperfections, mainly deviations from stoichiometry. The difficulty of small penetration of laser light continues to exist; to some extent this problem is offset by the common property of all materials under consideration to exhibit a reflection minimum in the middle-to-upper region of the visible spectrum, due to a plasma resonance[11–15].

The gold-colored samples used in this work were grown by K. Fischer, using a two-step process, i.e., directional freezing with post-annealing for recrystallization and homogenization. Tungsten ambules were used and special precautions were taken to avoid oxygen penetration. Most of the samples were single phase and stoichiometric. The present data were obtained from freshly cleaved [100] faces (typical dimensions 2x2x5 mm). Surfaces etched with syton produced similar spectra with somewhat increased background. The spectra were taken under vacuum at various temperatures (30 - 300 K). Backward scattering geometry along [100] was employed with either parallel or crossed polarization configurations (often the scattered radiation was kept unpolarized). Most of the Ar^+ and Kr^+ laser lines were used, including the uv line at 350.7 nm.

As expected, the spectra were very weak and observed through the use of either a single channel analyser or a multichannel analyser, connected to a Spex double monochromator and a photocounting system[19]. Two such spectra, obtained with different excitation lines, are shown in Fig. 1 (single channel collection). All the major peaks are reproduced fairly well. The spectra of Fig. 2 (multichannel collection) were taken at two different temperatures with 55 mW of the 476.5 nm line of an Ar^+ laser (below 150 cm^{-1} the background was too high to be shown). In spectra taken under different conditions some of the features were enhanced, like the ones around 100 and 500 cm^{-1}, while others were weakened, like the ones in the range 200 - 300 cm^{-1}. In general, the strength (but not the position) of the bands varied from sample to sample. In fact those samples which were characterized as stoichiometric exhibited no features at all.

This is consistent with the interpretation of the spectra as due to defects and/or deviations from stoichiometry.

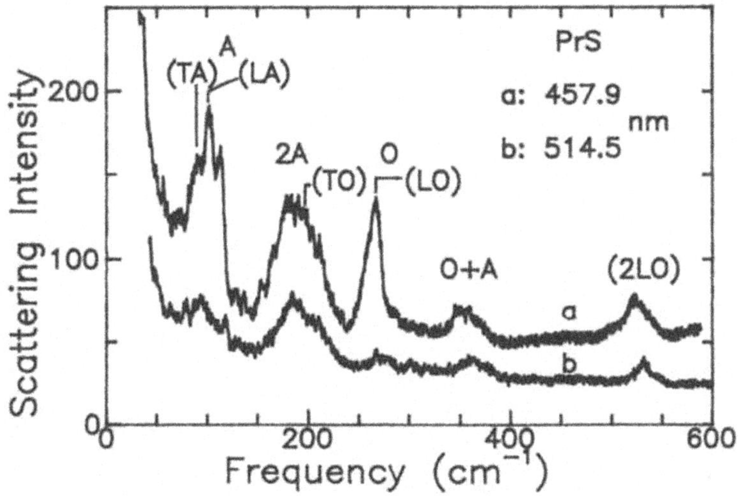

Fig. 1. Raman spectra of PrS taken at 300K with two different excitation frequencies. Acoustical (A) and optical (O) transverse (T) and longitudinal (L) phonon contributions are shown over the main spectroscopic features observed. Backward scattering along [001]. Incident polarization along [010]. Scattered light unpolarized. Scanning speeds 0.05 cm^{-1}/sec (a) and 0.02 cm^{-1}/sec (b).

The spectra in Figs. 1 and 2 exhibit striking similarities with those of GdS (Ref. 13) and other monochalcogenides[14]. Based on these similarities and the following general remarks, we have reached the assignments indicated in Figs. 1 and 2. In all these materials the main mechanism of scattering is related to lattice imperfections; it is reasonable, therefore, to expect sizeable DATA and DALA contributions at low frequencies, i.e., Defect-Activated scattering by TA and LA phonons, respectively, mainly from the L-point of the Brillouin zone[15]. The inset of Fig. 2 shows a detail recording, with the background removed, of the lowest band of the spectrum which includes scattering by acoustical phonons only. The band appears asymmetric, due to contributions from the TA and LA phonons. The dotted lines show the results of fitting of this band by two Gaussians, representing the two contributions.

The mass ratio between Pr and S is large enough (~ 4.4) to justify a distinct gap between the low frequency one-acoustic-phonon scattering (< 120 cm^{-1} and the rest of the spectrum (> 150 cm^{-1}. The latter includes one-phonon (LO, TO) and multi-phonon (overtone or combination) scattering. It is assumed that the broad asymmetric band around 200 cm^{-1} includes the overtones 2TA, 2LA, as well as first-order scattering from TO phonons. The frequency of the latter (195 cm^{-1}) has been determined from the fitting of the band with three independent Gaussians.

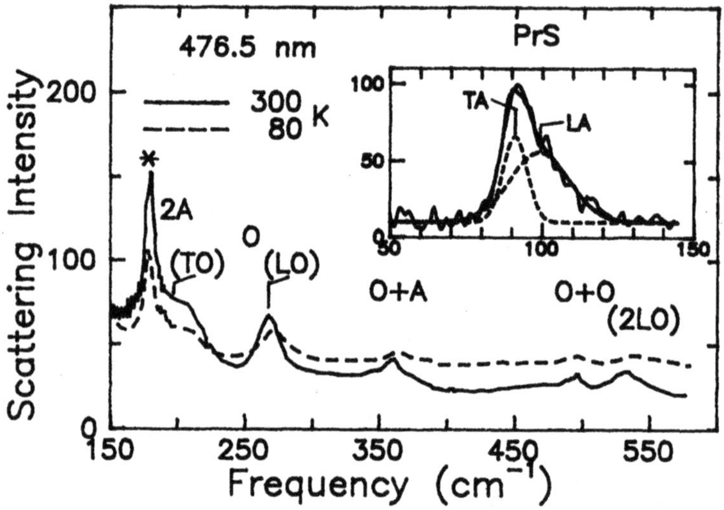

Fig. 2. Raman spectra of PrS at 80 and 300K. The polarization of the incident and scattered light was parallel to [010]. The inset represents first-order scattering by acoustical phonons, the dashed lines are the two Gaussians used for the fitting of the band, and specify the frequencies of the TA and LA phonon branches (Table I). The star signifies the presence of a laser plasma line on top of the 2A band.

As usual, the major contributions to the spectra of Figs. 1 and 2 arise from points of the Brillouin zone with high phonon density of states. Figures 1 and 2, therefore, represent a spectroscopic mapping of the corresponding density of states for acoustical and optical phonon branches. Strictly speaking, these results should be cross-checked through systematic measurements under variable temperature. Because of the weak signals involved, our measurements under variable temperature were not conclusive, although we observed a definite tendency of the peaks around 250 - 350 cm^{-1} towards higher frequencies, as the temperature was lowered (Fig. 2). In spite of these difficulties, the present data provide a starting point for considering the phonon dispersion curves of PrS. In this sense, the assignments of Figs. 1 and 2 are tentative and subject to confirmation from spectroscopic or other sources. There was no evidence of scattering by magnetic excitations.

In Table I we compile the frequencies for the first-order Raman spectroscopic features of PrS, as well as for other rare-earth monochalcogenides which have been studied in the past[13,20–24]

Table I. Acoustical and Optical modes (experimental values) of five representative, metallic paramagnetic (PrS), semiconducting ferromagnetic (EuS), metallic antiferromagnetic (GdS), semiconducting paramagnetic (YbS), and semiconducting diamagnetic (SmS) rare-earth chalcogenides, at 300K.

	PrS[a]	SmS[b]	EuS[d]	GdS[f]	YbS[g]
TA	91	70(L), 80(L)[c]	85	81	–
LA	99	105(L), 125(L)[c]	103(X), 140(L)	104	–
TO	195	170	178[e]	200	195(Γ)
LO	267	200	268[e]	285	272(Γ), 240(L)

[a] Present work, [b] Ref. 23, [c] Ref. 24, [d] Ref. 20, [e] Ref.21, [f] Ref. 13, [g] Ref. 22.

It may be worthwhile to undertake a comparative calculation of the phonon dispersion curves for this class of materials, based on well-accepted lattice dynamical models and whatever input information is available at present. Furthermore, neutron scattering measurements of phonon branches could be performed on crystals grown of rare-earth isotopes with low thermal-neutron capture cross section; unlike Raman spectroscopy, however, neutron spectroscopy would require large samples, a rather demanding crystal-growing task, at least for PrS.

We are thankful to Dr. Karl Fischer, Institute of Solid State Research KFA Julich, Germany, for providing the samples of PrS. Useful discussions with Prof. G. Güntherodt, are acknowledged with appreciation. Partial support was provided by the Greek Secretariat for Research and Technology, Greece and, during the early stage of this work, by Max Planck Institute for Solid State Research, Stuttgart.

REFERENCES

1. K. C. Turberfield, L. Passel, R. J. Birgeneau and E. Bucher, J. Appl. Phys. 42, 1746 (1974).
2. G. A. Smolensky, V. E. Adamjan, and G. M. Loginov, J. Appl. Phys. 39, 786 (1968).
3. R. P. Guertin, J. E. Crow, L. D. Longinotti, E. Bucher, L. Kupferberg, and S. Foner, Phys. Rev. B12, 1005 (1975).
4. E. Bucher, K. Andres, S. J. di Salvo, J. P. Maita, A. C. Gossard, A.S. Cooper, and G. W. Hull, Jr., Phys. Rev. B11, 500 (1975).
5. J. E. Schirber, H. T. Weaver, and D. S. Ginley, J. Appl. Phys. 49, 1469 (1978); B. Cornut, J. L. Genicon, R. A. B. Devine, and F. Holtzberg, Phys. Rev. B17, 4474 (1978).

6. C. Thomsen, R. Liu, M. Cardona, U. Amadov, and E. Moran, Solid State Commun. 67, 271, (1988).
7. A. A. Maksimov, I. I. Tartakovskii, and V. B. Timofeev, Physica C160, 249 (1989).
8. E. T. Heyen, G. Kliche, W. Kress, W. König, M. Cardona, E. Rampf, J. Prade, U. Schroder, A. D. Kulkarni, F. W. de Wette, S. Pinol, D. McK. Paul, E. Moran, and M. A. Alario-Franco, Solid State Commun. 74, 1299 (1990).
9. M. K. Crawford, G. Burns, G. V. Chandrashekhar, F. H. Dacol, W. E. Farneth, E. M. McCarron, III, and R. J. Smalley, Solid State Commun. 73, 507 (1990).
10. V. A. Serdyuk and V. L. Bekenev, Chemical Abstracts 91, 537 (1979), no. 99498.
11. L. A. Ivanchenco, Yu. B. Paderno, A. N. Pilyankevich, Chemical Abstracts 90, 508 (1979), no. 31283.
12. E. Anastassakis, H. Bilz, M. Cardona, P. Grünberg, and W. Zinn, in Light Scattering in Solids, eds. M. Balkanski et al. (Flammarion, Paris 1967), p. 367.
13. G. Güntherodt, P. Grünberg, E. Anastassakis, M. Cardona, H. Hackfort, and W. Zinn, Phys. Rev. B16, 3504 (1977).
14. A. Treindl and P. Wachter, Phys. Letters 64A, 147 (1977).
15. W. Spengler, R. Kaiser, and H. Bilz, Solid State Commun. 17, 19 (1975).
16. B. Morosin and J. E. Schirber, Phys. Lett. 73A, 50 (1979).
17. L. S. Palatnik, M. N. Naboka, A. P. Kulikov, V. Ya. Shevchenko, V. I. Sergienko, and S. S. Borisova, Inorg. Mater. (USA) 18 1531 (1982).
18. CRC Handbook of Chemistry and Physics, 52nd edition 1971-1972, pp. B-389, 363.
19. Some of the early measurements were taken at Max Planck Institute for Solid State Research, Stuttgart.
20. G. Güntherodt and R. Merlin in Light Scattering in Solids (Topics in Applied Physics, Vol. 54), Eds. M. Cardona and G. Güntherodt, Springer-Verlag, Berlin Heidelberg 1984, p. 243.
21. P. Grünberg, G, Güntherodt, A. Frey and W. Kress, Physica B89, 225 (1977).
22. R. Merlin, G. Güntherodt, R. Humphreys, M. Cardona, R. Suryanarayanan, F. Holtzberg, Phys. Rev. B17, 4951 (1978).
23. G. Güntherodt, R. Merlin, A. Frey and M. Cardona, Solid State Commun. 27, 551 (1978).
24. H. A. Mook, D. B. McWhan, F. Holtzberg, Phys. Rev. B25, 4321 (1982).

Dynamical Properties of Fast Ion Conducting Borate Glasses

M. Balkanski*, R. F. Wallis, J. Deppe
Department of Physics
University of California
Irvine, CA 92717, U. S. A.

M. Massot
Laboratoire de Physiques des Solides
Université Pierre et Marie Curie
4, place Jussieu, 75252 Paris Cedex 05, FRANCE

ABSTRACT

Experimental and theoretical work which leads to an improved understanding of the physics of ion diffusion in lithium doped borate glasses is presented. The matrix vibrations of undoped, i.e., B_2O_3, and doped, i.e. B_2O_3-xLi_2O, borate glasses are discussed in connection with Raman scattering and infrared reflectivity measurements. Addition of the modifier Li_2O is shown to cause a transformation of three- to four-coordinated boron atoms. The introduction of dopants, e.g. Li_2SO_4 or LiX (with X = F, Cl, Br, I) and their effect on the ionic conductivity are discussed.

1. INTRODUCTION

Fast ion conducting glasses have attracted considerable interest in recent years in view of their use as separators in solid state batteries. A solid state cell is formed of three components: an ion source, usually Li; a dielectric insulator but good ion conductor such as the lithium borate glass, for example, which acts as a separator between the anode or ion source and the cathode or electron exchanger; and an insertion compound acting as an electron exchanger. The requirements of the separator are to be shape adaptable in order to assure good contact with the ion source and the exchanger and to have a good ionic conductivity.

Another important characteristic is that the separator should be as perfect a dielectric as possible in order to prevent electronic leakage and assure a long shelf lifetime of the batteries. The borate glasses are one of the best materials among those known today to respond to all these requirements.

The boron oxide glass is a very good insulator and constitutes the glass former. Upon the addition of a modifier such as Li_2O, for example, the structure of the glass is significantly changed. In addition, some fraction of the Li_2O dissociates releasing free lithium ions and thus considerably increasing the ion conductivity of the glass. A further increase of the conductivity is achieved by adding a dopant containing the mobile ion, Li in this case. The effect of the dopant upon the concentration of free ions available for the conduction process has been investigated recently [1]. Spectroscopic investigations on the structure of borate glasses with different modifier concentrations and different dopants have also been extensively developed and a summary is given in a recent publication [2].

The structural modifications conditioning the fast ion conduction in borate glasses are inferred mainly from spectroscopic data which reflects the dynamical properties of the glass matrix. It was therefore important to develop theoretical studies on the vibrations of the glass in parallel with the experimental investigations. We shall present here simultaneously the theoretical and experimental results on the vibrational properties of the borate glass with the aim of reaching some conclusions on the conditions for increasing the ionic conductivity of the glass.

2. VIBRATIONS OF BORON OXIDE GLASS

The structure of the boron oxide glass has been studied by X-ray [3], Raman [4], infrared [5], nuclear magnetic resonance [6] and neutron diffraction [7] measurements. According to these observations, boron oxide glass is formed of an infinite network of boroxol groups B_3O_3. The six atoms inside the group form a planar ring and the B-O-B angle is 120°. These boroxol hexagons are interconnected by an oxygen bridge -0-, and the external B-O-B angle θ, less accurately determined, is about 130°. The relative geometry of two successive boroxol planes is not well known.

One of the characteristic features of the vibrational spectra of v-B_2O_3 observed by Raman scattering [2] and shown in Fig. 1 is the presence of a sharp peak at 800 cm^{-1} which is strong and well polarized [8]. Usually the vibrational spectra of amorphous solids are expected to

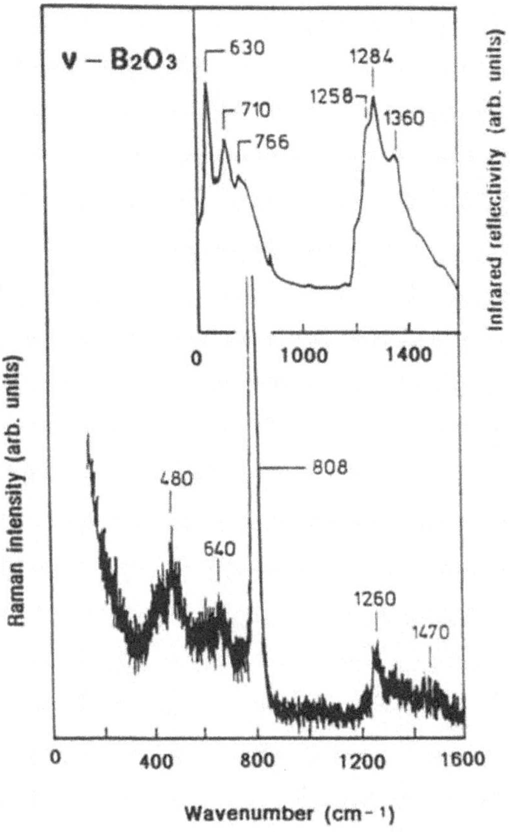

Fig. 1 Raman and infrared data for pure B_2O_3. The sharp peak at 808 cm^{-1} is attributed to the breathing mode of the six-membered boroxol ring.

consist of broad bands (see for example ref. 9). This is because the localized molecular mode responsible for a sharp peak couples with all other atoms and broadens when a network is formed. For example, Galeener et al. [8] have calculated vibrations of v-B_2O_3 using the Bethe lattice approximation (described in detail below) and obtained only broad bands due to extended modes.

There have been a few attempts to understand theoretically the possible existence of sharp features in the vibrational spectra. Kristiansen and Krogh-Moe [10] calculated the vibrational frequencies of the boroxol molecule with atoms of variable mass attached to the outer oxygen atoms and found that the mode at 800 cm^{-1} does not vary much when the mass is changed. Galeener et al. [11] explain the decoupling of rings from the rest of the network by a somewhat fortuitous cancellation of central and noncentral forces.

Galeener and Thorpe [12] have also discussed the introduction of 6-fold boroxol rings into a continuous random network model of v-B_2O_3. They conclude that the 800 cm^{-1} mode consists of a breathing motion of the intra-ring oxygen, and therefore does not couple to the outside network.

Kanehisa and Elliott [13] have demonstrated this decoupling explicitly using a cluster Bethe-lattice approximation.

3. EFFECT OF THE ADDITION OF MODIFIER Li_2O to v-B_2O_3: THE SYSTEM B_2O_3-xLi_2O

3.1 Experimental Results

The Raman spectra of the binary glass B_2O_3-xLi_2O are shown in Fig. 2. At low Li_2O concentration, x = 0.1, the 800 cm^{-1} band corresponding to the breathing oxygen motion in the boroxol ring, the ν_2 mode trigonal deformation of the ring shown in Fig. 3, is still narrow and strong. A new band appears at 780 cm^{-1} which corresponds to the trigonal deformation of six-membered borate rings with one or two BO_4 units [2] as shown in Fig. 4 for the tetraborate, triborate or diborate configuration. The band at 500 cm^{-1} corresponds to the breathing mode ν_3 (Fig. 3) where the boron and the bridging oxygen atoms move in phase.

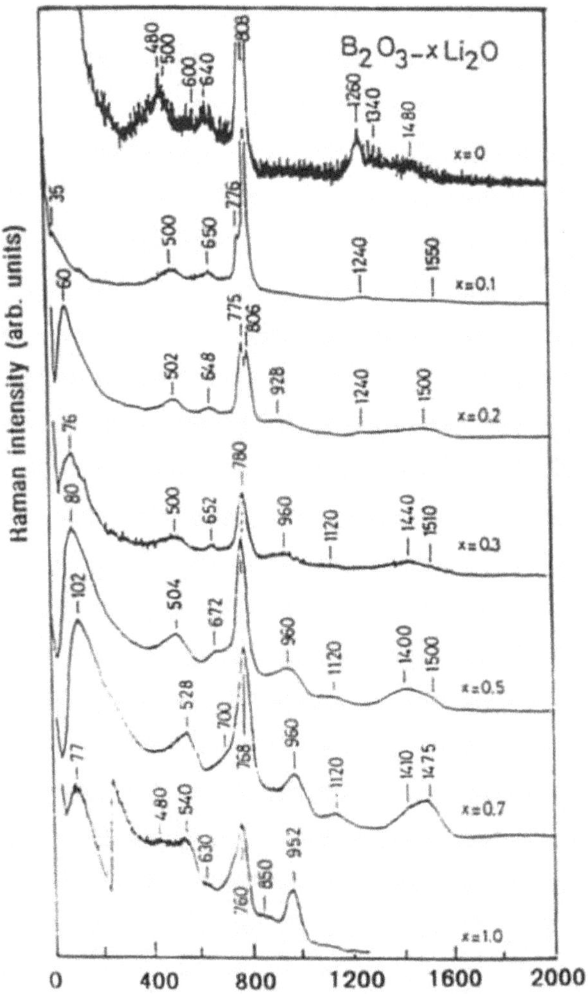

Fig. 2 Raman spectra of the binary glass system B_2O_3-xLi_2O.

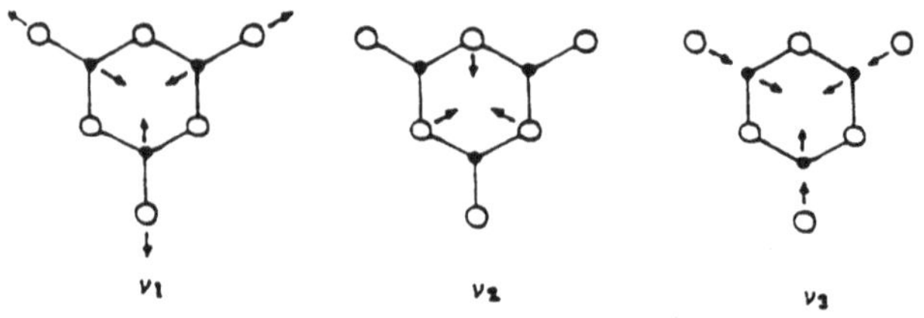

Fig. 3 The three symmetrical vibrations of the boroxol rings.

The mode ν_1, at 1240 cm^{-1}, is associated with the symmetric vibration of the same atoms. The band between 1400 and 1550 cm^{-1} is associated with the stretching vibrations of the B-O bonds in BO_3 units. The three regions associated with the stretching vibrations of the B-O bonds which link the boron-oxygen network will be modified upon addition of Li_2O, which leads to the transformation of trihedrally coordinated boron, B^{III}, to tetrahedrally coordinated boron, B^{IV}, according to the reaction

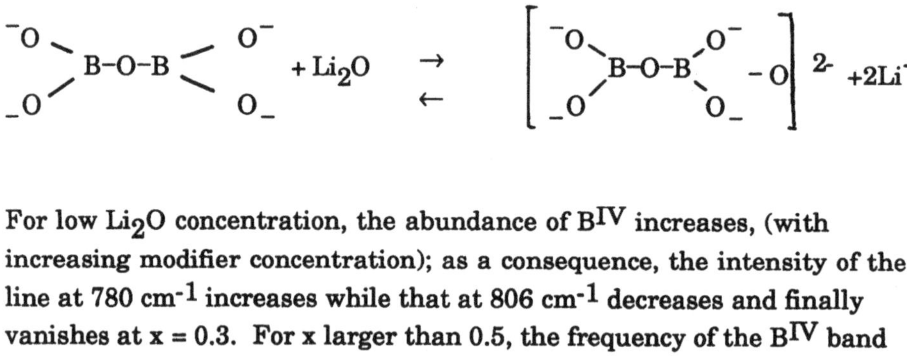

For low Li_2O concentration, the abundance of B^{IV} increases, (with increasing modifier concentration); as a consequence, the intensity of the line at 780 cm^{-1} increases while that at 806 cm^{-1} decreases and finally vanishes at $x = 0.3$. For x larger than 0.5, the frequency of the B^{IV} band decreases until it reaches 760 cm^{-1} in the metaborate glass $x = 1.0$.

At high Li_2O concentration ($x > 0.5$), the formation of non-bridging oxygen atoms is inferred from the relative intensity increase of the two bands at 960 and 1480 cm^{-1}.

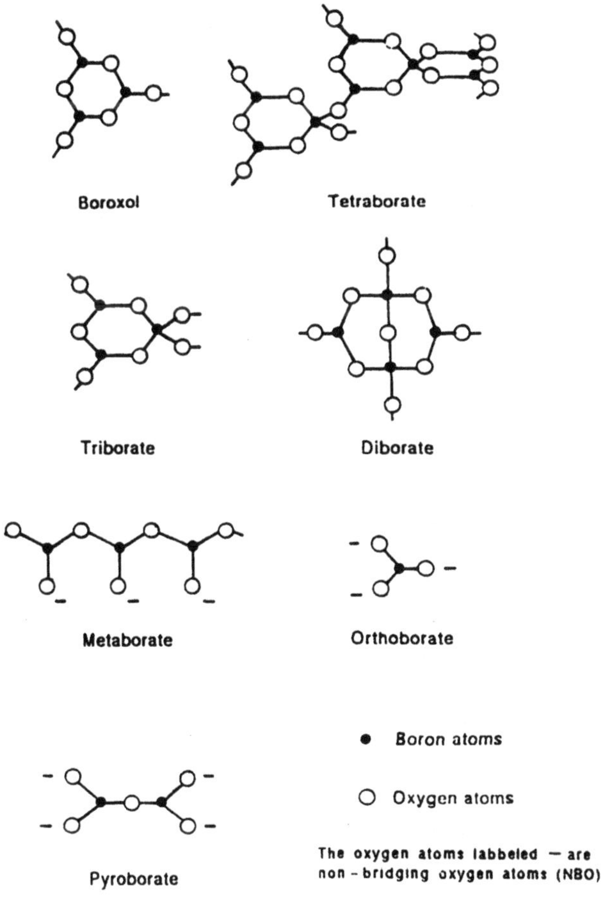

Fig. 4 Some typical borate groups.

Two stages of modification can be distinguished when the concentration of the modifier increases. During the first stage, the tetrahedrally coordinated boron units increase and lead to an increase of the linkage of the boron-oxygen network up to x = 0.5. Above this concentration, formation of BO_3 units with non-bridging oxygen atoms "opens" the borate units and consequently the degree of linkage of the network is decreased.

3.2 Theory

With the vibrational frequencies of groups of atoms in a given configuration deduced from spectroscopy data, it is of interest to calculate the vibrational spectrum with some specific short-range configuration. Toward this end, we have employed the network-dynamics theory and lattice dynamics in order to effect an indirect determination of the glass structure. Central force network-dynamics seems an ideal method to investigate the relation between local environmental changes and the modes of vibration, since the short range order can be easily modified.

3.3. Methods Of Calculation

3.3.1. Network dynamics

The network-dynamics method, introduced by Sen and Thorpe [14], considers only nearest-neighbor central forces in the continuous random network under consideration: the other forces, for example, long-range Coulomb and angle bending forces, are much smaller in many cases.

Thorpe and Galeener [15] have put the central-force model of Sen and Thorpe in a more general framework using a Lagrangian formulation.

As discussed in the previous section, we have seen that the addition of Li_2O to B_2O_3 is accompanied by the appearance of groups containing fourfold-coordinated borons. In addition, the concentration of these structures increases proportionally with the concentration of the modifier Li_2O up to a critical value x_c. We shall consider a network containing both threefold- and fourfold-coordinated borons. Indeed, for the composition B_2O_3-$0.5Li_2O$, one can envision a network of equal numbers of threefold- and fourfold-coordinated borons. Here we consider one such network, with each threefold-coordinated boron surrounded by fourfold-coordinated borons and vice versa. The equations of motion and their solution are presented in a separate publication [16].

3.3.2. Lattice dynamics

The lattice-dynamics calculations were carried out as an independent check on the network-dynamics results. As such, they were done in the nearest-neighbor central-force approximation. No disagreement was found between the lattice-dynamics and network-dynamics calculations. The two methods are quite complementary in examining the effect of varying various parameters, such as force constants, intermolecular bridging angles, or configurations on the vibrational spectrum of the solid.

3.4 Results And Discussion

In Fig. 5, the results of network- and lattice-dynamics calculations [16] are presented for $\beta/\alpha = 0.8$ where α and β are the force constants for four- and three-coordinated boron atoms, respectively. Note that the general features in the spectra, such as the narrowing of the middle frequency gap as compared to the data of pure B_2O_3, are found in the network-dynamics results. The widening of the high-frequency gap, presumably due to the difficulty of the modes associated with BO_3 groups to hybridize with those of the stiffer BO_4 groups, is also found. The narrow peak at 808 cm^{-1} in the Raman data for $x = 0$ is no longer visible, indicating that there are no longer any rings without at least one fourfold-coordinated boron. The new peak in the vicinity of 780 cm^{-1}, corresponding to the trigonal deformation of six-membered rings containing one or two BO_4^- units [17] increases in intensity, as can be seen by comparing Figs. 1 and 2. Note also that as the bridging angle θ approaches 90°, the modes decouple into a low-frequency singlet and high-frequency triplets characteristic of the threefold- and fourfold-coordinated molecules.

It is interesting to note that such simple calculations should yield results representing well the trends in the spectral distribution of different vibrational modes in the binary glass system as a function of modifier concentration. This result suggests that the force constants used are reasonable, and again underscores the importance of local order in understanding the vibrational response of covalently bonded glass.

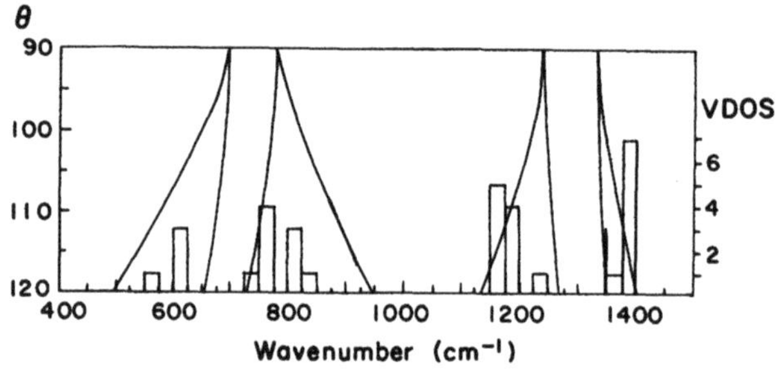

Fig. 5 Raman spectrum, network and lattice dynamics results for B_2O_3-$0.5Li_2O$. The ratio of the force constants of three-coordinated B-O bonds to that of four-coordinated ones is $\beta/\alpha = 0.8$. Histograms indicate the number of modes in the corresponding frequency band according to the lattice dynamics calculation for $\theta = 120°$.

4. THE EFFECT OF DOPANTS

4.1. Structural Modifications Induced By The Dopant

Two types of dopants which introduce additional free Li ions and increase the ionic conductivity of borate glasses are examined: the lithium halides LiX (X = F, Cl, Br, I), and lithium sulfate Li_2SO_4. A general observation is that the dopants do not significantly modify the glass structure which simply localizes the anion leaving as the only mobile species the small lithium cation. The larger the anion, the stronger is its interaction with the host network.

In the ternary glass B_2O_3-$0.57Li_2O$-yLiX, in which the modifier concentration is such that the glass does not contain boroxol rings, the

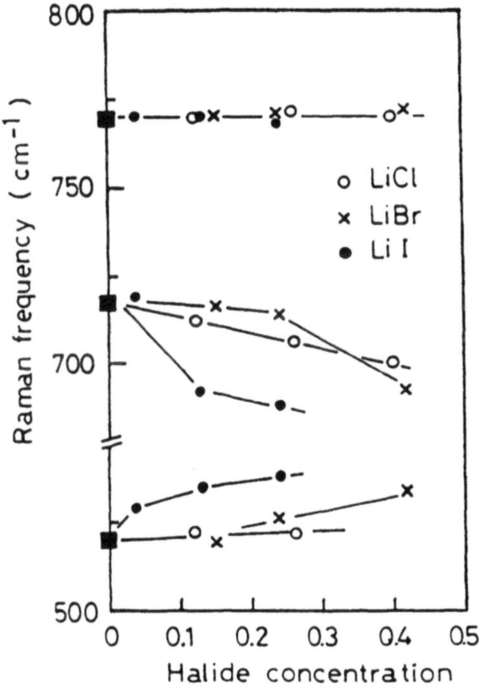

Fig. 6 The shift of the Raman frequencies versus the halide concentration.

modification due to the introduction of dopant salts is characterized in the Raman spectrum by the appearance of a shoulder at about 720 cm^{-1} on the 770 cm^{-1} band due to the vibration of four-coordinated boron atoms. For a given halide anion this new line shifts toward lower frequencies and its intensity increases with increasing dopant salt concentration, y, as shown in Fig. 6. At the same time the frequency of the band at 250 cm^{-1} due to the in-phase motion of the bridging oxygen and the boron atom increases. The stiffening of this mode, which is higher the larger the anion size, corresponds to a compression of the B-O-B band.

There is also an increase of intensity of the bands at 960 and 1480 cm^{-1} attributed to the vibrations of groups with non-bridging oxygen atoms. Sharpening of these bands and a shift toward higher frequencies is observed when going from LiCl to LiI with the same concentration.

These modifications of the boron-oxygen network of the ternary glasses have shown that in all cases the lithium halide addition results in an expansion of the network, the effect being most pronounced with LiI. These results agree with density and glass transition temperature measurements.

These observations are of prime importance for the understanding of the ion transport properties of the borate glasses. The BO_4 groups may be considered as negative ions with a large ionic radius which provide binding sites with small binding energy for the mobile lithium ions. At low modifier content the increase of ionic conductivity comes from the increase of the number of BO_4 units and of the number of free lithium ions. But for higher oxide content there is competition between two competing mechanisms: an increase in the number of free lithium ions provided by the dissociation of the dopant salt and conversion of the BO_4 units into BO_3 units with non-bridging negatively charged oxygen atoms which trap the lithium ions. In the range of large Li_2O modifier content the conductivity is less sensitive to the addition of the doping salt.

4.2 Effect Of The Dopant Li_2SO_4: B_2O_3-xLi_2O-yLi_2SO_4

Comparison of the Raman scattering spectra of the binary B_2O_3-0.7Li_2O and ternary B_2O_3-0.7Li_2O-0.42Li_2SO_4 systems, displayed in Fig. 7, shows that the spectrum of the Li_2SO_4 doped glass consists of the superposition of the spectra of B_2O_3-0.7Li_2O and Li_2SO_4. The peaks appearing at 456, 644, 1004 and 1100 cm^{-1} correspond respectively to ν_2, ν_4, ν_1 and ν_3 vibrations of the sulfate ion [18].

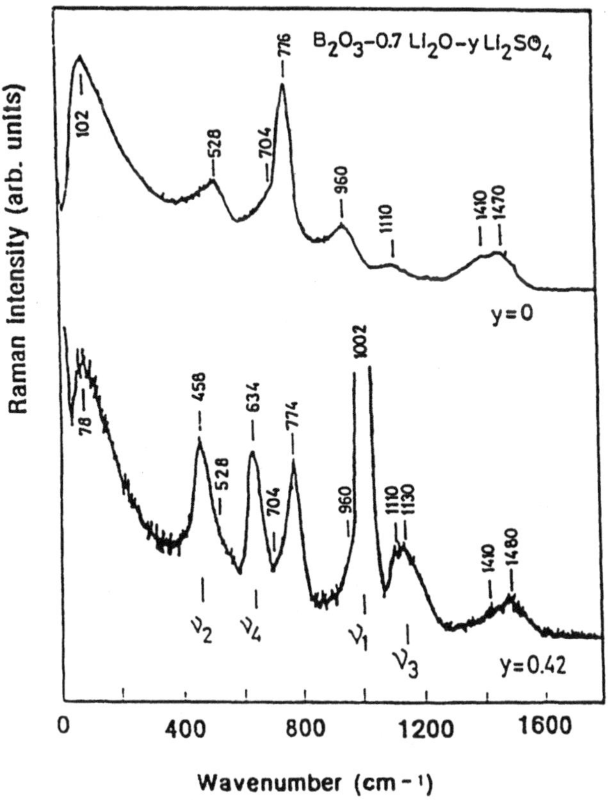

Fig. 7 Raman spectra of the ternary sulfoborate glasses B_2O_3-$0.7Li_2O$-yLi_2SO_4 versus the lithium sulfate content.

5. CONCLUSIONS

The effect of the addition of the modifier Li_2O, was investigated experimentally by Raman and infrared spectroscopy and theoretically by network dynamics and lattice dynamics calculations. The modifier was found to increase the number of free lithium ions which act as charge carriers, and also to significantly modify the host structure through the transformation of trihedrally coordinated boron atoms into tetrahedrally coordinated boron. The concentration dependence shows that this second effect goes through a maximum for x = 0.4 and then decreases as a result of the formation of BO_3 groups with non-bridging oxygen atoms which carry a negative charge. Compared with the experimental data, the results of network and lattice dynamics calculations seems to give an adequate picture of the structure and vibrational response of lithium doped borate glasses.

Introduction of dopants in the form of alkali halide salts produces by dissociation an additional free lithium concentration and also a modification of the host configuration dependent on the size of the anion. Flourine enters in the network in substitutional positions forming BO_3F and BO_2F_2 units. Chlorine, bromine and iodine enter interstitial positions which produces breaking of B-O-B links and formation of BO_4 units. This modification of the boron oxygen network results in a less linked network facilitating the free ion diffusion. The effect is more pronounced for LiI and decreases with the anion size.

Doping with Li_2SO_4 leads to analogous effects. The effect of Li_2SO_4 on the Raman spectra can be understood as simply a super-position of the vibrations of the host network and that of the SO_4 groups. The modifications of the host network with preferential formation of diborate groups tends to produce an expansion of the network in order to accommodate the sulfate ions and facilitate the lithium mobility.

References

*Permanent address: Laboratoire de Physique des Solides, Université Pierre et Marie Curie, 4 place Jussieu, 75252 Paris Cedex 05, FRANCE.

1. M. Balkanski, R. F. Wallis, I. Darianian and J. Deppe, Mater. Sci. Engin. B1, 15 (1988).
2. M. Massot and M. Balkanski in Festschrift for Sir R. J. Elliott edited by J. A. Blackman, Oxford University Press.
3. R. L. Mozzi and B. E. Warren, J. Appl. Crystallogr., 3, 251 (1970).
4. J. Goubeau and H. Keller, Z. Anorg. Chem. 272, 303, (1953).
5. J. Krogh-Moe, Phys. Chem. Glasses, 6, 46 (1965).
6. G. E. Jellison, Jr., L. W. Panek, P. J. Bray and G. B. Rouse, Jr., J. Chem. Phys., 66, 802 (1977).
7. R. N. Sinclair, J. A. E. Desa, G. Etherington, P. A. V. Johnson and A. C. Wright, J. Non-Cryst. Solids, 42, 107 (1980).
8. F. L. Galeener, G. Lucovsky and J. C. Mikkelsen, Jr., Phys. Rev. B, 22, 3983 (1980).
9. M. H. Brodsky, in Light Scattering in Solids, ed. M. Cardona, Springer, Berlin, 1975, p. 205.
10. L. A. Kristiansen and J. Krogh-Moe, Phys. Chem. Glasses, 10, 96 (1968).
11. F. L. Galeener, R. A. Barrio, E. Martinez and R. J. Elliott, Phys. Rev. Lett., 53, 2429 (1984).
12. F. L. Galeener and M. F. Thorpe, Phys. Rev. B28, 5802 (1983).
13. M. A. Kanehisa and R. J. Elliott, Mater. Sci. Eng. B3, 163 (1989).
14. P. N. Sen and M. F. Thorpe, Phys. Rev. B22, 3078 (1980).
15. M. F. Thorpe and F. L. Galeener, Phys. Rev. B22, 3078 (1980).
16. J. Deppe, M. Balkanski and R. F. Wallis, Phys. Rev. B41, 7767 (1990).
17. T. W. Brill, Phillips Research Rep., Suppl. 2 (1976).
18. E. Cazanelli and R. Fresh, J. Chem. Phys. 79, 2615 (1983).

Lattice Dynamics of γ-InSe Containing Intercalated Lithium

P. Gomes da Costa, M. Balkanski,* and R. F. Wallis
Department of Physics and Institute for Surface and Interface Science
University of California, Irvine, CA 92717

ABSTRACT

The lattice dynamics of the γ-polytype of the layer compound InSe has been investigated using a model containing short-range central forces and long-range Coulomb interactions. The normal mode frequencies and eigenvectors were determined by diagonalizing the dynamical matrix. The results have been correlated with the infrared and Raman spectra of γ-InSe. The effect of intercalated lithium atoms on the vibrational modes has been investigated both theoretically and with the aid of Raman spectra.

1. INTRODUCTION

In a number of solid state batteries currently under investigation, layered materials are used as electron exchange electrodes, since they can accommodate considerable amounts of intercalated lithium. For such an application it is essential that the electrode material have a sufficiently high ionic as well as electronic conductivity. A material that satisfies these requirements is InSe which can intercalate relatively large amounts of lithium with only a slight expansion of its lattice [1,2].

InSe is a layered compound having complex layers in which atomic layers of Se, In, In, and Se are bound together by covalent bonds with some ionic character. These complex layers are themselves bound to one another by weak Van der Waals interactions with a so-called Van der Waals gap between layers. There are four possible stacking arrangements of the complex layers leading to four polytypes designated $\beta, \varepsilon, \gamma$, and δ. Since the experimental results to be discussed are for the γ-polytype, we shall focus our attention on this polytype.

A microscopic understanding of both the electronic and ionic conductivity of lithium-intercalated γ-InSe requires a knowledge of the lattice dynamics of this system. Of particular importance is information concerning the forces of interaction between intercalated lithium atoms and nearby selenium and indium atoms.

In this paper we present the results of a theoretical investigation of the lattice dynamics of γ-InSe both with and without intercalated lithium. We develop the lattice dynamical model in section 2. The values of the force constants and calculated phonon dispersion curves for pure γ-InSe are presented in section 3. Similar results are presented in section 4 for lithium intercalated γ-InSe. Conclusions are presented in section 5.

2. LATTICE DYNAMICAL MODEL

The first step in setting up the lattice dynamical model is to summarize the pertinent information concerning the structure of InSe. The γ-polytype belongs to the trigonal system, and its space group is C_{3v}^{5} . The primitive translation vectors can be written as [3]

$$\vec{t}_1 = \left(\sqrt{3}/3\right)a_o\vec{i} + c_o\vec{k} \tag{1a}$$

$$\vec{t}_2 = -\left(\sqrt{3}/6\right)a_o\vec{i} + (1/2)a_o\vec{j} + c_o\vec{k} \tag{1b}$$

$$\vec{t}_3 = -\left(\sqrt{3}/6\right)a_o\vec{i} - (1/2)a_o\vec{j} + c_o\vec{k} \tag{1c}$$

where a_o is the lattice constant parallel to the layers, c_o is the lattice constant perpendicular to the layers, and $\vec{i},\vec{j},\vec{k}$ are unit vectors in the x, y, z directions, respectively. The values of a_o and c_o are [4] 4.00 Å and 8.44 Å respectively.

The primitive translation vectors of the reciprocal lattice can be obtained from $\vec{t}_1,\vec{t}_2,\vec{t}_3$ and are found to be

$$\vec{b}_1 = 2\pi\left[\left(2/\sqrt{3}a_o\right)\vec{i} + \left(1/3c_o\right)\vec{k}\right] \tag{2a}$$

$$\vec{b}_2 = 2\pi\left[-\left(1/\sqrt{3}a_o\right)\vec{i} + \left(1/a_o\right)\vec{j} + \left(1/3c_o\right)\vec{k}\right] \tag{2b}$$

$$\vec{b}_3 = 2\pi\left[-\left(1/\sqrt{3}a_o\right)\vec{i} - \left(1/a_o\right)\vec{j} + \left(1/3c_o\right)\vec{k}\right] \tag{2c}$$

Diagrams of the stacking of the layers of γ-InSe and its first Brillouin zone are shown in Fig. 1.

Since InSe is a polar semiconductor, the chemical bonds which bind neighboring atoms together have both covalent and ionic character. The interatomic forces should, therefore, have a short-range part associated with the covalent character of the bonds and a long-range Coulomb part associated with the ionic character of the bonds. We therefore write an element of the force constant matrix $\Phi_{\alpha\beta}(l\kappa;l'\kappa')$ as the sum of two contributions,

$$\Phi_{\alpha\beta}(l\kappa;l'\kappa') = \Phi^{C}_{\alpha\beta}(l\kappa;l'\kappa') + \Phi^{N}_{\alpha\beta}(l\kappa;l'\kappa') \quad , \tag{3}$$

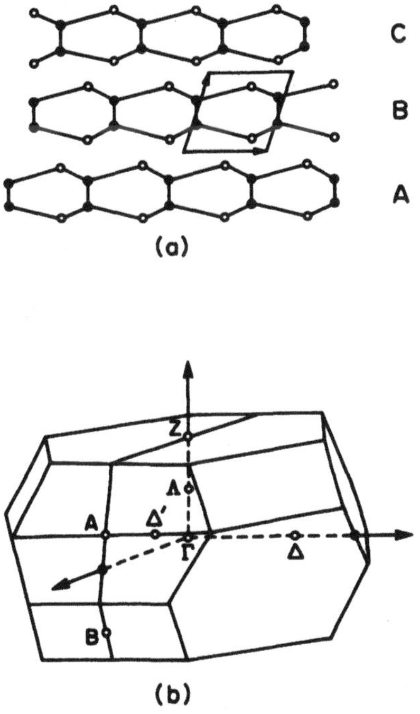

Fig. 1 (a) Stacking of layers and (b) first Brillouin zone of γ-InSe. In (a) the open circles are Se atoms and the closed circles are In atoms.

where $\Phi^{C}_{\alpha\beta}(l\kappa;l'\kappa')$ is the contribution from the long-range Coulomb interactions, $\Phi^{N}_{\alpha\beta}(l\kappa;l'\kappa')$ is the contribution from short-range non-Coulomb interactions, the index pair $l\kappa$ designates the κth atom in the lth unit cell, and α,β denote Cartesian coordinate components.

For the non-Coulomb interactions we adopted the central-potential model which Polian [5] found satisfactory for the related material GaS. With a central potential $\varphi(r)$ dependent only on the distance r between the interacting atoms, one can write the contribution to $\Phi^{N}_{\alpha\beta}(l\kappa;l'\kappa')$ for $l\kappa \neq l'\kappa'$ in the form

$$\Phi^N_{\alpha\beta}(l\kappa;l'\kappa') = -\varphi_{\alpha\beta}(l\kappa;l'\kappa') \tag{4}$$

where

$$\varphi_{\alpha\beta}(l\kappa;l'\kappa') = \left\{\frac{x_\alpha x_\beta}{r^2}\varphi''_{\kappa\kappa'}(r) - \frac{1}{r}\varphi'_{\kappa\kappa'}(r)\ \right] \quad + \frac{\delta_{\alpha\beta}}{r}\varphi'_{\kappa\kappa'}(r)\ \Big\}\Big|_{r=|\vec{x}(l\kappa;l'\kappa')|} \quad , \tag{5}$$

$$\vec{x}(l\kappa;l'\kappa') = \vec{x}(l\kappa) - \vec{x}(l'\kappa') \quad , \tag{6}$$

$\vec{x}(l\kappa)$ is the position vector of atom $l\kappa$ in the equilibrium configuration, and primes denote differentiation with respect to argument. A given interaction consequently involves two force constant parameters, $\varphi''_{\kappa\kappa'}(r)$ and $\varphi'_{\kappa\kappa'}(r)/r$, which will henceforth be designated by A and B, respectively.

The short-range model employed includes the following interactions:

1. In-In first neighbors
2. In-Se first neighbors
3. Se-Se interlayer
4. In-In second neighbors
5. In-Se second neighbors

The corresponding values of A and B are designated by A_i, B_i with i = 1,2,3,4,5.

The long-range Coulomb interactions are characterized by the effective charges of the In and Se ions. Since the In-Se bond is partly ionic and partly covalent, we expect the magnitudes of the effective charges to be less than the magnitude of the electron charge. From the symmetry and stoichiometry of γ-InSe and electroneutrality, we conclude that the effective charges of the In and Se ions are equal in magnitude and opposite in sign. The large anisotropy of the crystal leads us to introduce a tensorial effective charge of the form [5]

$$z^* e = \begin{bmatrix} z_1^* & 0 & 0 \\ 0 & z_1^* & 0 \\ 0 & 0 & z_2^* \end{bmatrix} e \quad . \tag{7}$$

This form for the effective charge accounts for the fact that the polarization produced by displacements parallel to the layers is different from that produced by displacements perpendicular to the layers.

The contribution of the long-range Coulomb interactions to the force constant matrix was evaluated using the Ewald technique by expressing the elements of this matrix in terms of a sum over the direct lattice and a sum over the reciprocal lattice. The elements of the Coulomb contribution to the dynamical matrix $D^C_{\alpha\beta}(\vec{k};\kappa\kappa')$ were then obtained using the relation

$$D^C_{\alpha\beta}(\vec{k};\kappa\kappa') = \frac{1}{\sqrt{M_\kappa M_{\kappa'}}} \sum_{l} \Phi^C_{\alpha\beta}(l\kappa;l'\kappa') e^{-i\vec{k}\cdot[\vec{x}(l\kappa)-\vec{x}(l'\kappa')]} \tag{8}$$

where M_κ is the mass of atom κ. The contribution of the sum over the direct lattice to $D^C_{\alpha\beta}(\vec{k};\kappa\kappa')$ is given by

$$D^C_{\alpha\beta}(\vec{k};\kappa\kappa')_d = -\frac{e_{\kappa\alpha} e_{\kappa'\beta}}{\sqrt{M_\kappa M_{\kappa'}}} \eta^3 \sum_l e^{-i\vec{k}\cdot[\vec{x}(\kappa)-\vec{x}(l\kappa')]} \quad H_{\alpha\beta}[\varsigma(l;\kappa\kappa')]$$

$$+ \frac{\delta_{\kappa\kappa'}}{M_\kappa} \eta^3 \sum_{\kappa''} e_\kappa e_{\kappa''} \sum_l H_{\alpha\beta}[\zeta(l;\kappa\kappa'')] \tag{9}$$

where $e_{\kappa\alpha}$ is the effective charge component of ion κ given by $z^*_{\kappa\alpha}e$, η was taken to have the value $\sqrt{\pi}/r_o$, r_o is the equilibrium nearest-neighbor separation,

$$\zeta(l;\kappa\kappa') = \eta|\vec{x}(\kappa) - \vec{x}(l\kappa')| \quad (10)$$

and

$$H_{\alpha\beta}[\zeta(l;\kappa\kappa')] \equiv H_{\alpha\beta}(\zeta)$$

$$= \frac{\zeta_\alpha\zeta_\beta}{\zeta^2}\left[\frac{3}{\zeta^3}Erfc(\zeta) + \frac{2}{\sqrt{\pi}}\left(\frac{3}{\zeta^2}+2\right)e^{-\zeta^2}\right]$$

$$-\delta_{\alpha\beta}\left[\frac{1}{\zeta^3}Erfc(\zeta) + \frac{2}{\sqrt{\pi}\zeta^2}e^{-\zeta^2}\right] \quad . \quad (11)$$

The reduced effective charge components $z^*_{In\alpha}$ and $z^*_{Se\alpha}$ satisfy $z^*_{Se\alpha} = -z^*_{In\alpha} = z^*_\alpha$ with z^*_α specified by Eq. (7) -- i.e., $z^*_x = z^*_y = z^*_1$ and $z^*_z = z^*_2$.

The contribution of the sum over the reciprocal lattice to $D^C_{\alpha\beta}(\vec{k};\kappa\kappa')$ is given by

$$D^C_{\alpha\beta}(\vec{k};\kappa\kappa')_r = \frac{4\pi}{v_a}\left\{\frac{e_{\kappa\alpha}e_{\kappa'\beta}}{\sqrt{M_\kappa M_{\kappa'}}}\left[\frac{k_\alpha k_\beta}{k^2}e^{-k^2/4\eta^2}\right.\right.$$

$$\left. + \sum_{\vec{G}\neq 0}\frac{(G_\alpha + k_\alpha)(G_\beta + k_\beta)}{|\vec{G}+\vec{k}|^2}e^{-|\vec{G}+\vec{k}|^2/4\eta^2}e^{i\vec{G}\cdot[\vec{x}(\kappa)-\vec{x}(\kappa')]}\ \right]$$

$$-\delta_{\kappa\kappa'}\frac{e_{\kappa\alpha}}{M_\kappa}\sum_{\kappa''}e_{\kappa''\beta}\sum_{\vec{G}\neq 0}\frac{G_\alpha G_\beta}{|\vec{G}|^2}e^{-|\vec{G}|^2/4\eta^2}e^{i\vec{G}\cdot[\vec{x}(\kappa)-\vec{x}(\kappa'')]}\Big\} \tag{12}$$

where the reciprocal lattice vectors $\vec{G}$ are given by

$$\vec{G}=m_1\vec{b}_1+m_2\vec{b}_2+m_3\vec{b}_3 \tag{13}$$

and m_1, m_2, m_3 are integers.

The full Coulomb contribution to the dynamical matrix is given by

$$D^C_{\alpha\beta}(\vec{k};\kappa\kappa')=D^C_{\alpha\beta}(\vec{k};\kappa\kappa')_d+D^C_{\alpha\beta}(\vec{k};\kappa\kappa')_r \quad . \tag{14}$$

Adding this result to the short-range contribution to the dynamical matrix $D^N_{\alpha\beta}(\vec{k};\kappa\kappa')$, where

$$D^N_{\alpha\beta}(\vec{k};\kappa\kappa')=\frac{1}{\sqrt{M_\kappa M_{\kappa'}}}\sum_{l'}\Phi^N_{\alpha\beta}(l\kappa;l'\kappa')e^{-i\vec{k}\cdot[\vec{x}(l\kappa)-\vec{x}(l'\kappa')]} \tag{15}$$

yields the full dynamical matrix $D_{\alpha\beta}(\vec{k};\kappa\kappa')$:

$$D_{\alpha\beta}(\vec{k};\kappa\kappa')=D^C_{\alpha\beta}(\vec{k};\kappa\kappa')+D^N_{\alpha\beta}(\vec{k};\kappa\kappa') \quad . \tag{16}$$

The normal mode frequencies and eigenvectors were obtained by diagonalizing the full dynamical matrix.

3. MODEL PARAMETERS AND PHONON DISPERSION CURVES FOR PURE γ-InSe

The central potential model which we employ contains twelve parameters to be determined: $A_1,\ldots.A_5$, $B_1,\ldots.B_5$, z_1^*,z_2^* . The experimental data that are available for γ-InSe, namely, infrared and Raman spectra, are insufficient to make a unique determination of these parameters. We have

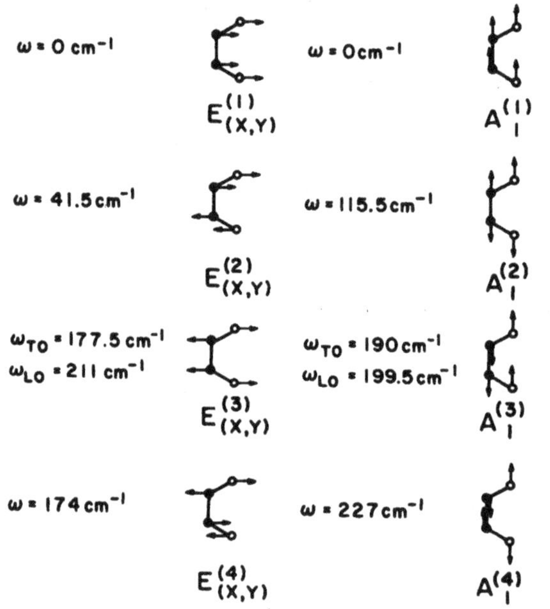

Fig. 2 Zone center normal modes and their corresponding frequencies for pure γ-InSe.

therefore taken as a starting point the values of the corresponding parameters for GaS obtained by Polian [5] by fitting phonon dispersion curves (determined by inelastic neutron scattering), elastic constants, and infrared and Raman data. We then fine-tuned the values of the parameters by adjusting them to reproduce the frequencies and polarizations of the optical modes of long wavelength derived from infrared and Raman data for γ-InSe [6] which are presented in Fig. 2. The values of the model parameters thus determined are given in Table 1.

Table 1. Short-range and Coulomb parameters for γ-InSe. The units for the A and B parameters are N/m.

i	1	2	3	4	5
A_i	78.35	32.17	0.5	7.0	3.4
B_i	4.65	31.0	-0.196	-8.584	0.9
z_i^*	0.71	0.43	---	---	---

The elements of the dynamical matrix were next evaluated using the tabulated values of the model parameters. Diagonalization of the dynamical matrix for various values of the wave vector $\vec{k}$ yielded the normal mode frequencies and eigenvectors. The dispersion curves were constructed for several high symmetry directions in the Brillouin zone and are presented in Fig. 3. Since there are four atoms per primitive unit cell in γ-InSe, there are

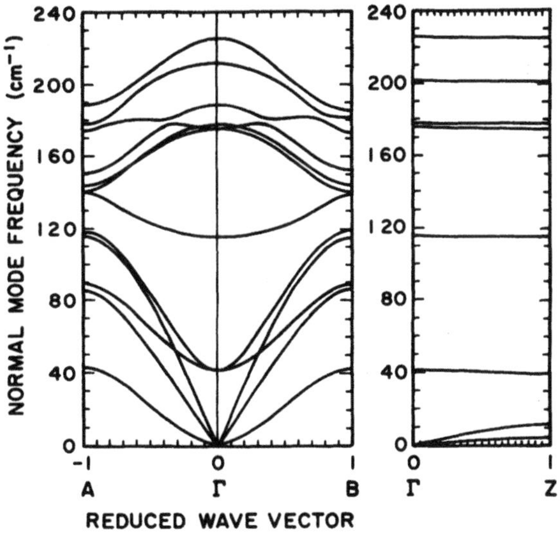

Fig. 3 Phonon dispersion curves for pure γ-InSe.

twelve normal modes for a given value of $\vec{k}$. There is considerable dispersion of the normal modes propagating in directions not perpendicular to the layers as a consequence of the strong intra-layer binding of the atoms. However, the weak interlayer binding (note the small values of A_3 and B_3 in Table 1) leads to nearly flat dispersion curves in the direction perpendicular to the layers, as evident in Fig. 3 for the ΓZ direction.

4. MODEL PARAMETERS AND PHONON DISPERSION CURVES FOR LITHIUM INTERCALATED γ-InSe

The intercalation of lithium into γ-InSe leads to new peaks [6] in the Raman spectrum at 92 cm^{-1} and 388 cm^{-1}. In addition, peaks associated with pure γ-InSe are modified by plasmon-phonon coupling [6]. We have made a theoretical investigation of the case in which there is one lithium atom per primitive unit cell and that atom occupies a site of lowest potential energy (A3-2 site [7]) in the Van der Waals gap between layers. We assume that there are central potential interactions between the lithium atom and the nearest selenium and indium atoms with interaction parameters A_6, B_6 and A_7, B_7, respectively. Since we have only two new normal mode frequencies and since B is typically small compared to A, we have neglected B_6 and B_7 and determined A_6 and A_7 by associating the frequencies 92 cm^{-1} and 388 cm^{-1} with normal modes dominated by lithium atom motion parallel and perpendicular to the layers, respectively. The resulting values for A_6 and A_7 are 1.61 N/m and 54.8 N/m, respectively. The large value of A_7 relative to that of A_6 is consistent with the facts that the indium atomic radius (0.76A) is larger than that of selenium (0.60A) [8] and that a lithium atom in the A3-2 site lies directly above the indium atom to which it is coupled. The restoring force on the lithium atom is therefore dominated by A_7 in the mode in which the lithium atom moves perpendicular to the layers (ω=388 cm^{-1}) and by A_6 in the mode in which the lithium atom moves parallel to the layers (ω=92 cm^{-1}.)

After incorporating the force constants A_6 and A_7 into the dynamical matrix, we calculated the normal mode frequencies and eigenvectors as functions of the wave vector $\vec{k}$ for $Li_{0.5}$ InSe. The dispersion curves thus obtained are plotted in Fig. 4. We note that the normal modes primarily

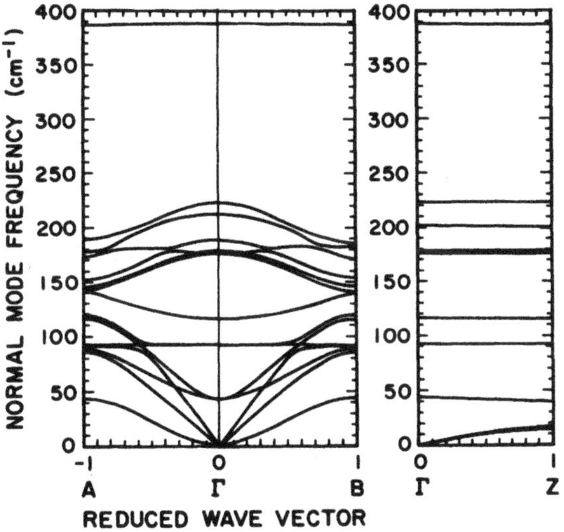

Fig. 4 Phonon dispersion curves for lithium intercalated γ-InSe.

associated with the InSe lattice are little affected by the presence of the lithium, since the mass of a lithium atom is much smaller than those of indium and selenium atoms. The normal modes primarily associated with lithium atom motion at 92 cm^{-1} and 388 cm^{-1} are essentially flat as a result of the weak Li-Se coupling and the lack of coupling between lithium atoms in different unit cells.

5. CONCLUSIONS

A lattice dynamical model has been developed for both pure and lithium-doped γ-InSe. The model contains short-range and Coulomb interactions. Phonon dispersion curves have been calculated for both the intercalated and non-intercalated cases. The results are consistent with observed infrared and Raman data.

ACKNOWLEDGEMENTS -- This work was supported in part by the U.S. Defense Advanced Research Projects Agency (DARPA) Contract No. N00014-88-K-0392. This research was also supported in part by the University of California, Irvine, through an allocation of computer time.

REFERENCES

*Permanent address: Laboratoire de Physique des Solides, Université Pierre et Marie Curie, 4 place Jussieu, 75252 Paris Cédex 05, France.

1. Julien, C. Hatzikraniotis, E. Paraskevopoulos, K. M., Chevy, A. and Balkanski, M., Solid State Ionics 18/19, 859 (1986).
2. Balkanski, M., Julien, C. and Emery, J. Y., J. Power Sources 26, 615 (1989).

3. Slater, J. C., "Quantum Theory of Molecules and Solids" (McGraw-Hill, New York, 1965).

4. Doni, E., Girlanda, R., Grasso, V., Balzaroti, A. and Piacentini, M., Il Nuovo Cimento 51 B(1),154 (1979).

5. Polian, A., Thèse de Doctorat d'Etat, Université Pierre et Marie Curie, Paris, 1981.

6. Burret, P. A., Thèse de Doctorat de l'Université Pierre et Marie Curie, Paris, 1989.

7. Kunc, K. and Zeyher, R., Europhys. Lett. 7, 611 (1988).

8. Pauling, L., "The Nature of the Chemical Bond" (Cornell University Press, Ithaca, 1940).

OPTICAL PHONONS IN BORON NITRIDE UNDER VERY HIGH PRESSURES

O. H. Nielsen[1] and K. Kunc[2]

(1) Laboratory of Applied Physics, Technical University of Denmark, Bldg. 307, DK-2800 Lyngby, Denmark

(2) Laboratoire de Physique des Solides associé au CNRS, Université P. et M. Curie, Tour 13, 4 pl. Jussieu, F-75252 Paris Cedex 05, France

Abstract

A calibration of diamond anvil high pressure cells may be obtained by monitoring, by Raman spectroscopy, the optical Γ-phonon in microcrystals placed in the pressure cell. This paper investigates the relation between the Γ-phonon and the equation-of-state for Boron Nitride. The BN crystal has the advantage that the Γ-phonon frequency is significantly smaller than that of diamond crystals, which constitute most of the high pressure anvils. We report the equation-of-state and Γ-phonon frequency of BN derived from self-consistent total-energy calculations within the local-density approximation. The results are parametrized for convenient perusal in experimental investigations.

1. INTRODUCTION

Development of the diamond anvil cell technique in the course of the last decade made it possible to study the properties of solids under very high pressures going to several hundreds of GPa. At the compressions of this order, the exact determination of the applied pressure is a non-trivial task and the commonly used "ruby scale" is not considered as reliable. In this work we are proposing a new way of calibrating the high pressures, *viz.* by using the data on the TO(Γ) mode in boron nitride calculated *ab initio*. The method of determination described below can very accurately predict the frequency of the Raman line under any compression. The volume is then converted to pressure by using the $p(V)$ equation of state, which can be determined either theoretically - as in the present work - or experimentally.

Using the cubic BN crystals (zincblende structure) as a gauge for calibrating high pressures was suggested by J. M. Besson [1]. There are several good reasons to support the proposal: Boron nitride does not show any phase transition until at least 50 GPa - and probably not until 1100 GPa [2]. On the other hand, both Raman scattering measurements of the TO(Γ) phonon and the Brillouin scattering measurements of the elastic constants c_{11}, c_{12}, c_{44} (as well as of their pressure derivatives) are possible, even on very small crystals. This means that the $p(V)$ equation of state can also be determined experimentally - eliminating thus the need for any extrapolations.

2. METHOD OF CALCULATION.

The reliability and precision of the *ab initio* calculations based on the determination of the total energy of solids using the Density Functional (DF) theory has been sufficiently demonstrated in the course of the last decade [3,4]. The present calculations are carried out within the local-density approximation (LDA) [5] for electron exchange and correlation. The Kohn-Sham equation is solved self-consistently, with a plane-wave basis set and with norm-conserving pseudopotentials, as e.g. in Ref. 6. We use the table of pseudopotentials given by Bachelet *et al.* [7]. Besides the total energy E_{tot}, we evaluate from the self-consistent solutions the atomic forces using the Hellmann-Feynman theorem, and the macroscopic stress (pressure) using the *stress theorem* [8].

The calculation starts by establishing static equilibrium. The calculated $E(V)$ equation of state is shown in Fig. 1, and the static quantities obtained by fitting the calculated points by the Birch equation of state are summarized in Tab. I. The 1 % disagreement between the calculated a_0 (the minimum of $E(V)$) and the experimental one is a well known consequence of using the LDA.

The optical TO(Γ) phonon frequency is determined by a "frozen phonon" calculation [9], applying a displacement $u = \pm\, 0.0002 \times (111)\, a$ to each of the sublattice atoms (a denotes the lattice constant). The magnitude of the anharmonicity of the TO(Γ) phonon was estimated by applying both positive and negative displacements along the (111) direction. We found a 0.5 % difference of the phonon frequency at the two displacements, indicating a negligible anharmonic contribution.

We have tested the convergence of these calculations with respect to the number of **k**-points sampled in the irreducible Brillouin zone. At a plane wave cutoff of 60 Ryd we performed calculations of the TO(Γ) phonon frequency and of the equation-of-state using both 2 and 10 special **k**-points [11]. We found, for the 2-point calculation relative to the

TABLE I : The equilibrium lattice constant a_0 (in Å), bulk modulus B_0 (in Mbar), and its pressure derivative B_0', as derived by fitting the calculated energies with the Birch equation of state (Eq. 2). Numbers in brackets indicate the fitting standard deviations.

	Present work	Experiment	Other calculations[a]
a_0	3.582(3)	3.615[b]	3.606
B_0	3.94(8)	---	3.67
B_0'	3.57(4)	---	---

a - Ref.2; *b* - Ref. 10.

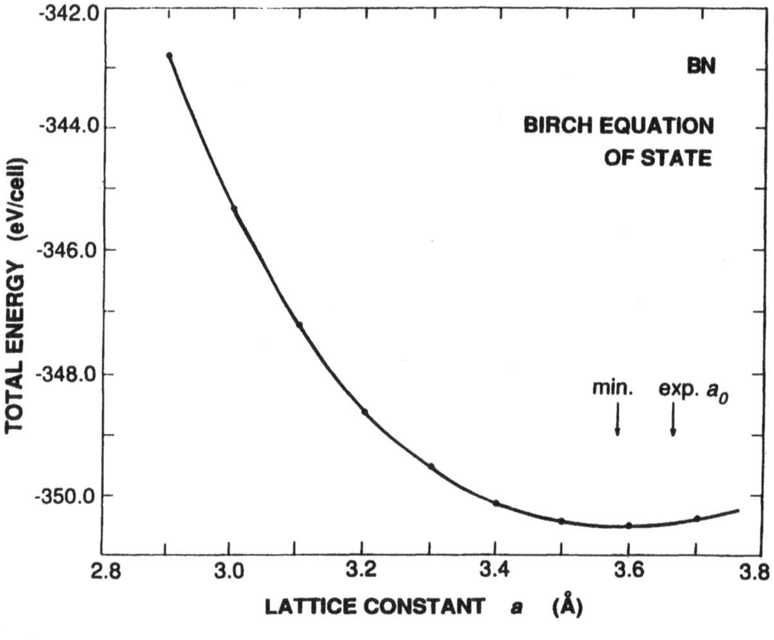

FIG. 1: Total energy of BN (zincblende structure), evaluated at different lattice constants, determines the $E(V)$ equation of state. The calculated points (full circles) are fitted by the Birch equation of state, the equilibrium lattice constant a_0, bulk modulus B_0 and its pressure derivative B_0' obtained in this way are summarized in Tab. I. The 1 % disagreement between the calculated a_0 (the minimum of $E(V)$) and the experimental one is a well known consequence of using the local density approximation. Cutoff $E_{PW} = 90$ Ryd; 2 special points sampling.

10-point calculation, the phonon frequency to be accurate to within 1 %, the lattice constant a_0 to within 0.6 %, and the bulk modulus B_0 to within about 10 %. We conclude that for our purposes a sufficient accuracy is obtained using only 2 special **k**-points.

We also tested the convergence with respect to the cutoff on plane waves included in the basis set. The plane wave cutoff, denoted as E_{PW}, is the maximum kinetic energy of the plane waves to be considered in the calculation. The calculated TO(Γ) phonon frequency converges to its asymptotic value at a cutoff of $E_{PW} = 90$ Ryd (Fig. 2); this corresponds to about 1300 plane waves at the equilibrium volume. This rather large number of basis functions is necessary for two reasons. Firstly, the pseudopotentials given by Bachelet *et al.* [7] have a rather small core radii, and are consequently quite accurate at the expense of an increase in the computational effort. Secondly, the Nitrogen atom with its five valence electrons has an inherently deep pseudopotential.

The equation-of-state (Figs. 1 and 3) is similarly converged at this cutoff value as the following arguments will show: The equation-of-state, i.e. $E_{tot}(a)$ versus a, generally differs from the results obtained directly in the DF formalism for pressure $P(a)$ [8], even at the large cutoff of $E_{PW} = 90$ Ryd, in the sense that $-\partial E_{tot}/\partial V$ deviates from P. This is a well-known problem due to incomplete basis sets [6], and is in principle similar to the problems encountered when calculating atomic forces with atom-centered basis set orbitals. When using plane-wave basis sets the atomic forces do not suffer from this problem, as the basis set remains constant even when the atom is displaced, but another complication appears. When calculating pressure, the plane-wave version of the stress theorem assumes a *constant number* of plane waves as the volume changes [8]. However, it is generally recognized [6] that a correct equation-of-state requires using a *constant kinetic energy cutoff* E_{PW} for the plane waves as the lattice constant a changes. It is fortunately straightforward to calculate a pressure which explicity refers to a constant E_{PW}, in the spirit of Ref. 6.

The pressure P is defined as the total derivative $-dE_{tot}(V)/dV$. With a constant E_{PW} (rather than with a constant number of plane waves N_{PW}) one must account explicitly for the change of the basis set as the volume V is varied. Taking derivatives under the condition of a constant number of plane waves, the corrected pressure becomes

$$P_{corrected} = \frac{dE_{tot}(V, E_{PW})}{dV} = -\frac{\partial E_{tot}}{\partial V} - \frac{\partial E_{tot}}{\partial E_{PW}} \frac{\partial E_{PW}}{\partial V} =$$

$$= P + (2\, E_{PW} / 3V)\, (\partial E_{tot} / \partial E_{PW}) \tag{1}$$

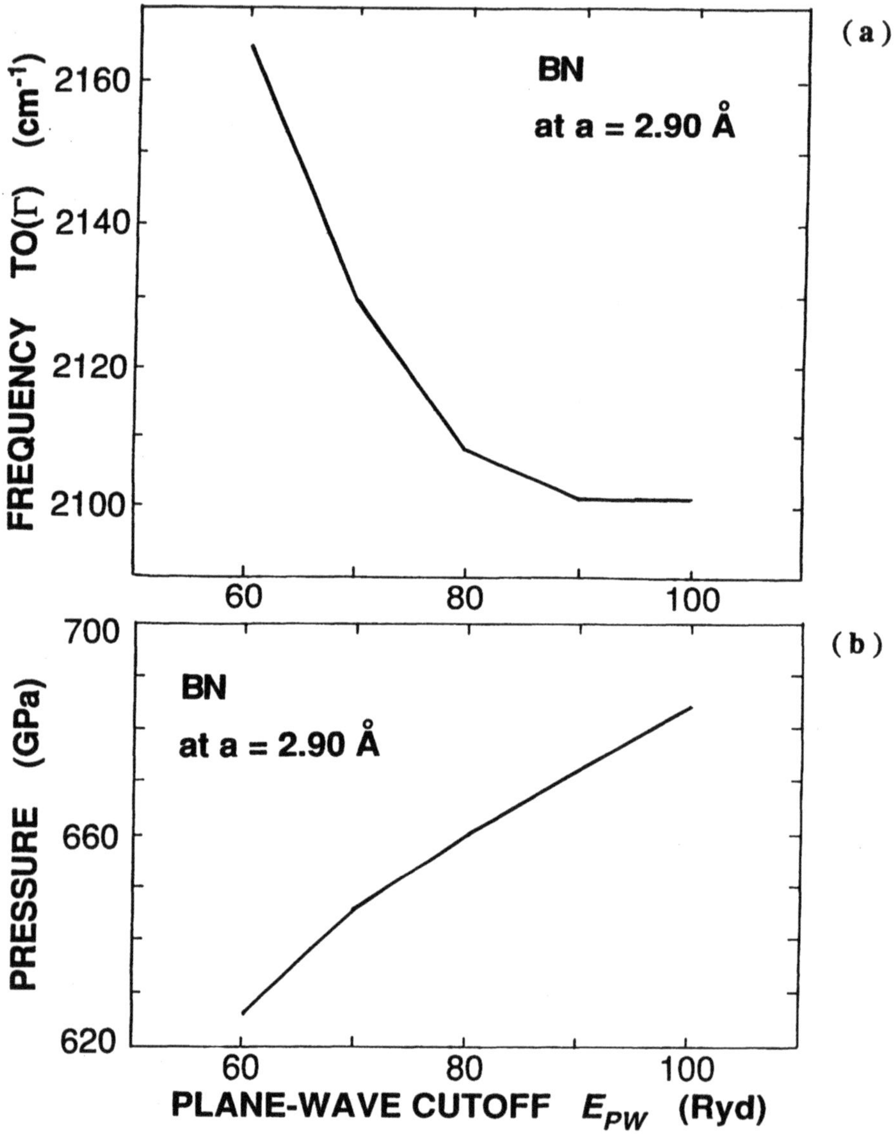

FIG. 2: Convergence test for the plane-wave cutoff E_{PW}: the TO(Γ) frozen phonon frequency (**a**) and pressure P provided by the stress thecrem (**b**) are calculated with different numbers of plane-waves. Convergence of the phonon frequency is achieved at $E_{PW} = 90$ Ryd. The case of P is discussed in the text; in what follows, $P \equiv -dE_{tot}/dV$ will be used for pressure (rather than the values plotted in this figure).

Here, the P is the pressure obtained from the stress theorem [8] in the DF calculations, and we took into account that E_{PW} varies as $V^{-2/3}$ when using a constant number of plane waves.

We evaluated $\partial E_{tot}/\partial E_{PW}$ at $E_{PW} = 90$ Ryd for the lattice constants $a = 2.90$ Å and $a = 3.50$ Å, and interpolated linearly for other lattice constants. The correction term in Eq. 1 is quite large, amounting to 300 - 700 kbar over the range $a = 2.90 - 3.50$ Å. Comparing the result of Eq. 1 with the derivative $-dE_{tot}/dV$ we find almost perfect agreement. The corretion term in Eq. 1 accounts for about 90 % of the difference between $-dE_{tot}/dV$ and P apparent in Fig. 3, which consequently agree to within 30 kbar over the entire range of a's. In addition, we find that $P_{corrected}$ is converged to within 30 kbar of its asymptotic value at $E_{PW} = 90$ Ryd.

In summary, the equation-of-state as well as the TO(Γ) phonon frequency are well converged at a cutoff of $E_{PW}= 90$ Ryd. We expect an overall calculational accuracy of 1 % in a_0 and a few percent in B_0. The phonon frequency should be determined to better than 1 % at each value of the lattice constant.

3. RESULTS

3.1. Equation-of-state

The total energy, pressure and frozen phonon frequencies for Boron Nitride were calculated for a range of lattice constants a between 2.90 - 3.70 Å with the calculational parameters described in the section above. Firstly, the equation-of-state is derived by fitting the $E_{tot}(a)$ values by the Birch equation [13]

$$P = (3/2)\, B_0\, (X^7 - X^5)\, [\, 1 + (3/4)\, (B_0' - 4)\, (X^2 - 1)\,] \tag{2}$$

($X \equiv a_0/a$) integrated to yield the total energy. Here a_0 and B_0 denote the equilibrium lattice constant and the bulk modulus, respectively. The $B_0' \equiv \partial B_0/\partial P$ is the pressure derivative of bulk modulus. The results derived with Eq. 2 are summarized in Table I. We find good agreement with the experimental value for a_0 as well as with the values for a_0 and B_0 calculated previously by Wentzcovitch *et al.* [2].

Alternatively, we employed the equation-of-state form suggested by Vinet *et al.* [14]

$$P = 3B_0\, [\, (1 - a/a_0) \,/\, (\, a/a_0\,)^2\,]\ \ \exp\{\, (3/2)\, (B_0' - 1)\, (1 - a/a_0)\, \} \tag{3}$$

The quality of the fit turned out to be almost as good as with the Birch equation, the a_0 and B_0 were virtually identical, B_0' differed by only 2 %. This is not surprising, since it is possible to show that, to first order in $(1 - a/a_0)$, the Birch and Vinet equations-of-state have identical values for a_0, B_0 and B_0' [15]. The Murnaghan equation [16] was also tested, but did not fit the numerical data as well as the above two forms.

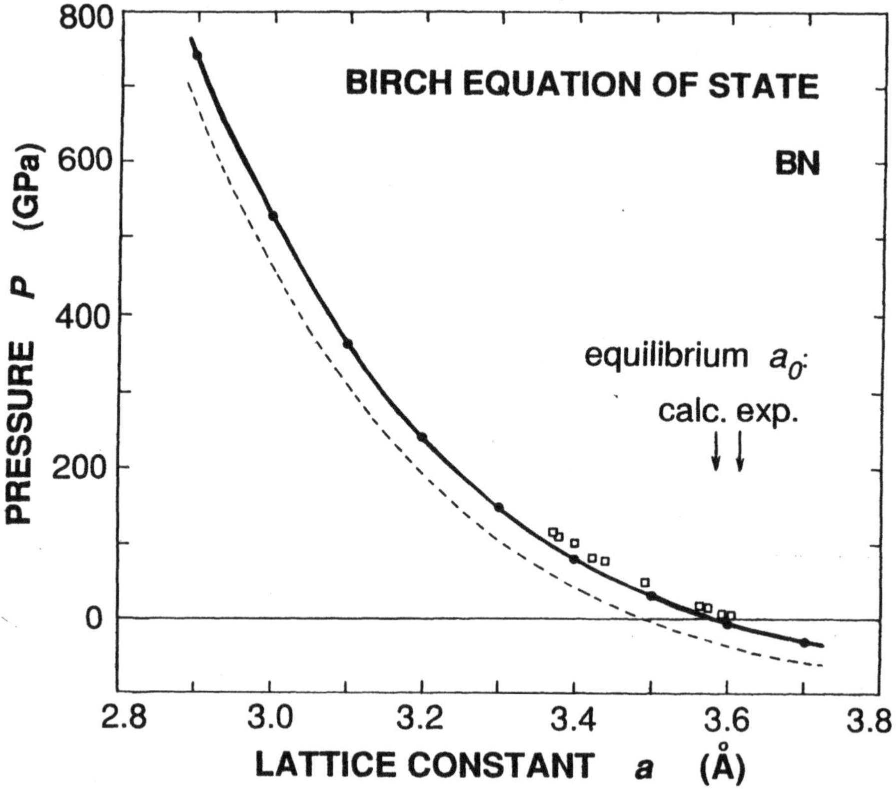

FIG. 3: Pressure calculated, on one hand, directly from the density-functional formalism using the stress theorem (dotted line) and, on the other hand, as a derivative of the *E(V)* curve in Fig. 1 ($-dE_{tot}/dV$, solid line).The difference between the two curves is due to the finite number of plane-waves - see the text. Experimental data (empty circles) from Ref. 12 follow the calculated *P(V)* variation (full circles) - the apparent disagreement is consequence of the 1 % difference between the experimental and calculated lattice constant. Cutoff E_{PW} = 90 Ryd; 2 special points sampling.

TABLE II : The optical $\nu(TO(\Gamma))$ phonon frequency and pressure P derived from Eq.2, as function of lattice constant a.

a (Å)	Frequency $\nu(TO(\Gamma))$ (cm^{-1})	Pressure P (GPa)
2.90	2101	741.8
3.00	1904	526.4
3.10	1728	363.5
3.20	1571	240.9
3.30	1427	149.2
3.40	1293	81.0
3.50	1168	30.7
3.60	1049	- 5.9
3.70	943	-32.3

3.2. Optical phonon frequency

The optical TO(Γ) phonon frequency $\nu(TO(\Gamma))$ was calculated along with the equation-of-state, and we show in Table II and Fig. 4 the frequency as function of lattice constant. At low and moderate pressures our results can be compared with the experimental data from Ref. 17 - the minor disagreement apparent in Fig. 4 is consequence of the -1 % error in the theoretical determination of the lattice constant a_0 ; for any practical purpose, the calculated and measured frequencies are to be compared at a/a_0. The phonon frequency may be parametrized as suggested by Hanfland *et al.* [18] by the Birch-like equation linking the pressure P with the phonon frequency ν,

$$P = (3c/2)\,(X^7 - X^5)\,[\,1 + b\,(X^2 - 1)\,] \qquad (4)$$

where b and c are parameters and $X \equiv \nu/\nu_0$, with ν_0 being the phonon frequency at zero pressure. This relation and other ones were discussed for the case of phonons in diamond crystals by Nielsen [15]. Finally, the combined results for P versus ν are displayed in Fig. 5, and the parameters of Eq. 4 given in Table III. One might wish to apply the Grüneisen relationship

$$\nu = \nu_0\ (\,a/a_0\,)^{-3\gamma} \qquad (5)$$

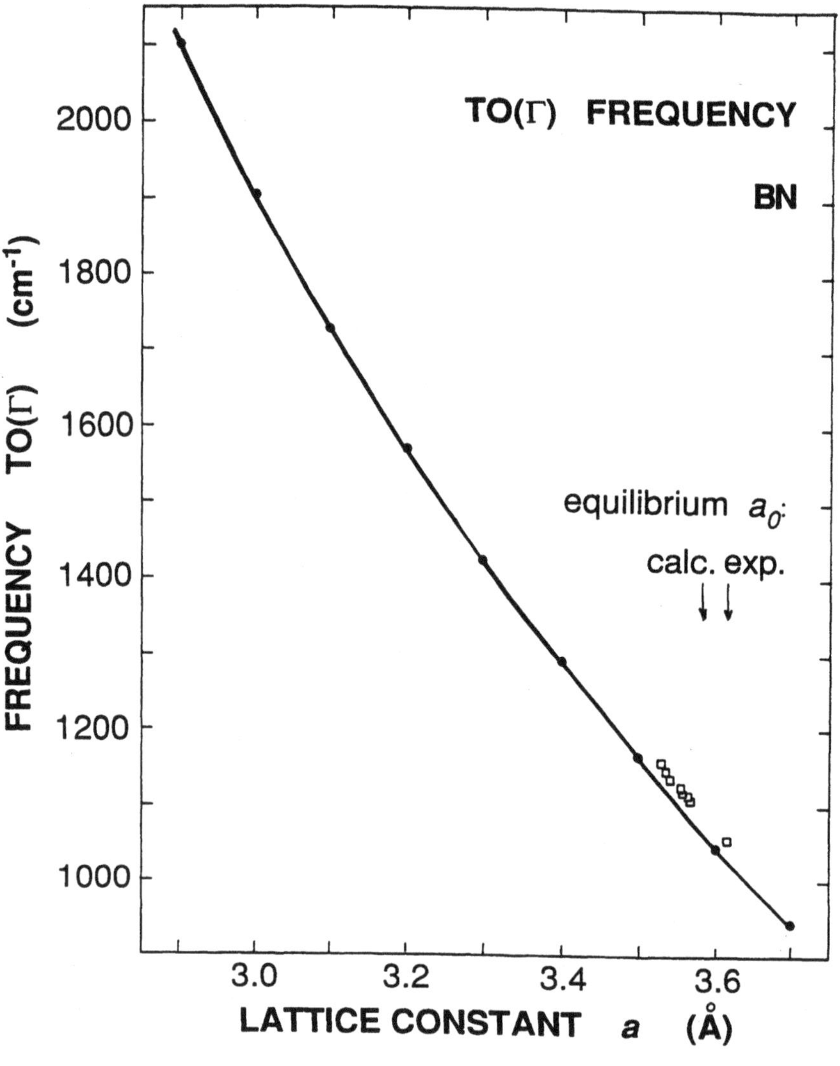

FIG. 4: The TO(Γ) frozen phonon frequency calculated at different compressions of the lattice (full circles), compared to the experimental data (empty squares) from Ref. 17. As in Fig. 3, the discrepancy between the calculation and the experiment disappears if one accounts for the -1 % error in the theoretically determined lattice constant. Cutoff E_{PW} = 90 Ryd; 5 special points sampling.

TABLE III : Parameters appearing in Eq. 4. Numbers in brackets indicate the fitting standard deviations.

ν_0 (cm^{-1})	b (Mbar)	$3c$ (Mbar)
1075(2)	0.117(12)	3.078(37)

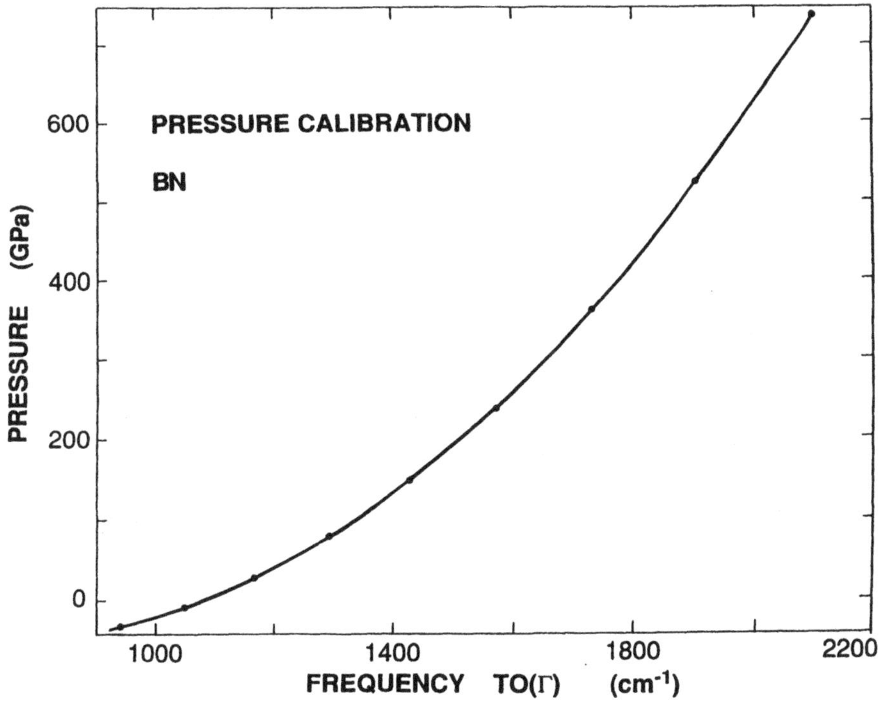

FIG. 5: Pressure P as derived from Eq. 2, plotted as a function of phonon frequency $\nu(TO(\Gamma))$. The curve displays the functional form of Eq. 4, and the points are the actually calculated values.

where γ is the Grüneisen parameter. This yields the value of $\gamma = 1.17$, which compares favourably with the experimental $\gamma = 1.188 \pm 0.002$ obtained in Ref. 19. However, we find the relation (5) valid only for a very close to a_0 and we recommend, for general values of a, to use the Eq. 4. For small pressures, the linear term in eq. (4) is equivalent to

$$\nu = \nu_0 + (\nu_0 / 3c) P \tag{6}$$

where by Table III, $\nu_0/3c$ = 3.49 cm^{-1}/GPa, in good agreement with the experiment by Sanjurjo et al. [20] who found the value 3.39 cm^{-1}/GPa.

4. CONCLUSION

The calculated $P(\nu)$ relationship (Fig. 5) can be used for measuring very high hydrostatic pressures. The available experimental data shown in Figs. 3 and 4 confirm the validity of the present calculations at low and moderate pressures and, in turn, they support the validity of the *ab initio* predictions at the extreme compressions. The present calculations assumed static lattice, i.e. the influence of the zero-point vibrations has been neglected.

ACKNOWLEDGMENTS

We want to thank J. M. Besson for the impetus to this work and for numerous discussions during its progress. Several references to experimental data were pointed out to us by A. Polian. The CRAY-2 computational facilities were provided by the Scientific Committee of the CCVR (Centre de Calcul Vectoriel pour la Recherche), Palaiseau, France. This work was supported in part by a NATO grant, and in part by a grant from the Danish Natural Science Research Council.

REFERENCES

[1] J. M. Besson, private communication, 1986.

[2] R.M. Wentzcovitch, M.L. Cohen, and P.K. Lam, Phys. Rev. B **36**, 6058 (1987).

[3] For a review see e.g. K. Kunc in *Electronic Structure, Dynamics and Quantum Structural Properties of Condensed Matter* (ed. by J. T. Devreese, P. E. Van

Camp), Plenum Press, New York, 1985, pp. 227 - 312.

[4] See e.g. K. Kunc, J. Chimie Physique (Paris) **86**, 647 (1989).

[5] W. Kohn and L.J. Sham, Phys. Rev. **140**, A1133 (1965) ; L.J. Sham and W. Kohn, *ibid.* **145**, B561 (1966).

[6] P. Gomes Dacosta, O.H. Nielsen and K. Kunc, J. Phys. C **19**, 3163 (1986).

[7] G.B. Bachelet, D.R. Hamann, and M.L. Schlüter, Phys. Rev. B **26**, 4199 (1982).

[8] O.H. Nielsen and R.M. Martin, Phys. Rev. Lett. **50**, 697 (1983) ; Phys. Rev. B**32**, 3780 (1985).

[9] K. Kunc and R.M. Martin, Phys. Rev. B **24**, 2311 (1981).

[10] *Landolt-Börnstein: Numerical Data and Functional Relationships in Science and Technology - New Series;* Group III: Crystal and Solid State Physics, Vol. 17: Semiconductors, Subvolume a (ed. by O. Madelung), Springer, Berlin, 1982.

[11] H.J. Monkhorst and J.D. Pack, Phys. Rev. B **13**, 5188 (1976).

[12] E. Knittle, R. M. Wentzcovitch, R. Jeanloz, and M. L. Cohen, Nature **337**, 349 (1989).

[13] F. Birch, J. Geophys. Res. **57**, 227 (1952).

[14] P. Vinet, J. Ferrante, J.R. Smith, and J.H. Rose, J. Phys. C **19**, L467 (1986).

[15] O.H. Nielsen, Phys. Rev. B **34**, 5808 (1986).

[16] F.D. Murnaghan, Proc. Natl. Acad. Sci. USA **50**, 697 (1944).

[17] I. V. Aleksandrov, A. F. Goncharov, S. P. Stishov, and E. V. Yakovenko, JETP Letters **50**, N° 3, 127 (1989) [original: Pis'ma Zh. Eksp. Teor. Fiz. **50**, N° 3, 116 (1989)].

[18] M. Hanfland, K. Syassen, S. Fahy, S.G. Louie, and M.L. Cohen, Phys. Rev. B **31**, 6896 (1985).

[19] I. V. Aleksandrov, A. F. Goncharov, E. V. Jakovenko, and S. P. Stishov, Proceedings of the AIRPAT Conference, Paderborn, July 1989; in print.

[20] J. A. Sanjurjo, E. Lopez-Cruz, Phys. Rev. B **28**, 4579 (1983).

EXCITONS AND OTHER OPTICAL PROBLEMS in SEMICONDUCTOR SYSTEMS OF REDUCED DIMENSIONALITY

by
John M. Worlock
Bellcore
Red Bank, NJ, USA

FOR ROGÉRIO CÉSAR DE CERQUEIRA LEITE

Since our honoree is so demonstrably a man of many dimensions, it might seem appropriate to celebrate his birthday with a short note in which dimensionality is highlighted. Indeed, within the scientific lifetime of Rogerio and his contemporaries, dimension has become one of the most important variables in physical thought. In cosmology, it can now be speculated that a multidimensional universe has somehow collapsed and knotted tightly all the dimensions but the three familiar cartesian ones. More mundanely, in condensed matter, critical phenomena at phase boundaries are sorted into universality classes which depend on the number of dimensions in which the interactions operate. And in our own realm of semiconductor physics we have achieved not only a wide variety of extremely useful devices based on quasi-two-dimensional electrons, but also discovered such unexpected new phenomena as the integral and fractional quantum Hall effects.

In this note, I deal with some recent achievements in producing, using, and studying the optical properties of semiconductor structures of low dimensionality. Optical properties of semiconductors are determined by transitions between valence and conduction bands. These transitions are governed by matrix elements and joint densities of states (occupied for emission, unoccupied for absorption). The principle role of dimensionality is in modifying the densities of states - the matrix elements being properties of single unit cells and not affected by their geometrical arrangement. Thus one of the additional requirements is that the structure must provide in the same region both valence and conduction band confinement, i.e. the joint density of states has to be spatially joint.

Figure 1 contrasts the band-structure diagram for a simple Metal-Oxide-Semiconductor Field-Effect-Transistor (MOSFET) with that of a quantum-well (QW) heterostructure. Each figure depicts the bands' spatial variation along the perpendicular to a layer in which the bands are flat and in which there are thus two dimensions of complete freedom. Hence we call it two-dimensional, or 2D, behavior, even though the third dimension has not shrunk to zero, and needs to be kept in mind. In the MOSFET, it is possible to achieve confinement of 2D electrons in the triangular well near the oxide-semiconductor interface, or alternatively confinement of 2D holes. The quantum well, having discontinuities in both valence and conductions bands, can confine *simultaneously* electrons and holes. Its optical properties depend on this joint confinement.

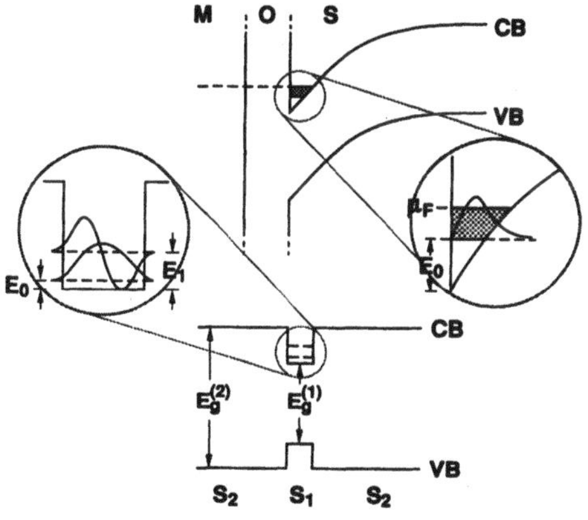

Figure 1. MOSFET and QW band structures, showing confinement of electrons only in the MOSFET, and some quantum-confined states of electrons in the QW. E_i indicates a confinement energy, the bottom of a 2D subband.

Lives are spent and books are written perfecting and explaining the techniques of epitaxial growth by which these deceptively simple structures are produced. The techniques are relatively new, and still developing. Here, until we want to go below two dimensions, we simply assume their existence. Likewise we will not dwell on the important contributions of microscopic techniques such as cross-sectional transmission electron microscopy, in finding and measuring suspected structures, and in diagnosing problems of growth and processing.

Figure 2 shows how the optical absorption spectrum of 3D GaAs near the fundamental band edge (Fig 2a) compares with that of one or a sequence of identical thin 2D GaAs layers grown within a matrix of AlGaAs (Fig 2b). The GaAs layers would be like the quantum well in Figure 1. To understand these spectra, we need the concept of the Wannier exciton, the semiconductor physicist's hydrogen atom. It is formed as a bound state of an electron with its counterpart in the valence band, the hole, at an energy below the single-particle gap energy E_g by its binding energy, E_b. In the spectra shown, the peaks correspond to exciton states, the excitons being created directly from annihilated photons. Similarly, the inverse process gives us luminescence spectroscopy, which has proved to be an indispensible tool for the characterization of newly grown quantum-well layers.

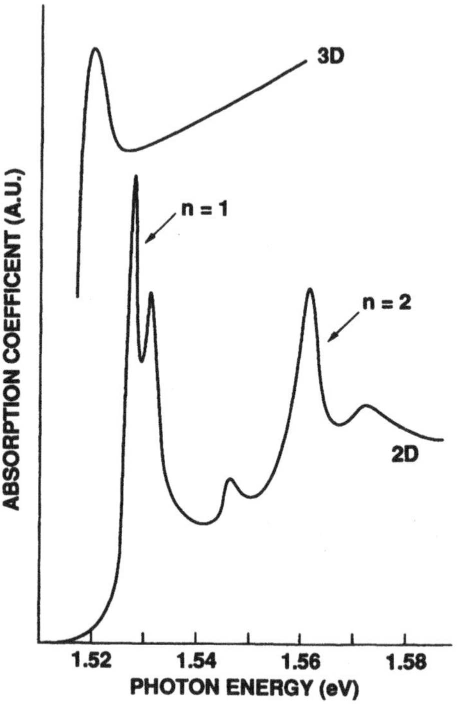

Figure 2. Low-temperature optical absorption spectra of a) 3D GaAs and b) a sequence of 19 nm quantum wells of GaAs embedded in AlGaAs.

Each peak in the spectra in Figure 2 corresponds to an exciton, and each exciton belongs to a unique conduction-valence band pair or band-gap. In 3D there is a single gap and so a single exciton peak. In the 2D spectra the multiplicity of excitons betrays a multiplicity of pairs of what are called subbands, each being a two dimensional continuum with origin simply determined by the energy of quantum confinement in the well depicted in Figure 1b. The 2D excitons come in pairs, identified by the quantum number n which is the number of (approximate) half-wave lengths required to span the quantum well. Of each pair, the lower in energy is an exciton bound to a heavy hole, the higher involving a light hole. Thus the quantum confinement removes the heavy-light hole degeneracy at the Γ-point.

What's so special about these 2D excitons? Let's make a list.

1. Because they are two-dimensional, their binding energies are enhanced. We can suppose that this is a consequence of the 2D density of the

subband states from which their hydrogenic wave-functions are constructed. The 2D density of states, above its origin in energy, is flat and constant, rather than rising from zero as in 3D, and as a result has more states at low kinetic energy available for constructing a bound state. This enhancement of binding energy never quite reaches the true 2D limit (a factor 4) because a) the quantum well never can be less than one cell thick, and b) if it did, the exciton would be in fact mostly outside the well, and would look like a 3D exciton weakly bound to a plane.

2. Because their binding energies are enhanced, 2D excitons are more robust, more resistant to thermal ionization. As a result, some of their spectral properties survive at room temperature, which is a strict requirement for a multitude of optoelectronic applications.

3. Accompanying the enhancement of binding energy is a reduction in the Bohr radius and consequently greater electron-hole overlap. This results in enhanced oscillator strength for the absorptive and radiative transitions.

4. Quantum confined Stark effect. Since the Coulomb binding of hole to electron operates only within the 2D plane, an electric field perpendicular to that plane will not destabilize the exciton. Consequently, it is possible to tune the resonance by substantial Stark shift. This effect is then potentially useful in electro-optic modulators and switches.

5. A consequence of the quantum-well confinement is a shift in the resonance, as depicted in Figure 2. Thus, within limits the resonant energy for the exciton can be freely chosen, since it is determined primarily by the thickness of the quantum well. Then it can be smoothly shifted as above.

6. The fact that the exciton wave-function can be confined in a specific layer allows it to be used as a *chromophore* in resonant Raman scattering, to probe the vibrational properties of the local region.

7. One of the properties of 2D excitons that I think we do not understand is their strong radiative efficiency. Perhaps the fact that they are confined in quantum wells means that they can not diffuse to the external surfaces, where large concentrations of non-radiative recombination centers lie in wait. But there may also be more to the question, as I believe that the process of radiative recombination of free excitons in 3D is itself a bit obscure.

8. Inhomogeneous broadening, localization, and all that. The exciton linewidth has come to be a standard measure for quantum well excellence, since

fluctuations in well thickness translate directly into inhomogeneous linewidth: narrow line, good sample. Localization is also an interesting application, since the neutral exciton's behavior is not complicated by the competition between interaction and localization experienced by charged wave-particles, and since localization is supposed to be a dominant phenomenon in two dimensions. On the other hand, the finite lifetime of the exciton makes its localization properties perhaps difficult to determine precisely. In addition, there appears to be an interplay between localization and radiative lifetime: more localized, longer lifetime. Stay tuned for further results.

The preceding paragraphs outline the salient properties of excitons in quantum wells. The *laser* community has also found interest in quantum wells, for reasons that are related. Many high-performance semiconductor lasers these days have quantum wells in the active region to assist with the attempt to achieve high electron-hole densities at low currents. The 2D densities of states appear to help in establishing low thresholds. Additionally, the emission energy can be chosen, within limits, by designing the quantum well thickness, without changing other design parameters or material specifications.

If you believe all that, you are doubtless ready to suppose that we ought to be pushing on to develop and study 1D and zero-D structures. Right. So we are. And by so doing we expect to discover the unexpected and to find, as in 2D, new phenomena and new uses.

But wait, you might say, we have been studying zero-D excitons for years, in the form of excitons bound to point defects. An example might be the isoelectronic trap, N, in GaP. It's a valid point, and we'll have to work hard indeed to do as well in a structure grown or produced by design. Nevertheless we will try.

One interesting approach that I mention only in passing is the precipitation of colloidal semiconductor particles from solution, either in liquid suspension or in glass matrices. Many interesting phenomena related to exciton confinement and nonlinear properties have already been investigated, although it is difficult as yet to produce a truly satisfactory uniformity of particle sizes, and hence the sharpness of resonance that might be desired.

There remain two basically different approaches to the problem. One is to begin with a high quality 2D layer and carve it up or otherwise modify it so as to provide lateral confinement for the excitons in one or both of the originally free dimensions. The second is to make use of the peculiarities of epitaxial growth on patterned or structured substrates.

In a number of large laboratories around the world, carving or etching on a very fine scale has been developed to the point where quantum wires and quantum dots have been produced with demonstrable optical properties. In this technique, optical or electron beams are used to write patterns in "resist" polymers spread on surfaces above good buried 2D quantum wells. Subsequent development and etching leaves typically a macroscopic array of microscopic free-standing ridges or posts which contain ribbon- or disk-shaped regions of the original quantum wells. The lateral confinement is provided by free space. A major problem often encountered with these structures is nonradiative recombination: as the structures become smaller, the surface-to-volume ratio becomes larger, and the radiative efficiency is degraded. Additionally, we suspect that the etching processes themselves must produce some internal damage, further degrading optical properties. Optical spectroscopy on such structures has succeeded in demonstrating the additional blue shifts that are expected from the quantum confinement in the lateral dimensions, as well as some polarization anisotropy and enhanced magnetic shifts. The confinement energy involved is on the order of meV for GaAs when the lateral confinement is 100nm or smaller. Resolution of these shifts requires not only small structures but also fine linewidths in the starting material.

Another promising technique which has not been so widely practiced is the use of inhomogeneous stress to provide lateral confinement, without leaving any exposed edges or surfaces. The deformation potential of a typical semiconductor being on the order of 10 eV, it is evident that band-gap shifts of $\sim$ 100 meV can be achieved with rather modest strains. Excitons should be almost as easily confined in such strain-generated wells as in those provided by heterointerfaces. How, then, do we provide the strain? A variety of different thin-film overlayers are candidates, getting stress from thermal expansion, lattice-mismatched epitaxy, or simply from the compressive stresses involved in deposition, as in diamond-like carbon films grown by plasma discharge. Such a thin film, uniformly deposited, will transfer strain to the substrate macroscopically, by bending it. But if it can be finely patterned, i.e. removed by lithographic processes analogous to those described above, then it can transfer strain to the underlying semiconductor in interesting inhomogeous patterns. Figure 3 shows a cross section of contours of dilational strain in the region beneath such a stressor. Strain penetrates well into the substrate, to a distance comparable to the dimensions of the stressor. Experiments making use of the natural compression in carbon films, patterned by electron beam lithography, have successfully shown that quantum wires and dots can be achieved with this technique.

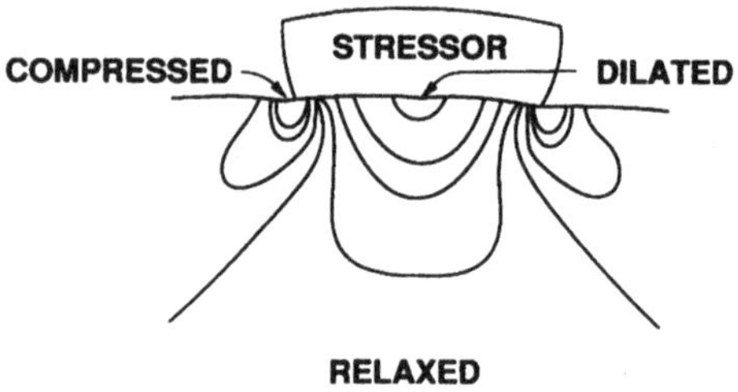

Figure 3. Calculated strain contours (dilatation) in cross-section beneath the surface of a semiconductor wafer with a compressed stressor attached. The band gap shift is proportional to the strain, and can confine excitons beneath the stressor.

As the different processes of epitaxy become more deeply understood, properties such as surface mobility of atomic species better defined, and experience with growth on surfaces of differing index more widespread, we have been able to grow some varieties of lateral microstructures directly, without the need for post-growth patterning. Generally, these techniques make use of nonplanar and high index surfaces and the properties of atomic steps on terraced and intentionally patterned surfaces.

So let's start with atomic steps. A perfectly planar <100> surface has no steps, but if it is tilted a few degrees toward <011> it will have a series of atomic steps which *could be* both straight and periodic. If periodic, it could be used as a template for growing half-monolayers in a kind of vertical superlattice, by filling in half a terrace at a time. Another idea takes advantage of different free-energies for growth of the different layer materials to sweep the steps into arrays, which are then somewhat more periodic than the original steps. Quantum wires can then be grown on the step-arrays. A third idea uses periodic grooves on such a misoriented substrate to cluster the steps at the grooves, thus ensuring better periodicity.

Another trick of epitaxy takes advantage of the different ways that GaAs and AlGaAs grow on the walls of macroscopic grooves in a < 100 > surface. The AlGaAs favors the high-index sidewalls, and so sharpens the apex, while the GaAs tends to fill in the bottom of the groove, leading to rather thicker regions in the bottom of the valley. Such crescent-shaped objects have been successfully incorporated into laser structures, thus verifying the predicted superior low-threshold properties of quantum-wire lasers. As in the quantum-well lasers, the superiority comes from the ability of the quantum wire to compress the charge carriers to a higher density at low current input, i.e. it is a direct consequence of the dimensionality.

Coming completely off of the vicinal-100 surface, it has been shown to be possible to grow high quality overlayers on the < 110 > cleavage planes of GaAs using molecular beam epitaxy. This feat opens the possibility of growing at the edge of a previously-grown superlattice or heterostructure. Generally, attempts to regrow or to cover exposed surfaces by epitaxy have failed because of contamination, but in the case of cleavage and regrowth, the newly exposed surface need never have been exposed to contamination.

Optical studies of this variety of structures of low dimensionality have until now been largely diagnostic and confirmatory. They have demonstrated the achievements as well as the shortcomings of the structure-makers. In general, it has been difficult to make arrays of structures with sufficient uniformity to be taken seriously in device design. Lateral confinement has not yet been achieved with the same atomic precision that is available in high quality planar epitaxial growth. But in time, with diligence and intelligence, we can be confident of having dense arrays of optically active structures. By then, the world may be clamoring for more information storage and computing capability and we shall all be heroes.

ACKNOWLEDGMENTS - While the opinions and judgments in this note are entirely mine, very few of the ideas and accomplishments are. I therefore give grateful thanks to all whose work is recapitulated here, and beg forgiveness for mentioning no names and giving no references. The avid reader who wants more information is assured of finding plenty in the published literature.

Evolution of Epitaxial III-V Semiconductor Materials

R.E.Nahory
Bellcore
Red Bank, NJ 07701, USA

In the years around 1970 I had the opportunity to work in several scientific collaborations with Dr. Rogerio Leite. It was during that time that high quality epitaxial GaAs was first becoming available, with impurity concentrations well below 10^{16} cm^{-3}. We were studying the optical properties of these new layers and seeing sharp line photoluminescence spectra, then a new event for GaAs. Previous to that time, GaAs generally available had room temperature carrier concentrations greater than 1×10^{16} cm^{-3} and narrow line PL spectra could not be resolved. An interesting comparison which I remember making at the time is depicted in Figure 1a and 1b. Fig. 1a shows the well known free exciton absorption spectrum of M.D.Sturge[1] from 1962. This measurement was carried out using bulk GaAs which had been mechanically thinned sufficiently to allow transmission of light in the bandgap vicinity at absorption coefficient values in the 10^4 cm^{-1} range. Note the width of the exciton absorption peak. In Fig. 1b is a photoluminescence (PL) spectrum as measured in 1969 using the newly available epitaxial GaAs.[2] This sample, a layer with a thickness of 69 μm, was grown using vapor phase epitaxy techniques by Ken Lawley. It had a background carrier density just below 1×10^{15} cm^{-3}, the best GaAs material I had seen up to that time. The low-temperature mobility reached a value as high as 50,000. The PL spectrum shows several resolved lines, although they are still broadened compared with what was to come shortly afterwards. The spectral features are related to bound exciton recombination associated with different impurities as well as to free exciton recombination at the high energy end of the range. The entire spectrum of lines has a width in energy which fits inside the exciton absorption peak in Fig. 1a. The early absorption spectrum was considerably broadened, perhaps telling of the effects of the built in impurities and internal strains characteristic of the bulk material which was state-of-the-art in 1962. Furthermore, Sturge[11] was careful to discuss the possible strain and damage effects involved in the preparation and mounting of the thin samples needed for transmission measurements. The epitaxial layer measured in Fig. 1b was one of the beginnings of a new era of semiconductor materials, which has by now changed the face of opto-electronic device technology.

With the observation of these new PL lines came plenty of discussion and thought about their origin. The idea of bound excitons was well worked out already, based on previous studies of wide bandgap materials, principally GaP among the III-V compounds along with various wide bandgap II-VI compound semiconductors.[31] I remember fondly many technical discussions on the identities of these

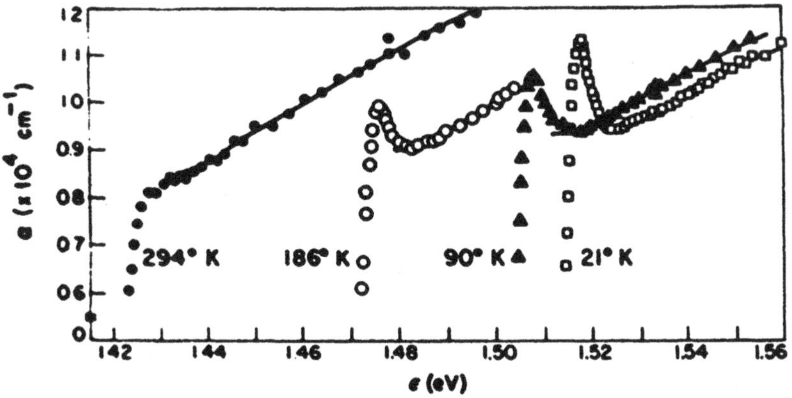

Fig. 1a. Exciton absorption spectrum for bulk GaAs in 1962. (Sturge)

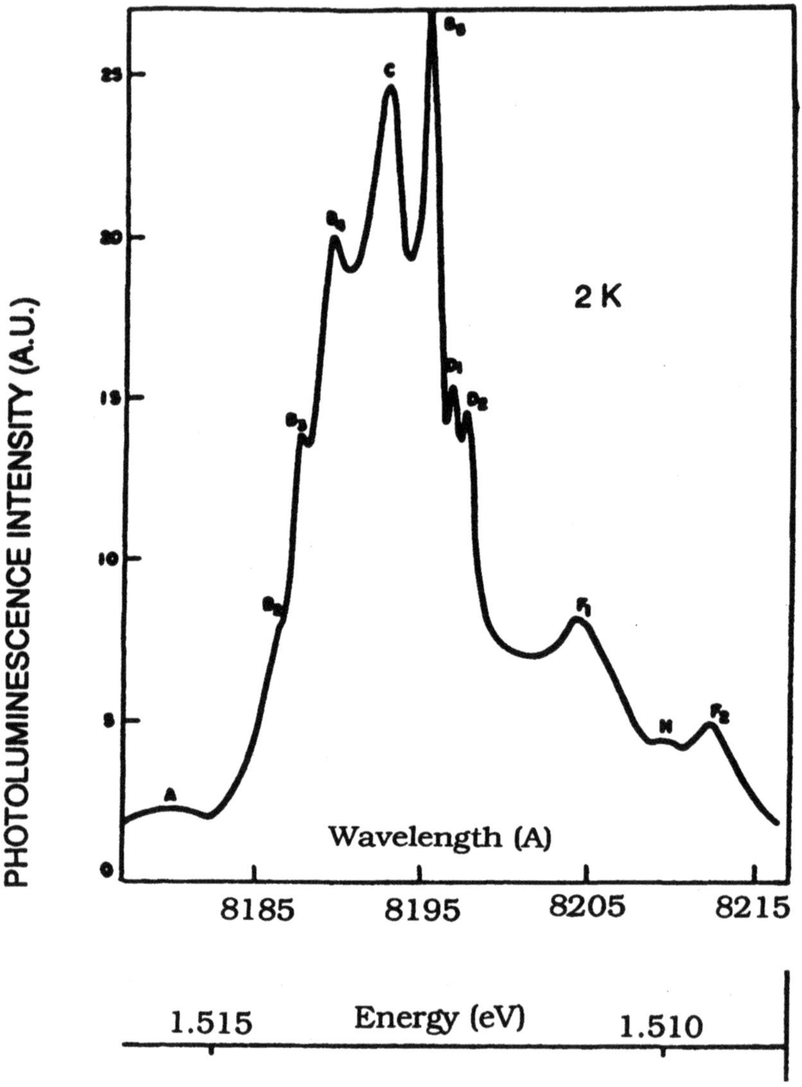

Figure 1b. Low temperature photoluminescence spectrum of a high purity epitaxial GaAs layer (Shah, Leite and Nahory, 1969).

new GaAs bound excitons, typically among Rogerio Leite and Jagdeep Shah and myself. In particular I fondly remember Rogerio at the black board describing his latest "proof" of a certain bound exciton model. He had a favorite trick, drawing a schematic of an impurity with bound electron and hole, and then arguing the energetics by removing either electron or hole from the impurity off to an infinite distance away. This simple artifice allowed all kinds of estimates of energies and enabled discussion of particular assignments of types of bound excitons. These discussions were both enlightening and intellectually pleasing.

These early epitaxial GaAs layers were only the beginning of a long line of advances still taking place today in the materials science of compound semiconductors. These advances have been made possible through the development of epitaxial growth techniques, including both liquid phase and vapor phase sources. Progress in the last two decades has been astounding, with device structures today not dreamed possible in 1970. Thinking fundamentally, at the root of these advances are several issues of importance. Among them are impurity control, alloy materials and the concept of heterostructures. In the remainder of this article I will discuss some of the advances which have followed over the past two decades from the beginnings of high quality epitaxial layers.

Very rapid advances were made in obtaining high purity GaAs epitaxial materials by various growth techniques in the early 1970's, by both liquid phase and vapor phase epitaxy. Low temperature electron mobilities in excess of 100,000 $cm^2/V/s$ came quickly, followed almost immediately by mobilities greater than 200,000 for GaAs epi-layers of the order of 10 μm in thickness.[4] Recently the record value of just over 300,000 was achieved[5] by low pressure metal-organic chemical vapor deposition techniques. These latter layers have carrier concentrations in the 10^{13} cm^{-3} range. They have sharp line photoluminescence spectra revealing bound exciton transitions as well as several features of the free exciton transitions as shown in Fig. 2. The three highest energy peaks in the spectrum were attributed to the lower and upper polariton branches as well as to an excited state of the free exciton. Again note the energy range of all the observed near-bandedge features, which is less than 8 meV. Such impurity control has improved by about 3 orders of magnitude in carrier and impurity densities since the days of bulk GaAs in the 60's, an outstanding achievement. Although these are special cases, routine purity control has become sufficiently good that intentional doping can now control carrier concentrations over the required ranges for virtually all practical device applications. This situation gives stability to the materials supply for devices, and in my experience it provides for better uniformity from wafer to wafer as well as across a given wafer.

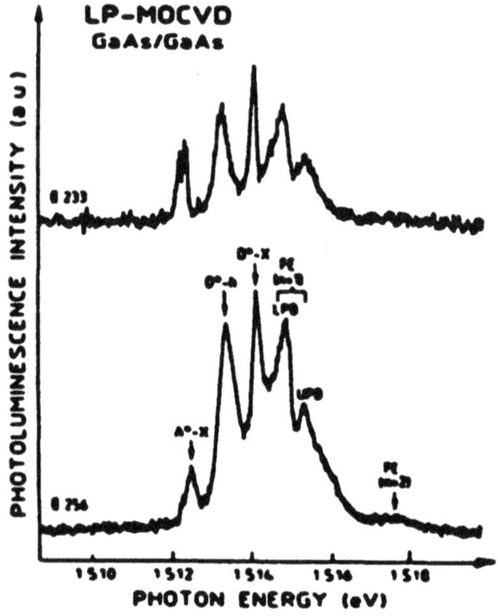

Figure 2. Photoluminescence spectra at 4.2 K of two ultrapure GaAs epitaxial layers grown by LP-MOCVD techniques in 1989. (Rezeghi et al)

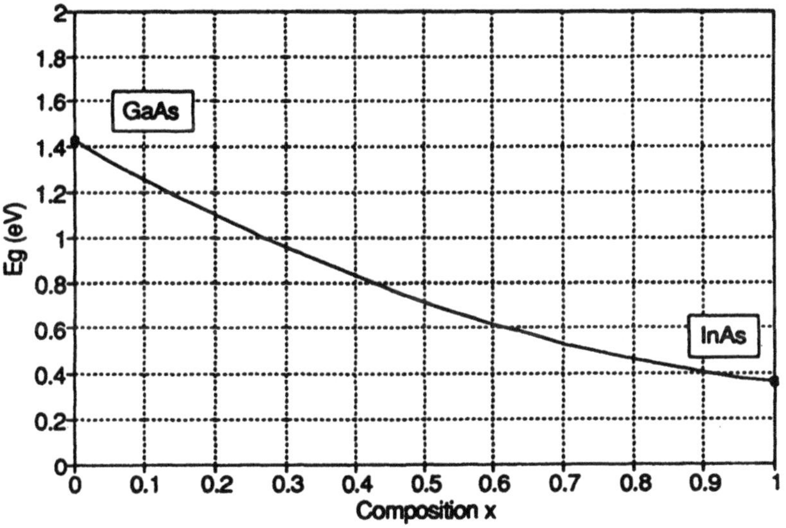

Figure 3a. Bandgap vs composition x for the $In_xGa_{1-x}As$ alloy, 300K.

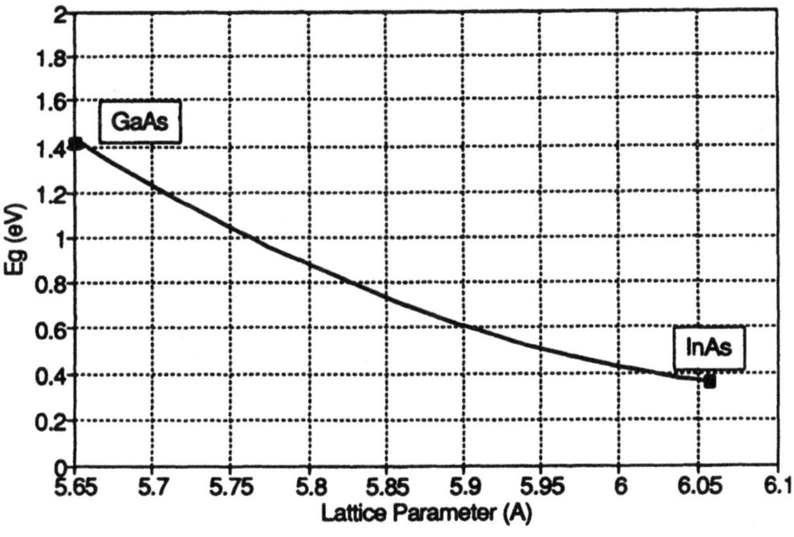

Figure 3b. Bandgap vs lattice parameter for the $In_xGa_{1-x}As$ alloy, 300K.

Another improvement in materials over the last two decades has been in the type of materials available, with the prominent impact[6] coming from ternary and quaternary alloys of various III-V compounds. One of the early alloys investigated was a mixture of GaAs and InAs to form the ternary $In_xGa_{1-x}As$. As shown in Fig. 3a, which shows bandgap vs composition x for $In_xGa_{1-x}As$, one can then engineer the bandgap of the resulting material between the binary extremes by choosing the appropriate composition. For this example the bandgap can be varied continuously from deep in the infrared at 0.36 eV (InAs) up to the near infrared at 1.42 eV (GaAs). This concept of bandstructure engineering is one of the most important developments over the past two decades for modern opto-electronic devices. It was established early that growth conditions allow very uniform compositions over large areas.

It was learned immediately however that the concomitant variation in lattice parameter with composition was a matter of concern. The range of variation in lattice parameter for this system is represented in Fig. 3b, where the bandgap energy is plotted vs lattice parameter in Angstroms. Actually it has been shown that this measure of lattice parameter is only an average of the microscopic behavior, where in reality the Ga-As separation is quite different from the In-As atomic separation.[7] For practical purposes, the values in the Fig. 3b are useful for determining the required lattice match, from layer to layer or between layers and substrate, for practical device structures. For reliable devices, especially minority carrier devices such as lasers, it was found that epitaxial layers must be lattice matched to each other and to their substrates in order to obtain reliable, long-lived devices. Tolerable degrees of mismatch are $\delta a/a < 10^{-3}$.

Alloying was extended to a variety of III-V materials during the 70's, but the compound found to be most useful for optoelectronics was

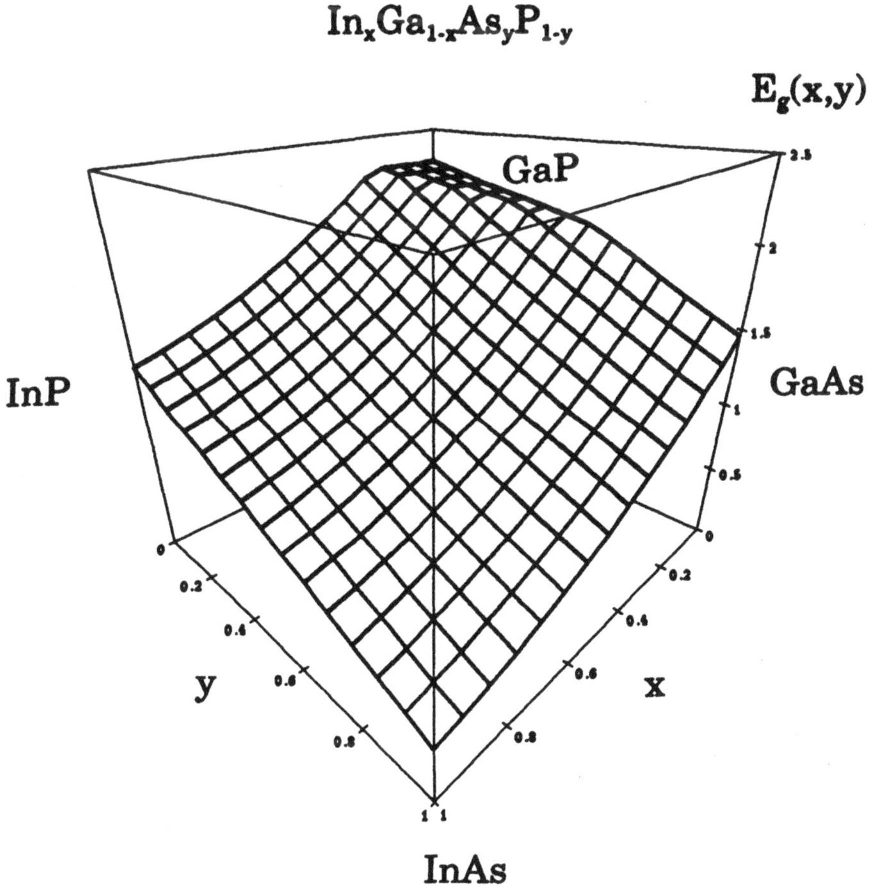

Figure 4. Bandgap surface Eg(x,y) for $In_xGa_{1-x}As_yP_{1-y}$ vs x and y. The four corners of the surface correspond to the four constituent binary compounds. The bandgap is direct except in the flattened region near the GaP corner.

the quaternary $In_xGa_{1-x}As_yP_{1-y}$ lattice matched to InP substrates. The bandgap of this system, $In_xGa_{1-x}As_yP_{1-y}$, which is made up of the four constituent binaries GaP, GaAs, InAs and InP, is illustrated in Fig. 4, which shows the bandgap surface plotted vs both x and y in a 3-dimensional representation.[8] The four corners of the surface represent the binaries, while the lines at the four sides represent the constituent ternaries. The flattening of the surface at higher energies occurs because of the change from Gamma conduction band to satellite conduction band valleys lying lowest, so that the bandgap is indirect at the highest energies. A more convenient plot for the ternary is shown in Fig. 5, which gives bandgap energy vs lattice parameter. On this graph, one obtains a fixed lattice parameter along any vertical line. Thus it is easy to see that InP occupies a unique position for use as a substrate material for a large range of lattice matched quaternaries extending from InP on one end to $In_{.47}Ga_{.53}As$ at the other extreme. It is this range of alloys that provides the source of materials for practical opto-electronic devices for use in modern lightwave tele-communications systems. Lasers are constructed using $In_xGa_{1-x}As_yP_{1-y}$ at compositions vertically below InP (Fig. 5) which give emission at wavelengths of either 1.3 μm (E_g = 0.95 eV) or 1.55 μm (E_g = 0.8 eV). Photo-detectors are fabricated from $In_{.47}Ga_{.53}As$ at the lower end point of this vertical line, for which all energies above 0.75 eV are detectable. Another way of plotting data from this same vertical line is given in Fig. 6, which shows the bandgap of $In_xGa_{1-x}As_yP_{1-y}$ plotted vs composition y under the constraint that the material is lattice-matched to InP. Looked at in this way, the lattice-matched quaternary is as easy to comprehend as a ternary alloy.

The alloying of different epitaxial binary compounds has been extended to a plethora of materials by now. A view of several of the

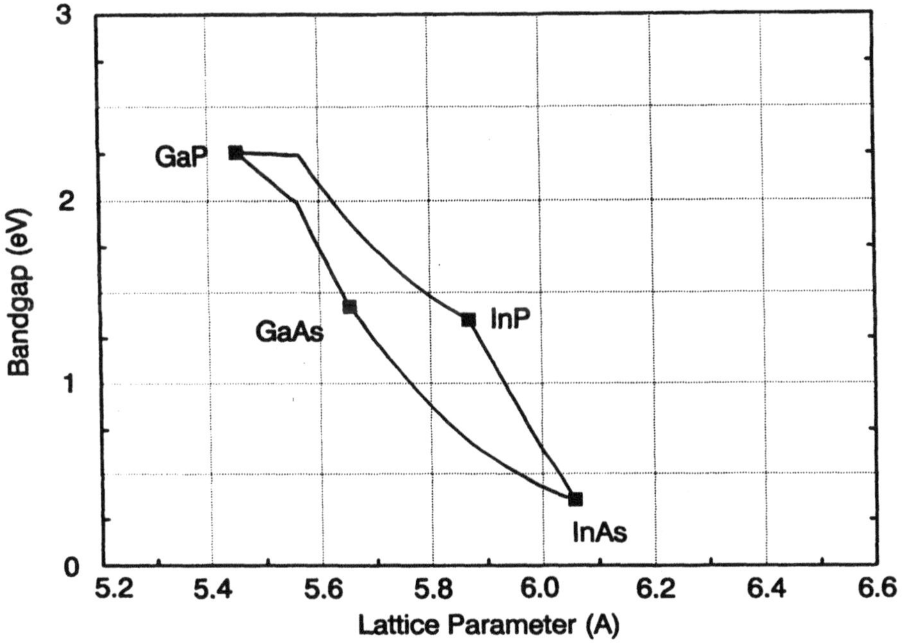

Figure 5. Bandgap vs lattice parameter for $In_xGa_{1-x}As_yP_{1-y}$.

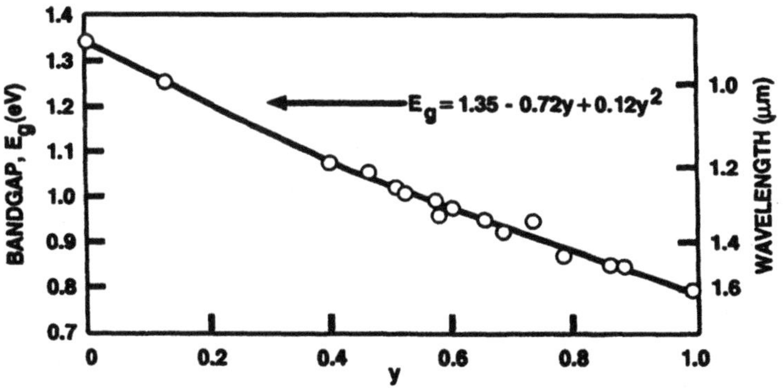

Figure 6. Bandgap vs composition y for $In_xGa_{1-x}As_yP_{1-y}$ which is lattice-matched to InP. Data is from reference 8.

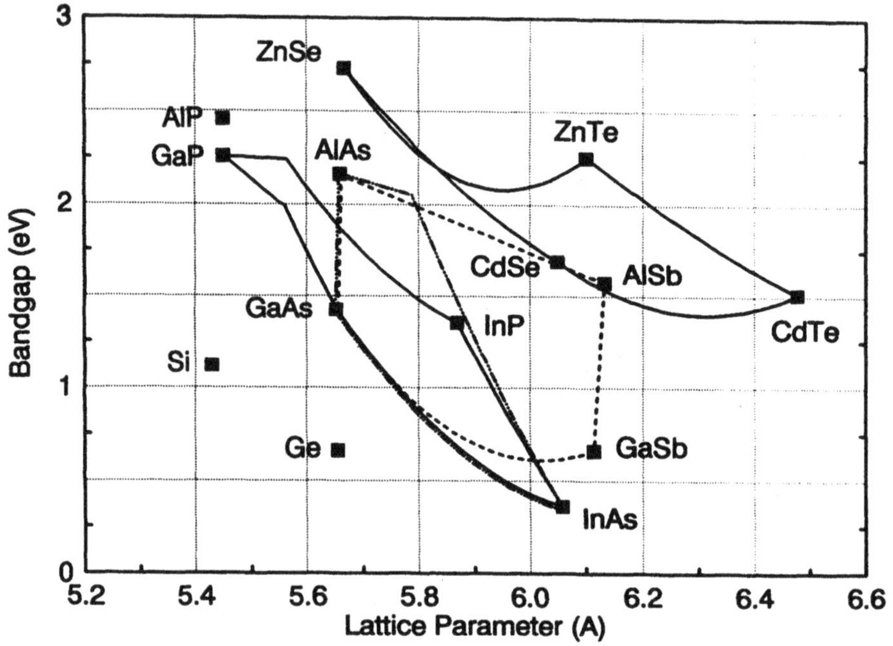

Figure 7. Bandgap vs lattice parameter for four different quaternary alloy systems. The lines represent the ternary alloy boundaries, with the area inside representing the quaternary alloy parameters.

possibilities, although not all, is shown in Fig. 7. Four quaternary systems are represented, including $In_xGa_{1-x}As_yP_{1-y}$, $Al_xGa_{1-x}As_ySb_{1-y}$, $Al_xIn_yGa_{1-x-y}As$ and $Zn_yCd_{1-y}Se_{1-x}Te_x$. The concept of alloying has indeed advanced, with all kinds of possibilities of materials and substrates which can be chosen from this chart alone. By choosing specific materials and alloy compositions one can engineer band structures to provide the properties needed for many different applications.

By stacking various alloy layers on top of each other one can form lattice-matched heterostructures which control the location of both charge carriers and photons. Combining the concepts of band structure engineering and heterostructures, all kinds of optoelectronic devices have been designed and demonstrated. In this paper I will illustrate some of the ideas by following the evolution of semiconductor laser structures as an example of the science and art of this type of device design. Again the materials capabilities of the various epitaxial processes have been the enabling technology.

Useful semiconductor lasers are pn-junction diode devices. In general, application of forward current gives rise to injection of carriers, such as electrons from the n-side into the p-side of a diode, where radiative recombination gives a source of photons. By use of cleaved mirror facets an optical cavity is readily formed, enabling laser action.

In this geometry, carrier diffusion lengths determine the size of the inversion and threshold current densities are necessarily large,

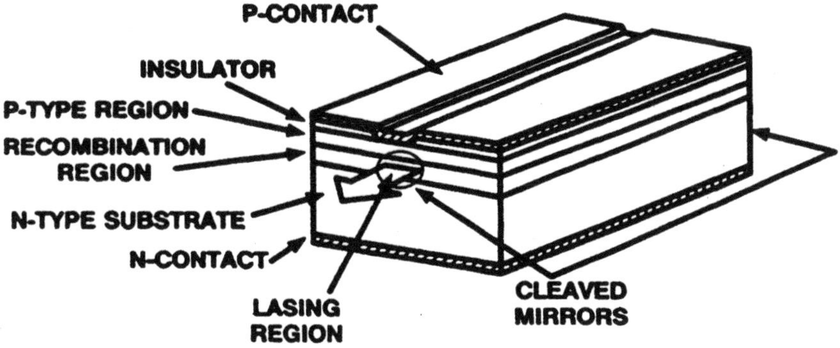

Figure 8. Schematic of a simple stripe geometry double heterostructure laser.

greater than 10^4 Amps/cm^2 at room temperature. These values were reduced to values near 10^3 Amps/cm^2 through use of double heterostructure (DH) geometries,[6] illustrated in Fig. 8 for general materials. In this case both electrons and holes are injected from the large bandgap n-InP and p-InP, respectively, into the relatively small bandgap quaternary alloy InGaAsP. The location of electrons and holes is then controlled by the thickness of this layer, typically 0.2 μm for a simple DH laser, and radiative recombination is very efficient. In almost all cases, smaller bandgap materials have higher indices of refraction, thus producing a waveguide for light in the smaller bandgap layers. This also contributes to the improved efficiency of the DH. Room temperature laser threshold current densities are near 1000 Amps/cm^2, corresponding to carrier densities of about 1×10^{18} cm^{-3}. As shown in Fig. 8, current was supplied through a stripe contact on the epilayer surface, but current spreading internally limited the efficiency of these simple DH lasers.

Increased sophistication was required in the materials preparation to control this current spreading. Fig 9a shows three well known ways in which internal currents are controlled. All of these configurations accomplish lateral carrier confinement and wave guiding. Without going into detail which is beyond the scope of this paper, all three structures are prepared by sequential growth, etching, and then a further growth step. The lateral patterns are thus formed, giving a lateral step in refractive index which confines light to the central region. Furthermore, the growth is such that a narrow stripe pn-junction structure is formed, surrounded by pnpn structures, thus restricting current to the central stripe. With these features all 3 of these structures have improved efficiency, with thresholds around 20 milliamps. These types of lasers form the basis of modern lightwave

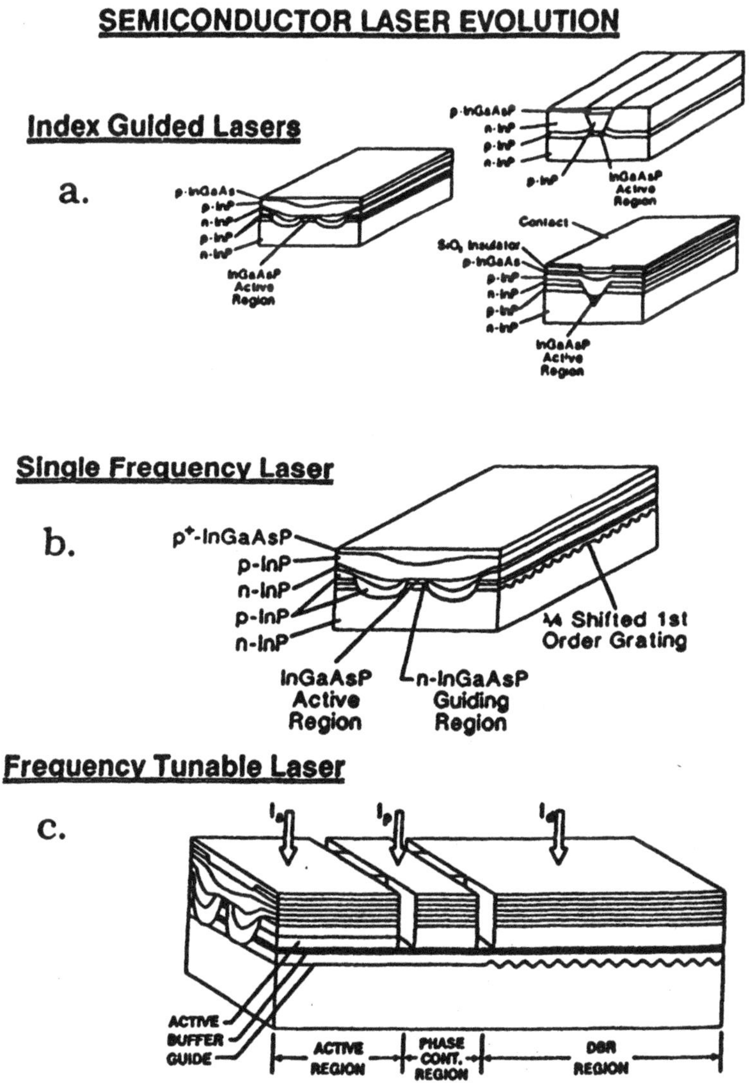

Figure 9. Schematics of several laser structures, with increasing sophistication from top to bottom. (a.)Three index guided lasers. (b.)Single frequency laser with distributed feedback provided by the internal grating. (c.)Multi-section frequency tunable laser with Bragg reflector tuning section.

telecommunications systems, with the InGaAsP active layers engineered through composition choices to emit light at either 1.3 μm or 1.55 μm where glass fiber waveguides work with lowest losses.

Further advancements are developing rapidly in many research labs. In Fig. 9b the so-called single frequency laser is illustrated. It is basically one of the structures from Fig. 9a, called the double channel planar buried heterostructure or DCPBH, with the addition of an extra waveguide layer and a fine etched grating. In this structure the light in the waveguide senses the periodicity of the grating, which provides a distributed reflectivity enhancement for a particular wavelength obeying the Bragg condition. Overall, it is this wavelength which dominates the active cavity, producing a narrow wavelength output light beam. Line widths for this type of laser are very narrow, typically of the order of 1 MHz or less.[9-12]

Taking this a step further we come to the tunable semiconductor laser of Fig. 9c.[13-20] This is a marvel of technology and design. Such lasers are actually 4 terminal devices. Light is generated and amplified in the left hand section only. The other sections contain passive waveguides along with the grating reflector. Briefly this laser operates as follows: current in the left hand section generates light and is used to turn the laser on and off, allowing pulsed output which can encode information for transmission. Light emerges from the cleaved facet at the left. The light generated at the left is guided by the waveguide layer to the grating region, which reflects a specific wavelength back for amplification. This provides narrow band laser output at a specific wavelength. The refractive index n in the waveguide is determined by the material and also by the concentration of charge carriers, with the latter controlled by current flow in the right hand section of the laser.

The wavelength reflected is lambda$_i$ = lambda$_o$/n where subscripts i and o refer to internal and external wavelengths. Thus by controlling the charge density, accomplished by adjusting current in the grating section, one can vary the index n and thereby the output wavelength lambda$_o$ of the laser. Wavelength tunability over a range of 10 nm has been demonstrated for 1.55 μm lasers. This is adequate for multi-wavelength systems with 10's of different wavelengths. Thus one laser can couple many different wavelengths into one optical fiber. Switch times of 1 nsec from one wavelength are possible, with pulse code modulation rates of 1 Gbit/sec.[20] Applying this technology to future telecommunications systems suggests wavelength-multiplexed, fast Gbit/sec transmission.

There is more to this story as research is directed toward lower and lower threshold lasers through use of quantum confinement effects. For example, by deposition of layers in a V-groove by organo-metallic chemical vapor deposition techniques, very thin active layer lasers have been prepared with the so-called quantum wire structure as the active region.[21] With injected electrons and holes quantum confined in both vertical and horizontal directions, carrier densities of states are changed. Such lasers have operated at threshold currents of about 1 milliamp, with predictions of even lower values to be sought after.[22]

Finally, multilayer structures of 600 epilayers have been prepared as surface emitting lasers.[23] In this case multilayer structures form high reflectivity "mirrors" above and below an active layer to form a vertical optical cavity which gives laser output perpendicular to the surface. By etching such structures into small cylinders a few μm in diameter large arrays of surface emitting lasers have been formed, with device densities of 2 million per cm^2. Such arrays, although in their infancy, hold promise for new technologies, such as future optical processing applications.

I suppose the point is now made, that epitaxial materials advances have indeed been astounding over the last two decades or so. The sophistication of present day semiconductor laser structures is more than one might have dreamed. Ideas based on epitaxial materials, band structure engineering and heterostructures have given possibilities which still have not been plumbed to their full depths. We work in a situation where we are limited more by our imagination than by the materials technologies. Rogerio, what a distance has been traveled!

REFERENCES:

1] Sturge, M.D., Phys. Rev. **127**, 768 (1962)

2] Shah, J., Leite, R.C.C. and Nahory, R.E., Phys. Rev. **184**, 811 (1969)

3] Hopfield, J.J., in II-VI Semiconducting Compounds, edited by D.G.Thomas, p. 786, (Benjamin, 1967)

4] Stillman, G.E., Wolfe, C.M. and Dimmock, J.O., in Semiconductors and Semimetals, Vol. **12**, p. 169 (Academic Press, 1977).

5] Razeghi, M., Omnes, F., Nagle, J., Defour, M., Acher, O. and Bove, P., Appl. Phys. Lett. **55**, 1677 (1989)

6] Casey, H.C.Jr. and Panish, M.B., Heterostructure Lasers (Academic Press, 1978)

7] Mikkelsen, J.C. Jr. and Boyce, J.B. Jr., Phys. Rev. B28, 7130 (1983)

8] Nahory, R.E., Pollack, M.A., Johnston, W.D.Jr. and Barns, R.L., Appl. Phys. Lett. **33**, 659 (1978)

9] Zah, C.E., Caneau, C., Menocal, S.G., Gozdz, A.S., Lin, P.S.D., Favire, F., Yi-Yan, A., Lee, T.P., Dentai, A.G. and Joyner, C.H., Electr. Lett. **25**, 650 (1989)

10] Imai, H., Proceedings of Conf. on Quantum Electronics and Laser Science, Baltimore, MD (1989)

11] Kitamura, M., Yamada, H., Takano, S., Sasaki, T., Henmi, N. and Mito, I., Proceedings of the Optical Fiber Communications Conf., paper WB2, Houston, TX (1989)

12] Yamazaki, H., Sasaki, T., Kida, N., Kitamura, M. and Mito, I., Proceedings of the Optical Fiber Communications Conf., paper PD-33, SanFrancisco, CA (1990)

13]Murata, S., Mito, I. and Kobayashi, K., Electr. Lett. **23**, 403 (1987)

14] Kobayashi, K. and Mito, I., IEEE J. Lightwave Techn. **6**, 1623 (1988)

15] Kotaki, Y., Matsuda, M., Ishikawa, H. and Imai, Electr. Lett. **24**, 503 (1988)

16] Koch, T.L., Koren, U., Gnall, R.P., Burrus, C.A. and Miller, B.I., Electr. Lett. **24**, (1988)

17] Lee, T.P., International Electron Devices Meeting, Proceedings, p. 35.1.1 (1989)

18] Koch, T,L. and Koren, U., IEEE J. Lightwave Techn., 18, 8, 274 (1990)

19] Pan, X., Olesen, H. and Tromborg, B., IEEE J. Quantum Electr. **QE-24**, 2423 (1988)

20] Chapuran, T.E., Kobrinski, H., Vecchi, M.P., Zah, C.E., Caneau, C., Menocal, S.G., Lin, P.S.D., Gozdz, A.S. and Favire, F., Proceedings of the Optical Fiber Communication Conf. paper TUE5, SanFrancisco, CA (1990)

21] Kapon, E., Hwang, D.M. and Bhat, R., Phys. Rev. Lett. **63**, 430 (1989)

22] Yariv, A., Appl. Phys. Lett. **53**, 1033 (1988)

23] Jewell, J.L, Scherer, A., McCall, S.L., Lee, Y.H., Walker, S., Harbison, J.P. and Florez, L.T., Electr. Lett. **25**, 1123 (1989)

DYNAMICS OF FORMATION AND RELAXATION OF INTRINSIC EXCITONS AND RELAXATION AND THERMALIZATION OF EXCITON-SPIN IN GaAs QUANTUM WELLS

(To Rogerio C.C. Leite)

by

Theo C. Damen

AT&T Bell Laboratories

Holmdel, N.J. 07733

Optical properties of excitons in quantum wells have been very fashionable in recent years [1-5 and references therein]. Rogerio does not know it, but he started this surge with exciton line-width measurements [6] in GaAs in the year 1969. Since then Rogerio reaped the fruits of other endeavors such as politics, philosophy, the musical and editorial arts, he even studied the behavioral science of various animals including humans. In the meantime the physical sciences did not stand still. On the contrary, the sophistication of high technology allows us to investigate the formation and relaxation of intrinsic excitons, and exciton-spin following excitation of electron- hole pairs close to the bandgap and at the exciton resonance using subpicosecond laser pulses, fancy luminescence up-conversion equipment, and a streak camera. Rogerio always liked the use of the maximum number of senses for a communication, and since a picture is worth more than a thousand words, the results of the measurements will be presented primarily in pictorial form. We were fortunate to have nearly perfect samples as shown in figure 1. Note that the luminescence and luminescence excitation spectra of the sample overlap very well, indicating an absence of a Stokes shift.

The absence of a Stokes shift ensures measurements of intrinsic excitons rather than localization effects so dominant in earlier studies [7-9]. The two exciton luminescence lines are a result of a multiple quantum well (MQW) stack with two independent and distinct regions. The bottom 35% of the MQW-stack was grown with a slightly narrower well-width.

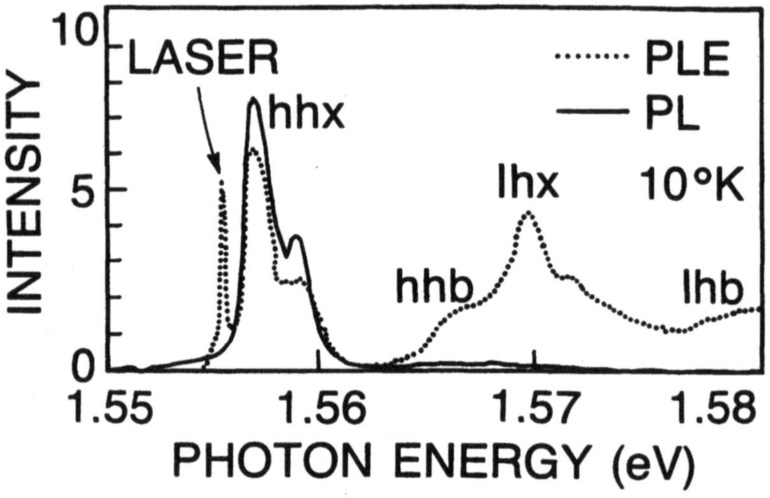

Figure 1.

When we examine the time evolution of exciton luminescence intensity at the spectral peak for various temperatures, shown in figure 2, we find that the maximum is reached in about 300 to 400 ps at 10 K. On the other hand the electron-hole pairs relax to the bottom of the band in about 1ps, even at the very low densities used in our experiments. As the lattice temperature is increased a quasi equilibrium amongst the electron-holes and excitons is reached much sooner (<5ps at 60 K). So..., where are the excitons?

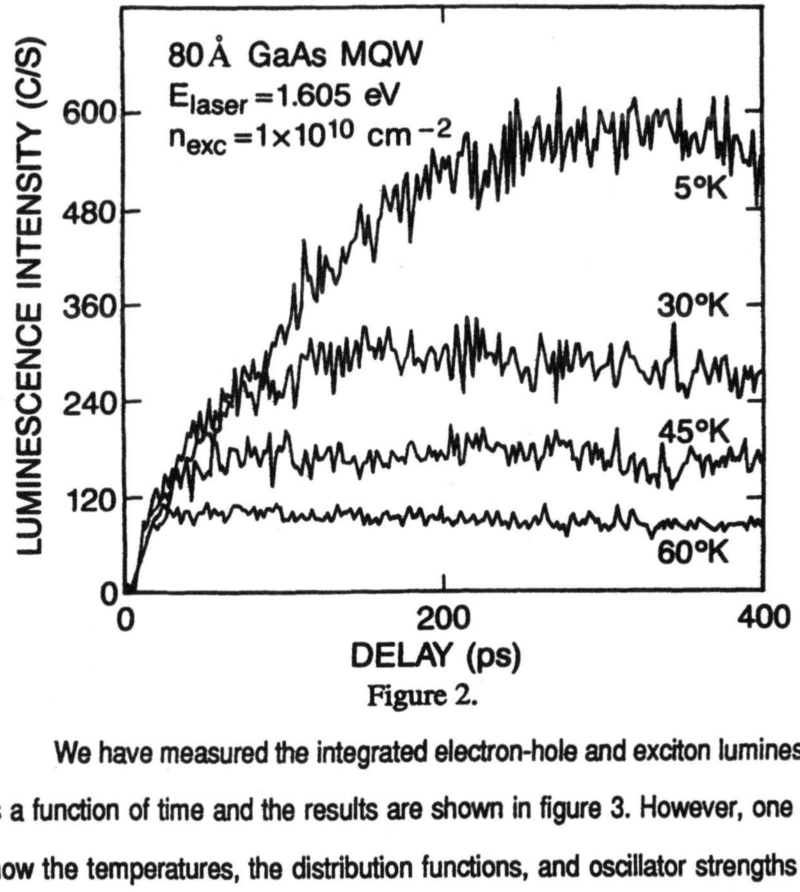

Figure 2.

We have measured the integrated electron-hole and exciton luminescence as a function of time and the results are shown in figure 3. However, one has to know the temperatures, the distribution functions, and oscillator strengths of the carriers at short times for quantitative determination. We therefore determine the electron-hole density by measuring the temporal evolution of the homogeneous linewidth of excitons, which is strongly influenced by the presence of electron-hole pair density [6,10]. We consider the exciton lineshape to be a convolution of a Gaussian lineshape observed at very long times with a Lorentzian whose width is adjusted to give a good fit to the low energy side of the spectrum, since the lineshape is asymmetric at short times due to band-to-band recombination at the high energy side.

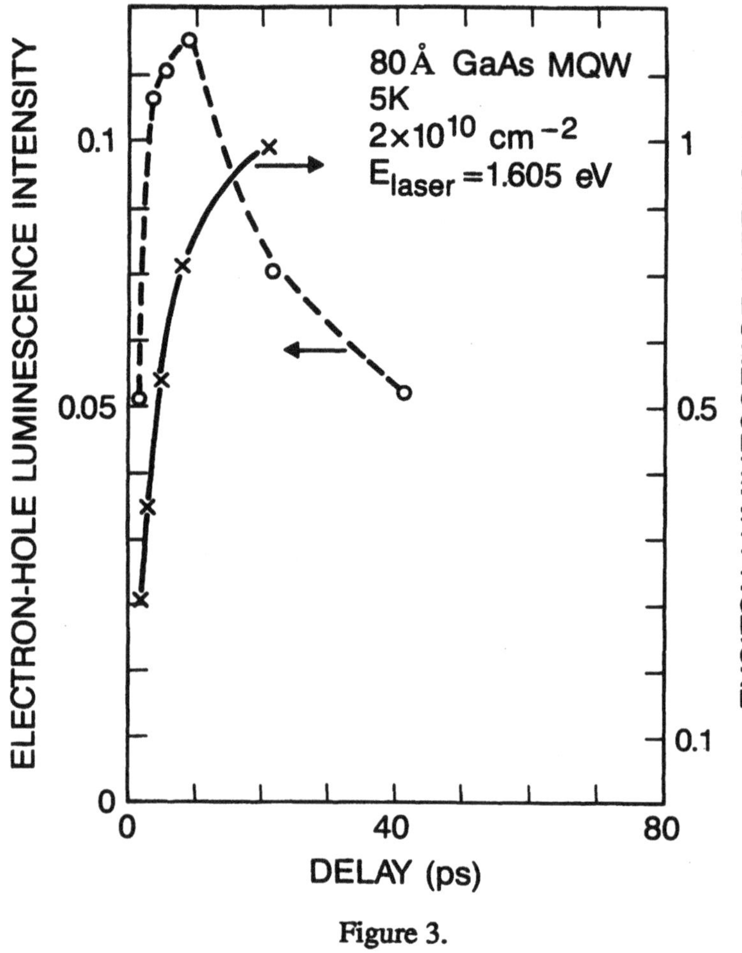

Figure 3.

This procedure is illustrated in the following figure 4 where (a) are typical time resolved spectra and (b) is the lineshape analysis of an observed spectrum at 20ps. The solid curve is experimental, the dotted curve is a

Gaussian-Lorentzian fit and the dashed curve is twice the difference between the two, which represents the band-to-band recombination.

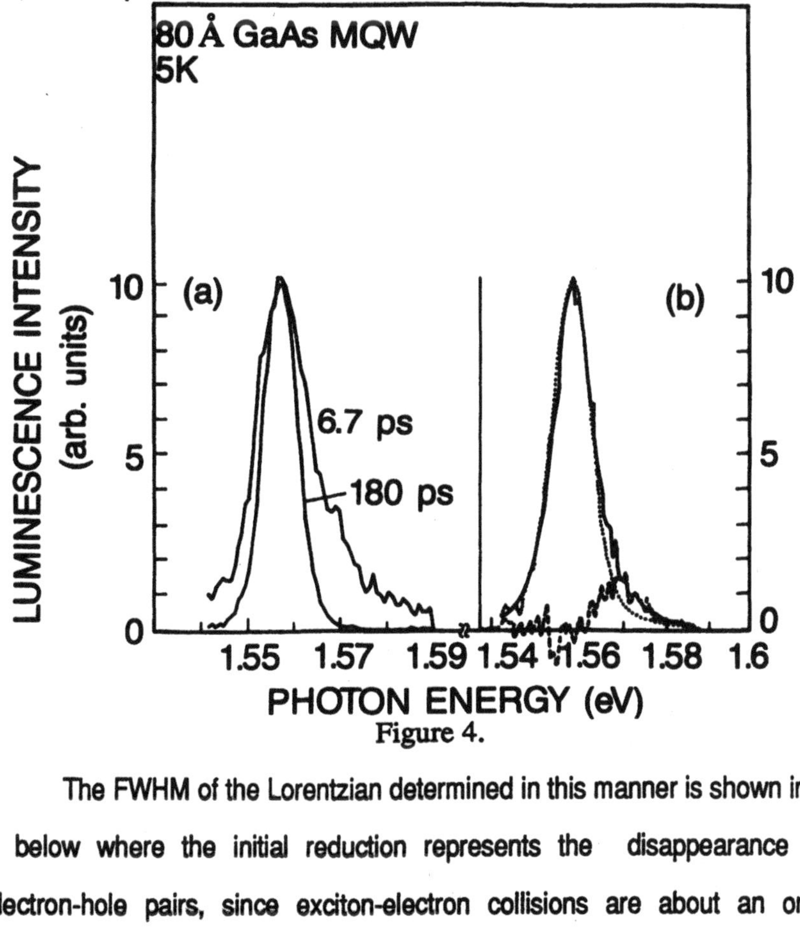

Figure 4.

The FWHM of the Lorentzian determined in this manner is shown in figure 5 below where the initial reduction represents the disappearance of the electron-hole pairs, since exciton-electron collisions are about an order of magnitude more efficient [9] than exciton-exciton collisions [8]. The disappearance of the electron-hole pairs is considerably shorter than the rise of the luminescence. We therefore presume that the electron-hole pairs have formed large K excitons (about 20 ps) and that the long risetime of exciton luminescence is caused by the slow relaxation of large K excitons to K=0 excitons.

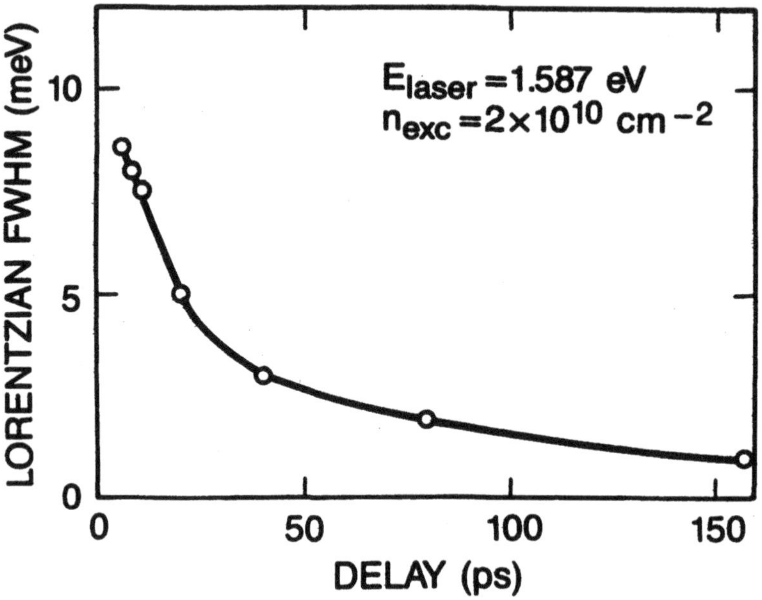

Figure 5.

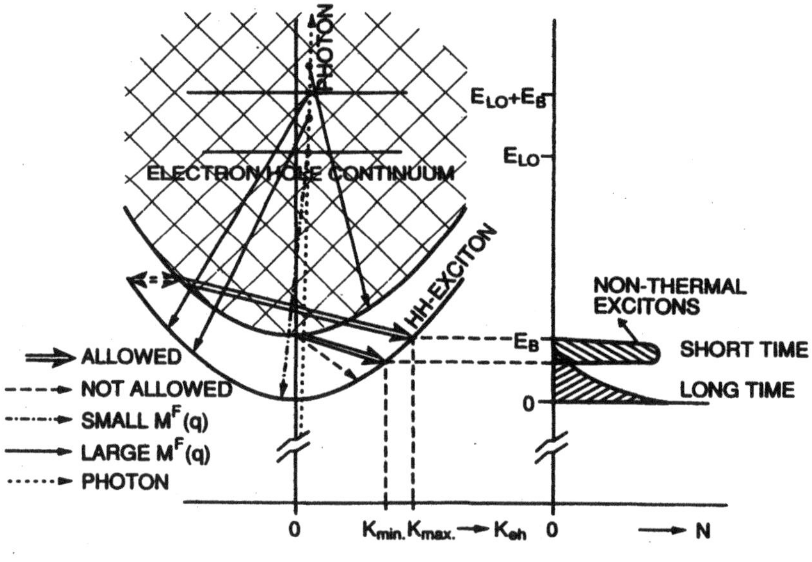

Figure 6.

We explain the above premise by considering the energy and momentum conservation requirements for the interaction of free electron-hole pairs with the acoustic phonons as illustrated in figure 6. This forces the majority of the formed excitons to occupy large K states, which can not radiate, hence the weak luminescence intensity at short times. Although the homogeneous linewidth of excitons is large at early times, a simple estimate shows that the excitons are primarily formed at energies larger than 6 meV and the exciton binding energy is 9 meV. Thus these excitons do not auto-ionize and the exciton distribution is **highly non-thermal** which is consistent with the slow luminescence risetime.

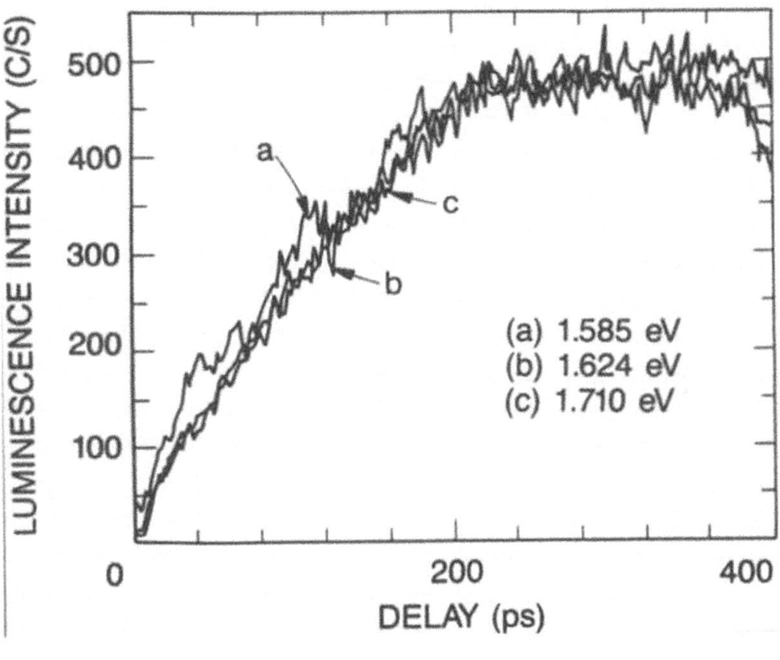

Figure 7.

With this model we now can explain a more fascinating observation that the luminescence risetime is independent of the excitation energy, which is illustrated in figure 7 for three different excitation energies. The results were obtained on a sample with a larger splitting between the heavy and light hole excitons, so that we could vary the excitation energy through the light-hole and LO-phonon resonances. In our case the light hole state is degenerate with the electron-hole continuum, which interaction might dominate over that of the light-heavy hole exchange interaction, resulting again in large K excitons. The LO-phonon-exciton Froehlich interaction is only strong when the wavevectors involved are about the inverse of the exciton radius and vanishes at $q=0$. All the above is also illustrated in figure 6 presented earlier. Therefore, no resonance is observed, consistent with our model.

We can also explain the density dependence of the luminescence rise shown in figure 8 which shows that the relaxation of large K excitons to those that can radiate takes place via exciton-exciton collisions and exciton-accoustic phonon collisions. We find that at high densities the exciton-exciton interaction dominates resulting in a decrease of the luminescence risetime [3,4] and when the exciton density is below $4.10^9 cm^{-2}$ per well the intrinsic exciton-accoustic phonon collisions drive the relaxation, which results in a limiting risetime of about 400ps. If we imagine collisions between hard discs, the collision time is inversely proportional to the exciton density and velocity. Hence, at a temperature of 10 K and at the transition density where exciton-exciton interactions become negligable, we estimate that the effective exciton-accoustic phonon collision time is approximately 30 ps.

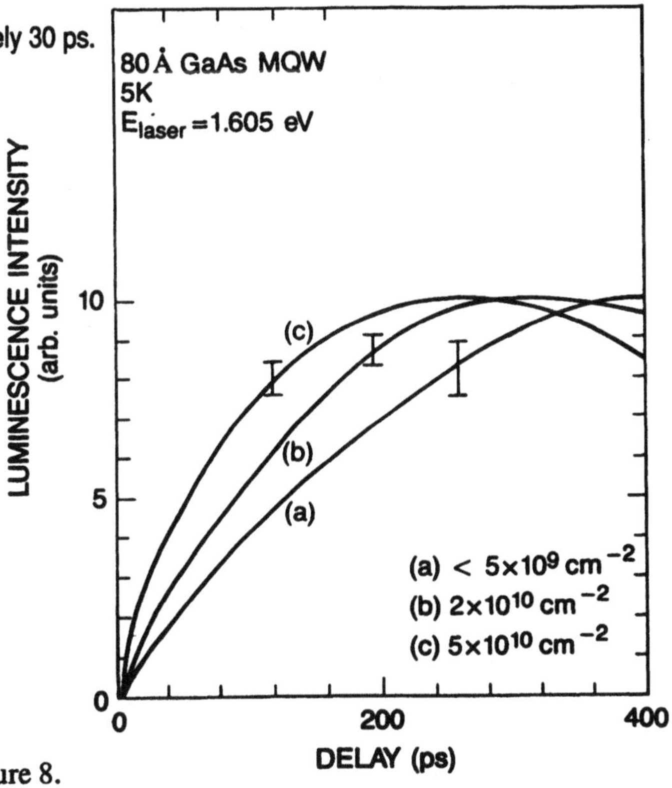

Figure 8.

Although exciton-acoustic phonon interactions related to exciton linewidth have been investigated [14-16], relaxation of excitons within the exciton branch remains unexplored, and no comparison with theory is possible at this time.

Lastly, we have measured the dynamics of spin relaxation and thermalization of these excitons in GaAs quantum wells under resonant as well as non-resonant excitation [15]. Polarization measurements of luminescence processes under cw conditions [16,17] have been performed and various theories [18-20] have been proposed to explain these results. Only two studies with linearly [8] and circularly [21] polarized light have been reported in GaAs quantum wells. Stolz et al [8] have also recently measured the depolarization times of excitons in GaAs quantum wells using linearly polarized light. Although their measured times are similar, their samples showed large (approx 8 meV) PL linewidths and a rate equation analysis (including effects of exciton localization) was needed to deduce times which were quite close to the time resolution of the

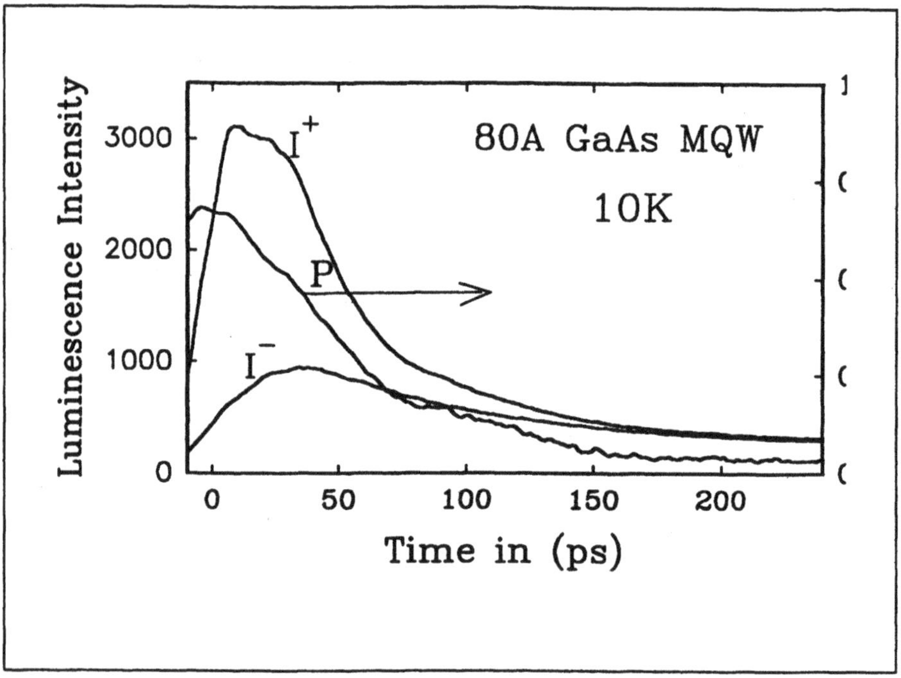

Fgure 9

experimental setup. Therefore, their results should not be directly compared with our results.

Time evolutions of the luminescence intensity at the exciton wavelength were measured with both σ^+ and σ^- excitations (and denoted by I^+ and I^- respectively). Figure 9 shows the time evolutions of I^+ and I^- for resonant smaller than I^+ but becomes equal to I^+ for long delay times. Polarization P = excitation

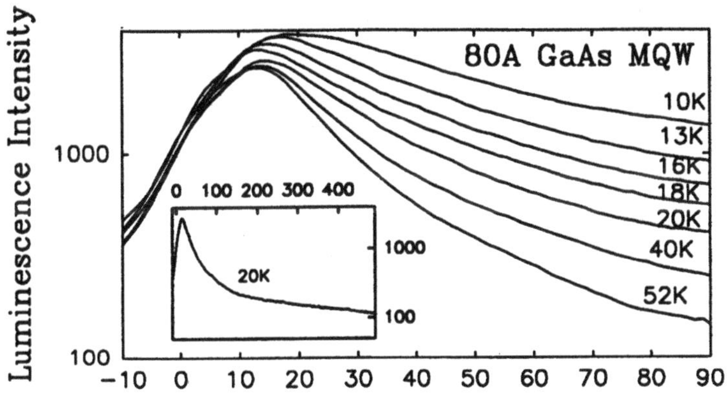

Figure 10

of the heavy hole exciton. The depolarized signal I^- is initially much$(I^+-I^-)/(I^++I^-)$ is calculated for each time and is also plotted in Figure 9, which shows that P decays with a time constant of approx 50 ps. For resonant excitation, the risetime of the luminescence is extremely fast (limited by our instrument response time), in contrast to the long (>100ps) risetime observed for non-resonant excitation [5,7]. For non-resonant band-to band-excitation, P is identically zero at all times even when the excitation is within a few meV of the bandgap. These measurements provide information about the spin relaxation of carriers as discussed below.

Information about the energy relaxation of excitons is contained in the time evolution of the total luminescence intensity $I=I^++I^-$ shown in Figure 10 for several different temperatures, for resonant excitation. The inset clearly shows that the total intensity has a fast decay followed by a slow decay and furthermore the initial decay becomes faster with increasing temperature.

Fig. 11 shows that the initial decay rate (i. e. the inverse of the initial fast decay time) increases nearly linearly with temperature and then saturates as it approaches the system time resolution.

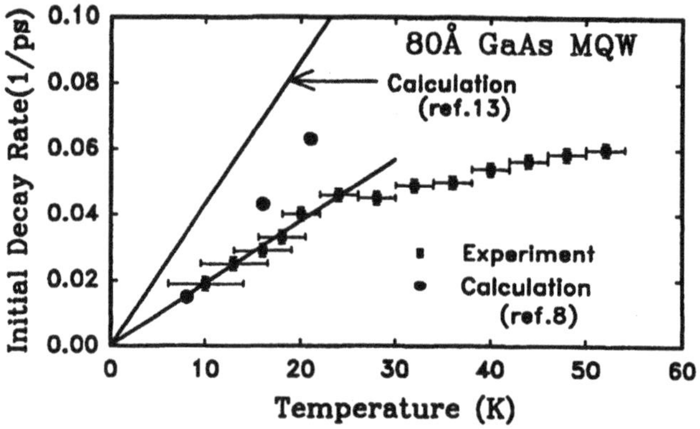

Figure 11.

We refer to Figure 12 (a) to discuss the data on polarization relaxation and to consider only the heavy hole states and no heavy-hole-light-hole coupling. For band-to-band absorption or emission, $\sigma+$ light couples only the -3/2 heavy hole state and the -1/2 electron state, whereas $\sigma-$ couples only the +3/2 heavy hole state with the +1/2 electron state. For excitons, $\sigma+$ couples to the exciton made up of +3/2 heavy hole state and -1/2 electron state, whereas $\sigma-$ couples to the exciton made up of -3/2 heavy hole and +1/2 electron. For both resonant and non-resonant $\sigma+$ excitation, the observation of $\sigma-$ polarization in luminescence is possible if both hole as well as electron spins have relaxed.

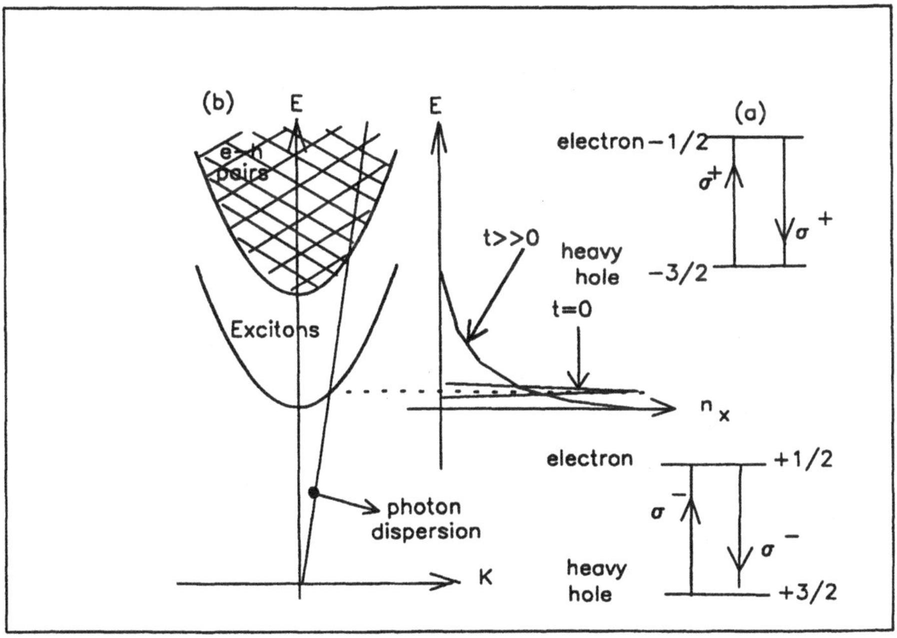

Figure 12

With the usual assumption [16] that the hole spins relax very fast, the polarization relaxation time measured might imply that the electron spins relax with a time constant of 50 ps. For non-resonant excitation very close to the bandgap, only heavy hole states are excited so similar arguments should be valid. Absence of any polarization implies that the electron spins relax in less than the time resolution of 10 ps. This conclusion is consistent with our cw polarization spectroscopy data [22] but in disagreement with the cw data of Miller et al [16] and Freeman et al [23]. The electron spin relaxation times deduced from our measurements are considerably shorter than the values for free electrons in bulk GaAs. The difference may arise from the fact that we investigate intrinsic excitons rather than free carriers.

We now turn our attention to the thermalization of the resonantly excited heavy hole excitons. Figure 12b schematically displays the relevant energy diagram in the two particle or exciton representation. The absorption of photons takes place only within the homogeneous width of the exciton and the PL linewidths in our sample are < 0.7 meV so that the homogeneous linewidth is less than kT even at 10 K. As the photoexcited cold excitons thermalize, their distribution becomes wider than the initial distribution so that fewer excitons remain within the homogeneous linewidth of the exciton with increasing time. Since only excitons within the homogeneous linewidth couple to light, this leads to a decrease in the luminescence intensity even though the total number of excitons has not decreased. This is the cause of the initial rapid decay of exciton intensity. The initial decay rate increases with increasing temperature because the phonon population and hence the exciton-acoustic phonon scattering increases with temperature. Furthermore, the homogeneous linewidth increases slowly with temperature so that the fraction of excitons in the homogeneous linewidth decreases with temperature. Therefore, the initial drop in the luminescence intensity is larger at higher temperature. This model explains all the results presented in Figure 11 and we deduce γ=2x10^9 (1/s-K) for T <25K from the exciton-acoustic phonon scattering rate W given by W(1/s)=γT(K). This is the first measurement of exciton-acoustic phonon interaction rate for thermalization of excitons in a semiconductor. Lee et al [13] as well as Stolz et al [8] have calculated exciton-acoustic phonon scattering rate [24]. Their calculated results are also plotted in Figure 11. The calculated value γ=4.45x10^9(1/s-K) [8] is a factor of 2.3 larger than the value determined from our data. Also, exciton-acoustic phonon scattering rate calculated by Stolz et al [8] (Fig.11) and the rate deduced from four-wave-mixing measurements [10] are larger than our results. This difference may be explained by considering that all exciton-acoustic phonon collisions contribute to linewidth broadening and exciton dephasing but only a subset of these collisions, which remove an exciton out of the homogeneous width of the exciton, contributes to the measured thermalization rate. Also, excitons excited out of this linewidth may scatter back into the homogeneous linewidth by subsequent collisions, thus reducing the net outgoing rate. Our measurements show that only about half the collisions are effective in removing photoexcited excitons from the optically coupled region. Although the agreement between the calculations and experiments is reasonably good, there is at present no calculation on the time evolution of the exciton distribution function. Calculations and experiments on this subject would be very interesting.

We have given Rogerio the new view of the dynamics of formation and relaxation of intrinsic excitons and exciton-spin in a semiconductor. We show that momentum and energy conservation require that the free excitons formed from the electron-hole pairs by emission of phonons are created at large K resulting in a **highly non-thermal** exciton distribution at short times. The relaxation of this non-thermal initial distribution to the thermalized distribution is responsible for the long risetime of the exciton luminescence. Creating K=0 excitons by emission of LO-phonons we find is very weak since the Frohlich matrix element is small for small wave vectors. At higher excitation energies the matrix element becomes large creating large K excitons. A number of fundamental questions concerning excitons in quantum wells have been adressed and a good qualitative picture for the formation and relaxation of exciton-spin has been developed. The Festschrift might entice Rogerio to go back to his initial state, i.e. **Physics.**

The above work has been done in collaboratin with Jagdeep Shah, Karl Leo, J.E.Cunningham, D.Y.Oberli, and J.M.Kuo.

We thank D. S. Chemla, S. Schmitt-Rink and L. J. Sham for useful discussions and K. Kaufman and W. Cieslik of Hamamatsu Corp. for the loan of their synchroscan streak camera.

REFERENCES

[1] Schmitt-Rink S., Chemla D.S., and Miller D.A.B., Advances in Physics, **38,** 1, 1989.

[2] Miller D.A.B., Chemla D.S., Damen T.C., Wood T.H., Burrus C.A., Gossard A.C., and Wiegmann W.W., IEEE J.Quan.Electronics **QE-21**, 1462 (1985)

[3] Damen T.C., Shah Jagdeep, Oberli D. Y. , Chemla D. S. ,Cunningham J. E. and Kuo J. M. , J. of Lumin. **45,** 181 (1990).

[4] Damen T.C., Shah Jagdeep, Oberli D. Y. Chemla D. S., Cunningham J. E. and Kuo J. M., Phys. Rev. B , Oct. 15, (1990)

[5] Damen T.C., Shah J., Oberli D.Y., Chemla D.S., and Cunningham J.E. SPIE-Proceedings of the International Conference on Quantum-Well and Supefast Laser Probe Phenomena in bulk and Microstructure Semiconductors III, edited by R. Alfano, 18-19 March 1990; p.110-120.

[6] Leite R.C.C., Shah Jagdeep and Gordon J.P., Phys. Rev. Letters, 23, 1332 (1969).

[7] Kusano J, Segawa Y., Aoyagi Y., Namba S. and Okamoto H. Phys. Rev.B **40**, 1685 (1989).

[8] Stolz H., Schwarze D., von der Osten W., and Weimann G. Superlattices and Microstructures, **6**, 271 (1989)

[9] Amand T., Lephay F., Valloggia S., Zoillot F., Brousseau M., Regreny A. Superlattices and Microstructures, **6**, 223 (1989)

[10] Honold A, L.Schultheis, J.Kuhl, and C.W.Tu, Proceeding of Ultrafast VI, Mt.Hiei,Japan, (1988); edited by T.Yajima, K.Yoshihara, C.B.Harris and S.Shionoya, Springer-Verlag, Berlin 1988; p.307.

[11] Note that holes are less effective in broadening the exciton since Γ is inversely proportional to the carrier mass.

[12] Takagahara T., Phys.Rev. **B32**, 7013 (1985)

[13]Lee Johnson., Koteles E.S., and Vassell M.O., Phys.Rev.B 33, 5512 (1985).

[14] Rudin S., and Reincke T.L., Superlattices and Microstructures, **3**, 137 (1987)

[15] Damen T.C., Leo K, Shah J. and Cunningham, J.E. to be published in Appl.Phys.Letters.

[16] Miller R.C. and Kleinman, D.A. J.Lumin. **30**, 520 (1985).

[17] Sooryakumar R., Pinczuk A., Gossard A.C., Chemla D.S. and Sham L.J., Phys. Rev. Letters **58**, 1150 (1987).

[18] Ruckenstein A. E., Schmitt-Rink S. and Miller R. C. Phys. Rev. Letters **56**, 504 (1986).

[19] Twardowski A, and Hermann C,. Phys. Rev. B 35, 8144 (1987).

[20] Uenoyama T and Sham L.J. Phys. Rev. Letters 64, 3070 (1990).

[21] Freeman, M.R., Awschalom, D.D., Hong, J.M., and Chang, L.L.,presented at ICPS 20, Thessaloniki, Greece, Aug.6-10, 1990; Paper MoP-57.

[22] Damen, T.C., K.Leo, Jagdeep Shah and J.E. Cunningham, to be published.

[23] Freeman et al [21] have recently shown that polarization of luminescence is sample dependent, for reasons that are not well understood at this time.

[24] For temperatures above 10 K, the average phonon energy involved in the scattering process is considerably larger than the homogeneous linewidth. Comparison of Fig. 7 and 8 of Ref. [8] leads to a similar conclusion. Therefore, it is more meaningful to compare our results to scattering rate rather than energy loss rate.

COUPLED INTERSUBBAND-CYCLOTRON RESONANCES IN QUANTUM-WELL STRUCTURES

R. Merlin

The Harrison M. Randall Laboratory of Physics
The University of Michigan
Ann Arbor, MI 48109-1120, U. S. A.

Magnetic fields combined with quantum confinement generally lead to coupling between the motions parallel and perpendicular to the confinement plane.[1-17] The tilt angle θ, defined by the direction of the field and the normal to the plane, determines the degree of admixture vanishing at $\theta = 0$. For various reasons, the spectrum of quasi two-dimensional electrons in tilted fields has been the object of intensive study in recent years.[4-17] In the case of free electrons, tilting results in subband-Landau-level hybridization and associated excitations of mixed character.[1-15] Experimentally, tilted-field-induced coupling with allowed excitations makes possible for infrared[4-12] and Raman[14,15] spectroscopy to probe nominally forbidden intersubband or cyclotron resonances. Tilted magnetic fields have also been applied to related studies of confined shallow impurities[14,16] and, more recently, they have helped elucidate the role of reversed spins in the fractional quantum Hall effect.[17]

In this work, we present a review of our recent Raman scattering study of intersubband-Landau-level coupling in quantum-well structures.[14,15] We also discuss results of finite-basis calculations of the tilted-field spectrum and derive selection rules for the coupling relying on the even symmetry of the confinement potential. The calculations and the Raman measurements agree substantially over a broad range of angles and fields. In addition, the finite-basis (and the experimental) data for the *lowest-lying* coupled levels show a remarkable agreement with predictions of the parabolic-potential model.[2,3] This is obviously relevant to recent infrared work on coupled-excitations in parabolic wells.[11,12]

Coupled intersubband-Landau-level excitations have been extensively discussed in the theoretical literature.[1-3] The Hamiltonian for non-interacting electrons confined in the (x,y) plane and magnetic field $\mathbf{B} = B\ (0,\sin\theta,\cos\theta)$ can be written as[1-3]

$$H = H_W + H_P + H_D + H_M \quad , \tag{1}$$

with

$$H_W = \frac{1}{2m_0}\, p_z^2 + V(z) \tag{2}$$

$$H_P = \frac{1}{2m_0}\, p_x^2 + \frac{1}{2m_0}\left(p_y + \hbar\cos\theta x/l^2\right)^2 \tag{3}$$

and

$$H_D = \frac{\hbar^2}{2m_0}\,(z^2\sin^2\theta/l^4) \tag{4}$$

$$H_M = (\hbar/m_0)(zp_x\sin\theta/l^2)\quad . \tag{5}$$

Here, $\mathbf{p} = (p_x,p_y,p_z)$ and m_0 are the momentum and the effective mass of the electron, θ is the tilt angle and $l = (c\hbar/eB)^{1/2}$ is the magnetic length. $V(z)$ is the confinement potential; for a quantum well of width L and barrier energy V_0, it is given by $V(z) = V_0$ ($V(z) = 0$) for $|z| > L/2$ ($|z| \le L/2$). The eigenenergies and wavefunctions of H can not be obtained analytically. Various methods giving approximate solutions which are adequate for some range of parameters are discussed in the following.

At $\theta = 0$, $H_D \equiv H_M \equiv 0$ and the motions along z and in the (x,y) plane (determined by H_W and H_P, respectively) trivially decouple. The eigenstates of H are of the form $\psi^{(0)}_{n,N,X}(x,y,z) = \phi^{(0)}_n(z)\ \chi_{N,X}(x,y)$, where $\phi^{(0)}_n(z)$ ($n = 0, 1,..$) is the nth-eigenstate of H_W and $\chi_{N,X}$ is a Landau state of energy $(N + 1/2)\hbar\omega_C$ ($N = 0, 1,..$); $\omega_C = eB/m_0c$ is the cyclotron frequency and X is the center coordinate. If θ is small, H_D and H_M can be treated as perturbations.[1-3,6] To lowest order in θ, *i.e.*, first-order in H_D and second-order in H_M, both terms lead to shifts of the energy levels $\propto \sin^2\theta$. For $B \to 0$, the H_M-contribution behaves as B^3 while that of the diamagnetic term H_D is $\propto B^2$ and, therefore, dominant. Accordingly, if $\hbar\omega_C$ is negligible compared with intersubband spacings, perturbation theory predicts a B^2-like increase for the eigenenergies (positive diamagnetic shift).[1] Moreover, transition energies are also expected to increase with field given that higher-lying wavefunctions generally show a larger spread, *i.e.*, larger $\langle z^2\rangle$. Note, however, that the shifts can have either sign if the contribution of the mixing term H_M becomes important.[3] H_M couples states with Landau indices differing by ± 1 and is primarily responsible for the crossing-avoiding behavior of levels at points of degeneracies.[1]

To compute the electronic states at arbitrary tilt angle, we follow the finite-basis formalism developed by Ando.[1] Here, H is represented in the basis $\psi_{n,N,X}(x,y,z) = \phi_n(z)\,\chi_{N,X}(x,y)$, where $(H_W + H_D)\phi_n = \varepsilon_n\phi_n$. From the eigenstates ϕ_n computed numerically, it is straightforward to calculate the matrix elements of H_M given by[1]

$$\langle nN|H_M|n'N'\rangle = (1-\delta_{n,n'})\,\frac{\hbar^2 z_{nn'}}{m_0 l^3}\,\sin\theta\,\cos^{1/2}\theta$$

$$\left[\left(\frac{N'+1}{2}\right)^{1/2}\delta_{N,N'+1} + \left(\frac{N'}{2}\right)^{1/2}\delta_{N,N'-1}\right] \qquad (6)$$

with $z_{nn'} = \langle\phi_n|z|\phi_{n'}\rangle$. In this expression, we have dropped the center coordinate X which remains as a good quantum number because H does not depend on y. Since $z_{nn'}$ vanishes unless $(-1)^n \neq (-1)^{n'}$ and the off-diagonal elements are non-zero only for $N = N' \pm 1$, the H_M matrix separates into two blocks corresponding to states with either $(-1)^{n+N} = 1$ or $(-1)^{n+N} = -1$. It follows then that no crossing is allowed for $(-1)^{n+N} = (-1)^{n'+N'}$ while states having $(-1)^{n+N} \neq (-1)^{n'+N'}$ can cross.[15]

For parabolic wells, *i.e.*, $V(z) = m_0\Omega^2 z^2/2$, Schrödinger's equation can be solved analytically for arbitrary θ.[2,3] The eigenvalues are of the form $(n_\alpha + 1/2)\,\varepsilon_\alpha + (n_\beta + 1/2)\,\varepsilon_\beta$, where n_α and n_β are integers and ε_α, ε_β can be shown to satisfy the relations

$$\varepsilon_\alpha^2 + \varepsilon_\beta^2 = (\hbar\Omega)^2 + (\hbar\omega_c)^2$$

$$\varepsilon_\alpha\varepsilon_\beta = \hbar^2\omega_c\Omega\cos\theta \qquad , \qquad (7)$$

which are appropriate for scaling purposes. In the comparison with experiments, we have taken $\Omega = E_{01}\hbar^{-1}$ with the intersubband energy $E_{01} = [\varepsilon_1^{(0)} - \varepsilon_0^{(0)}]$. Here, $\varepsilon_{0,1}^{(0)}$ are the eigenenergies of the ground and first excited state of H_W [Eq. (2)]. A few features of these solutions should be emphasized. First, the lower branch of the coupled modes tends to $E_{01}\cos\theta$ at large fields giving rise to a gap in the excitation spectrum.[8,14,15] This resembles, in some sense, the coupling between photons and infrared-active phonons (polaritons) leading to splitting into transverse and longitudinal modes.[18] In addition, the asymptote of the upper branch is $\hbar\omega_c$ indicating that B (as opposed to $B\cos\theta$) determines the position of the cyclotron resonance for $\hbar\omega_c \gg E_{01}$. However, the lower branch is $\cong \hbar\omega_c\cos\theta$ for $\hbar\omega_c \ll E_{01}$.[3]

The samples were GaAs-$Al_{1-x}Ga_xAs$ multiple quantum-wells grown by molecular beam epitaxy on (001) semi-insulating GaAs substrates. Three structures with GaAs-well thicknesses L = 700 Å, 460 Å and 380 Å were studied in detail. Values of E_{01} measured at zero field were 30, 65 and 80 cm^{-1}, respectively. These energies should be compared with $\hbar\omega_c \approx 100$ cm^{-1} for GaAs at B = 7 T; the latter being the largest field provided by our split-coil superconducting magnet.

Raman measurements were performed at T = 4K using a DCM dye laser pumped by an argon-ion laser. The photoelectrons were generated using the same beam. The DCM dye covers the region of the resonance with the critical point derived from the spin-orbit split-off gap of GaAs where electron (although not hole) scattering is enhanced.[19] Spectra with a resolution of ≈ 1 cm^{-1} were obtained in the $z(x',y')\bar{z}$ and $z(x',x')\bar{z}$ scattering configurations where, as before, z is normal to the layers and x', y' are along the [110] and [1$\bar{1}$0] directions. These geometries allow, respectively, scattering by spin-density and charge-density excitations.[19] Power densities were kept purposely low, in the range 0.5-2 Wcm^{-2}, to create an as small a population as possible and to avoid thermal broadening. It is well known that the carrier concentration can be gained from the positions of intersubband peaks in the two geometries (there is a depolarization shift for the charge-density, but not the spin-density scattering).[19] In our experiments, the power density was reduced to the above quoted values so that the Raman shifts in $z(x',y')\bar{z}$ and $z(x',x')\bar{z}$ could not be distinguished. From the limiting spectral resolution, we estimate an upper bound of 10^9 cm^{-2} for the photoexcited electron density. At this level, many-body corrections are negligible and, thus, our measurements give essentially the bare transition energies.

Figure 1 shows typical Raman data for the sample with L = 700 Å (θ is given in radians). The low-energy peaks labeled by arrows are hybrid excitations involving predominantly the states $|n,N\rangle = |1,0\rangle$ and $|0,1\rangle$. In the top trace (B =2.15 T, θ = 0.14), the feature at ≈ 80 cm^{-1} corresponds primarily to the transition $|0,0\rangle \rightarrow |2,0\rangle$ while that at ≈ 45 cm^{-1} is $|1,0\rangle \rightarrow |2,0\rangle$.

Data at a large angle ($\theta = \pi/3$) for the three samples are shown in Fig. 2. The most significant feature of these results is the anticipated large gap separating the lowest $\{|0,1\rangle,|1,0\rangle\}$ hybrid excitations (the lowest two branches in Fig. 2 (b)). In the range of interest, the spectrum of our quantum-wells is well approximated by that of an infinitely-deep potential and, at a fixed angle, the latter can be scaled for arbitrary L if one chooses E_{01} as the unit of energy. The calculations in Fig. 2 using the finite-basis approximation are for $V_0 = \infty$. Theory and experiment agree extremely well other than for the second branch from the top. The source of this discrepancy has not as yet been identified. The branch corresponds to the transition into a level resulting from the hybridization of mainly three states: $|0,2\rangle$, $|2,0\rangle$ and $|1,1\rangle$.

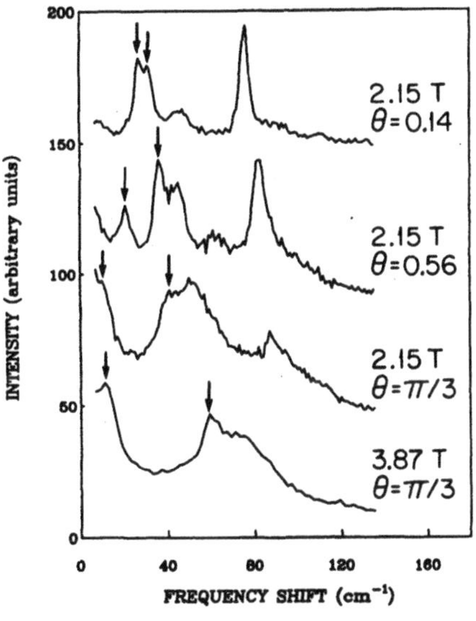

FIGURE 1. Electronic $z(x',y')\bar{z}$ Raman spectra of the L = 700 Å quantum well at various tilt angles (θ) and magnetic fields. Arrows denote transitions into coupled modes $|0,1\rangle$ and $|1,0\rangle$ (combined intersubband-cyclotron resonances). The temperature is T = 3.5 K and the laser energy is 1.873 eV.

For $\hbar\omega_c \geq E_{01}$, finite-basis calculations at large angles show (in most cases) strong mixing involving a large number of states with comparable weights. However, the lowest two branches contain nearly exclusi-

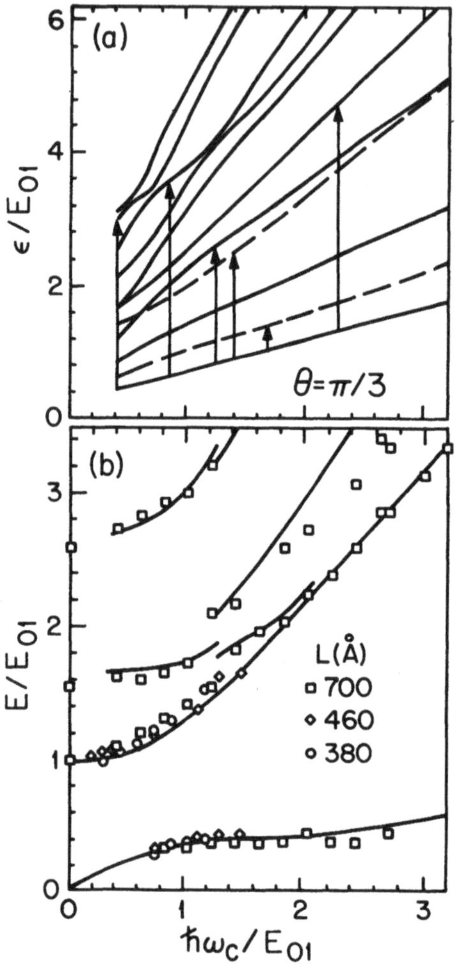

FIGURE 2. (a) Energy levels of an infinitely deep quantum well (arbitrary L) at $\theta = \pi/3$ as a function of cyclotron energy ($\hbar\omega_c$) calculated using the finite basis approximation discussed in the text. The scales are normalized to the energy separation between the two lowest quantum-well states at zero field, E_{01}. Solid (dashed) curves indicate even (odd) parity levels. Arrows denote transitions from the ground state observed in the experiments. (b) Comparison between calculated and measured transition energies ($\theta = \pi/3$). The second branch from the top at $\hbar\omega_c = 0$ is the intersubband excitation $|1,0\rangle \rightarrow |2,0\rangle$. Other branches correspond to the transitions labeled by arrows in (a).

vely a combination of $|0,1\rangle$ and $|1,0\rangle$. This and, in addition, the fact that the eigenfunctions $|0,0\rangle$ and $|1,0\rangle$ in parabolic and square wells are very similar,[14,15] provide a justification for the applicability of the parabolic-potential model to our problem. Actually, finite-basis and parabolic-well results[2,3] for the lowest two branches exhibit substantial agreement in a wide range of angles and fields. As shown in Figs. 3, our measurements also agree remarkably well with parabolic-potential predictions. The results for

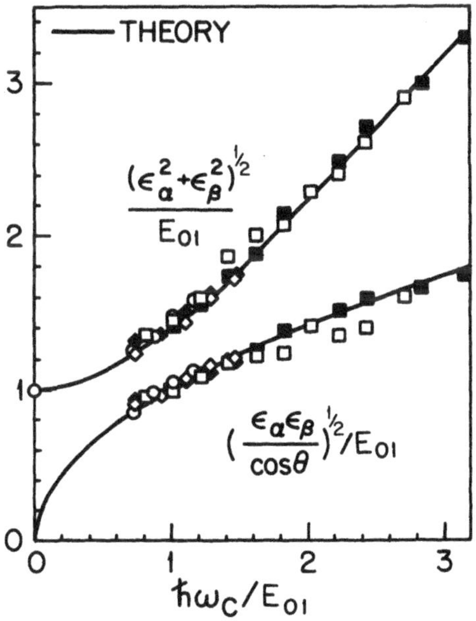

FIGURE 3. Scaled plot of experimental data for three samples (squares ≡ 700 Å, diamonds ≡ 460 Å and circles ≡ 380 Å) at two different angles. α and β denote the coupled intersubband-cyclotron modes ($|0,1\rangle$ and $|1,0\rangle$). Solid (open) symbols are for θ = 0.14 ($\pi/3$). The solid lines are results of the parabolic-well model [Eq. (7)].

the three samples and two different angles closely follow the scaled expressions [Eq. (7)] valid for parabolic wells. We remark that, *a priori*, there are no reasons for piecewise-constant potentials (including $V_0 = \infty$) to exhibit such scaling form.

Acknowledgements - The author wishes to thank R. Borroff and C. Tejedor for their contributions to much of the work discussed here. The preparation and characterization of the samples by P. K. Bhattacharya is gratefully acknowledged. This work was supported in part by the U. S. Army Research Office under Contract No. DAAL-03-89-K-0047 and the University Research Initiative Program Contract No. DAAL-03-87-K-0007.

1. See: T. Ando, A. B. Fowler and F. Stern, Revs. Mod. Phys. **54**, 437 (1982), and references therein.
2. J. C. Maan, in *Two Dimensional Systems, Heterostructures, and Superlattices*, ed. by G. Bauer, F. Kuchar and H. Heinrich (Springer, Berlin, 1984), p. 183.

3. R. Merlin, Solid State Commun. 64, 99 (1987).
4. W. Beinvogl and J. F. Koch, Phys. Rev. Lett. 40, 1736 (1978).
5. Z. Schlesinger, J. C. M. Hwang and S. J. Allen, Jr., Phys. Rev. Lett. 50, 2098 (1983).
6. G. L. J. A. Rikken, H. Sigg, C. J. G. M. Langerak, H. W. Myron, J. A. A. J. Perenboom and G. Weimann, Phys. Rev. B 34, 5590 (1986).
7. M. A. Brummell, M. A. Hopkins, R. J. Nicholas, J. C. Portal, K. Y. Cheng and A. Y. Cho, J. Phys. C 19 L107 (1986).
8. S. Huant, M. Grynberg, G. Martinez and B. Etienne, Solid State Commun. 65, 457 (1988).
9. A. D. Wieck, F. Thiele, U. Merkt, K. Ploog, G. Weimann and W. Schlapp, Phys. Rev. B 39, 3785 (1989).
10. K. Ensslin, D. Heitmann and K. Ploog, Phys. Rev. B 39, 10879 (1989).
11. M. Shayegan, T. Sajoto, J. Jo, M. Santos and H. D. Drew, Phys. Rev. B 40, 3476 (1989).
12. K. Karrai, H. D. Drew, M. W. Lee and M. Shayegan, Phys. Rev. B 39, 1426 (1989).
13. V. E. Kirpichev, I. V. Kukushkin, V. B. Timofeev and V. I. Fal'ko, JETP Lett. 51, 436 (1990).
14. R. Borroff, R. Merlin, R. L. Greene and J. Comas, Surface Sci. 196, 626 (1988).
15. R. Borroff, R. Merlin, J. Pamulapati, P. K. Bhattacharya and C. Tejedor, Phys. Rev. B 43, 2081 (1991).
16. N. C. Jarosic, E. Castano, B. D. McCombe, Y. C. Lee, J. Ralston and G. Wicks, Surface Sci. 170, 459 (1986).
17. R. G. Clark, S. R. Haynes, A. M. Suckling, J. R. Mallett, P. A. Wright, J. J. Harris and C. T. Foxon, Phys. Rev. Lett. 62, 1536 (1989); J. P. Eisenstein, H. L. Störmer, L. Pfeiffer and K. W. West, *ibid.* 62, 1540 (1989).
18. M. Born and K. Huang, *Dynamical Theory of Crystal Lattices* (Clarendon Press, Oxford, 1954).
19. See, e.g., G. Abstreiter and A. Pinczuk, in *Light Scattering in Solids V*, ed. by M. Cardona and G. Güntherodt, Topics Appl. Phys. 66 (Springer, Berlin, 1989), p. 153.

Wannier-Stark Quantization in Semiconductor Superlattices

Michel Voos

Laboratoire de Physique de la Matière Condensée de l'Ecole Normale Supérieure
24 rue Lhomond, 75005 Paris, France

ABSTRACT

The application of an external electric field parallel to the growth axis of a superlattice quantizes the energy continuum associated with the miniband dispersion into a Stark ladder of discrete energy levels, and transforms the extended Bloch waves into strongly localized wave functions. This Wannier-Stark quantization is clearly evidenced from electro-optical experiments performed in various semiconductor superlattices.

1. INTRODUCTION

Semiconductor superlattices (SL), which were first proposed by Esaky and Tsu[1], consist, as shown in Fig. 1, of periodic stackings of thin alternate layers of two different materials deposited on a suitable substrate by molecular beam epitaxy (MBE) or metalorganic chemical vapor deposition. In first approximation, they can be viewed as a series of quantum wells separated by potential barriers and coupled through resonant tunneling interaction. As a result of this coupling, the discrete energy levels, which would correspond to isolated quantum wells, broaden in the z-direction perpendicular to the plane of the layers and give rise to superlattice conduction and valence subbands or minibands, leading to band structures which are actually different from those of the corresponding host materials and can be tailored by changing parameters like the thickness of the layers for instance.

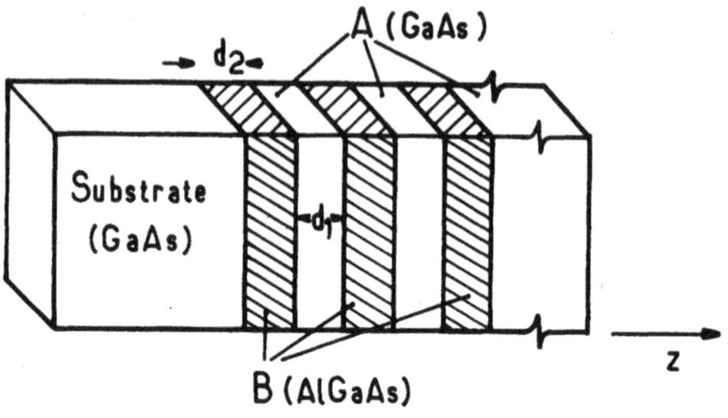

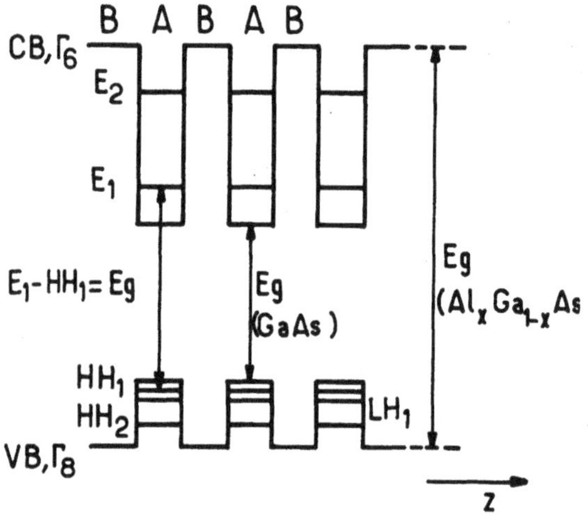

Fig. 1 : Schematization of a type I semiconductor superlattice. The conduction and valence bands are drawn in the z direction parallel to the growth axis, exhibiting quantum wells (A layers) and potential barriers (B layers). E_1, E_2..., HH_1, HH_2..., LH_1 are conduction, heavy-hole and light-hole subbands, respectively, with corresponding widths ΔE_1, ΔHH_1, for instance, in the z direction. The superlattice band gap is $E_g = E_1 - HH_1$. Γ_6 (Γ_8) is the symmetry of the conduction (valence) band.

Superlattices involving III-V compound semiconductors have been extensively investigated for the last fifteeen years about, and many new phenomena have been observed in these man-made heterostructures. Two striking examples are the quantum confined Stark effect[2] and the Wannier-Stark localization[3] which, depending on the layer thicknesses, result in a red and a blue shift of the band gap, respectively, when an electric field F is applied in the z direction.

We describe here the Wannier-Stark quantization (WSQ) which is interesting from the point of view of both basic and applied physics. This effect has been proposed recently[3] and observed independently in GaAs-AlGaAs SL's by Mendez et al[4] and Voisin et al[5], and also in GaInAs-AlGaInAs SL's by Bleuse et al[6].

2. THEORETICAL ASPECTS

The WSQ, which results from the application in the z direction of an electric field F to a semiconductor SL, has in fact been predicted and analyzed by Wannier[7] in the case of bulk materials, but never convincingly observed, probably because the required electric fields are much too large. The effect of an external electric field on the SL structure can be understood easily using a simple intuitive description. As already mentioned, a superlattice can be viewed as a series of quantum wells (QW) coupled through resonant tunnelling which results in a broadening of the QW discrete energy levels into a series of conduction and valence subbands of widths Δ_c and Δ_v, with $\Delta_c = \Delta E_1$ and $\Delta_v = \Delta HH_1$ if we consider only the ground conduction (E_1) and heavy-hole (HH_1) subbands. Besides, the band gap E_g^{QW} of the isolated QW is[3] larger than that of the SL, E_g^{SL}, by about $(\Delta E_1 + \Delta HH_1)/2$. Under an electric field F along z, the energy levels in adjacent QW's are misaligned by eFd, where d is the SL period (Fig. 2). Resonant tunnelling is thus inhibited and, for eFd $> \Delta E_1, \Delta HH_1$, the subband widths become equal to zero while the eigenfunctions are fully localized over each QW, leading to a two-dimensional (2D) situation. Since $E_g^{QW} = E_g^{SL} + (\Delta E_1 + \Delta HH_1)/2$, the fundamental optical transitions are, in these conditions, blue shifted by[3] $\Delta/2 = (\Delta E_1 + \Delta HH_1)/2$.

For F = 0, the eigenfunctions in a semiconductor SL are delocalized Bloch waves, which corresponds to a three-dimensional (3D) configuration and gives rise to a 3D absorption lineshape as illustrated in Fig. 3. In the large field limit (eFd $> \Delta_c, \Delta_v$) the SL behaves therefore as a series of uncoupled QW's with an E_1 wavefunction, for instance,

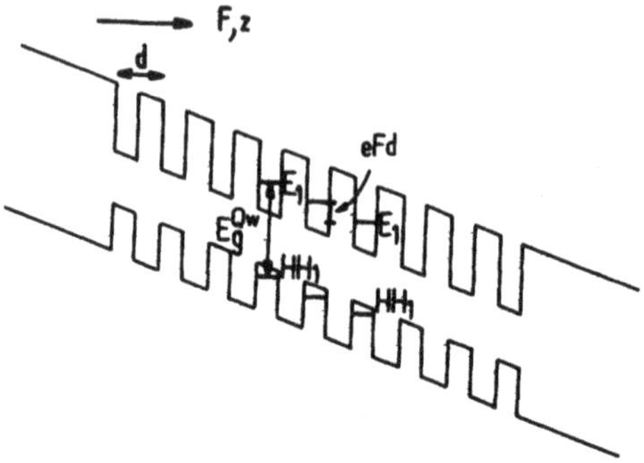

Fig. 2 : Band structure in real space of a type I superlattice when an electric field F is applied in the z direction. E_1 and HH_1 are the ground conduction and valence subbands; d is here the superlattice period, and E_g^{QW} is the band gap of isolated quantum wells.

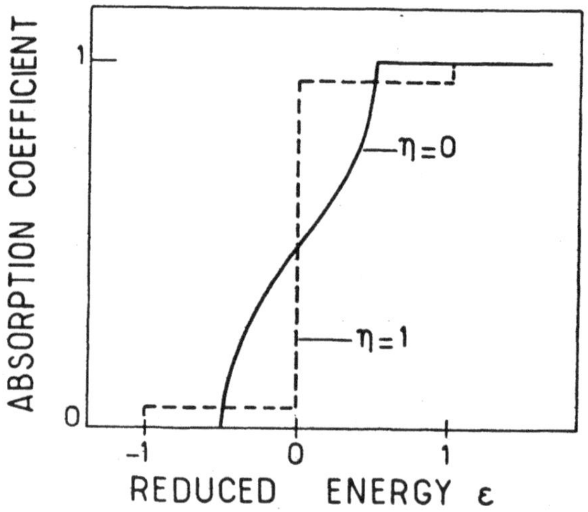

Fig. 3 : Absorption coefficient at zero and large reduced electric field η as a function of the reduced energy ε. As defined in the text, $\eta = eFd/\Delta$ and $\varepsilon = (h\nu - E_g^{QW}) / \Delta$..

localized over each quantum well, and it recovers a 2D step-like absorption spectrum (Fig.3). At intermediate fields, the wavefunctions extend over several quantum wells, and this leads to some overlap between conduction and valence eigenstates centered in different quantum wells. These overlaps should result in additional optical transitions which are indirect or oblique in real space, at the transition energies $E_g^{QW} \pm peFd$, where $p = \pm 1, \pm 2...$ and E_g^{QW} is the bandgap of the isolated quantum well (Fig. 4). This leads therefore to a quantization of the energy continuum associated with the miniband dispersion into a Stark ladder of discrete energy levels. When the electric field is swept at a fixed photon energy, these oblique transitions generate a new type of electro-optical oscillations periodic in 1/F, as illustrated in Fig. 5.

The one-band tight binding analysis of the envelope wavefunctions[3,8] provides a particularly convenient theoretical approach to this problem, and has the great merit of giving analytical expressions for the absorption coefficient. If edge effects are neglected (infinite SL), the energy spectrum corresponding to a given miniband (e.g. the ground conduction miniband) is exactly a Stark ladder :

$$E_\nu = E^{QW} + peFd \qquad (1)$$

The associated wavefunction X_ν is localized symmetrically in the vicinity of the νth quantum well and is expended as

$$X_\nu(z) = \Sigma_n J_{n-\nu}\left(\frac{\Delta_c}{2eFd}\right) \phi\,(z - nd) \qquad (2)$$

where $\phi(z)$ is the corresponding isolated quantum well eigenfunction, and $J_p(x)$ the Bessel function of integer index p. At large F (small x) all the J_p (x) but J_0 tend to zero, which means that the eigenstate $X_\nu(z)$ is entirely concentrated in the quantum well of index ν.

Assuming parabolic in-plane dispersion relations, the optical absorption coefficient is then given[9], in the case of type I SL like the GaAs-AlGaAs system, by :

$$\alpha(h\nu) = \frac{\alpha_0}{d} \Sigma_p J_p^2\left(\frac{\Delta}{2eFd}\right) Y\left(h\nu - E_g^{QW} + peFd\right) \qquad (3)$$

where $\alpha_0 \approx 5 \times 10^{-3}$ is the strength of the absorption for the HH_1-E_1 heavy-hole to

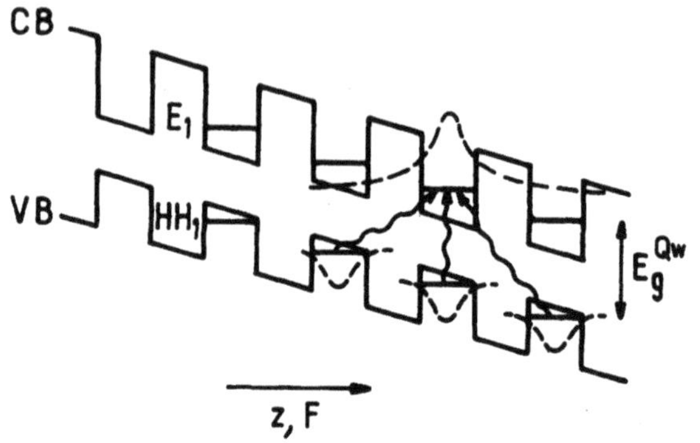

Fig. 4 : Schematization of oblique transitions in real space for intermediate electric fields F parallel to the superlattice growth direction z. The dashed lines give the behavior of the ground conduction (E_1) and valence (HH_1) wavefunctions.

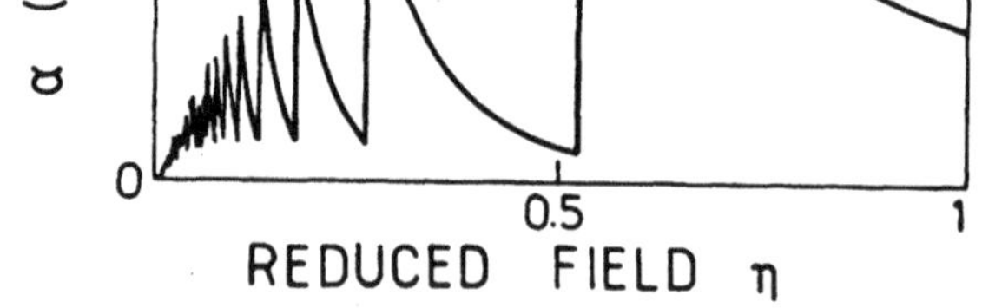

Fig. 5 : Absorption coefficient as a function of η, evidencing the electro-optical oscillations associated with the oblique transitions.

conduction transitions in a type I quantum well[8,10], $\Delta = \Delta_c + \Delta_v$ is the sum of the related conduction and valence miniband widths, and Y(x) is the step function at x=0.

From these expressions, it can be deduced that the absorption on the two-dimensional (2D) plateau (i.e. for a photon energy $h\nu >> E_g^{QW}$) saturates always to the same field-independent value, and that the absorption profiles depend on a single dimensionless parameter, the reduced field $\eta = eFd / \Delta$. In other words, the plots of α shown in Fig. 6 using the reduced field η and the reduced energy $\varepsilon = (h\nu - E_g^{QW}) / \Delta$ are universal. The absorption profile follows straightforwardly the joint density of state in both limits of zero field (quasi-3D miniband absorption) and large field (2D quantum well absorption). At large field, only the J_0^2 term survives, corresponding to electron and hole wavefunctions localized in the same well which give rise to vertical transitions, and the absorption is a single step at E_g^{QW}, like that of a series of uncoupled quantum wells. The transitions at $h\nu$ ($p \neq 0$) correspond to electrons and holes whose wave functions are centered in QW's distant from pd, which lead to the oblique transitions mentioned previously. Figure 3 compares the absorption profiles at $\eta = 0$ and $\eta = 1$, evidencing the $(\Delta_c + \Delta_v)/2$ blue shift of the absorption edge under electric field.

Fundamental optical transitions in type I semiconductor SL's involve usually free excitons which modify the optical absorption spectra. From this point of view, one should note that the WSQ results in an enchancement of the exciton binding energy and oscillator strength associated with a change in exciton dimensionality from 3D at zero field to 2D at large field.

3. EXPERIMENTAL RESULTS

These new electro-optical effects resulting from the WSQ have been first evidenced at low temperature in GaAs-AlGaAs[4,5], InGaAs-AlGaInAs[6], and InGaAs-GaAs[11] type I superlattices, using photocurrent, electro-reflectance and transmission spectroscopies. We present in Fig. 7 electro- absorption spectra obtained from direct low temperature optical transmission measurements[6] in a MBE-grown $Ga_{0.47}In_{0.53}As$-$Al_{0.24}Ga_{0.24}In_{0.52}As$ SL sandwiched between n-doped GaInAs layers to form a n-i-n structure. The thicknesses of the SL GaInAs and AlGaInAs layers are 50 and 55 Å, respectively.

The relatively small HH_1-E_1 excitonic peak occuring at 914 meV at zero bias (F = 0

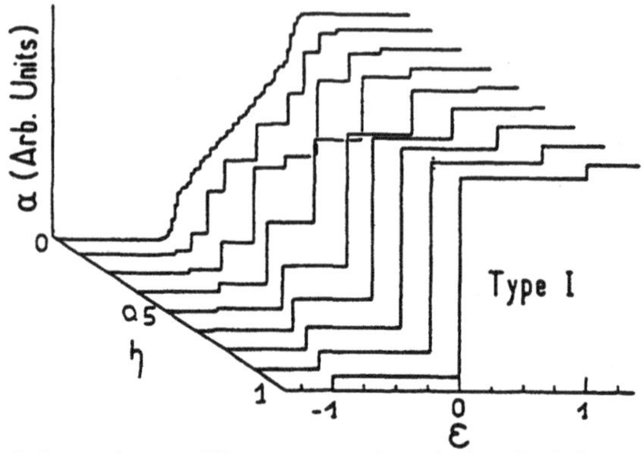

Fig. 6 : Plots of absorption profiles versus reduced electric field η and reduced energy ε for a type I superlattice.

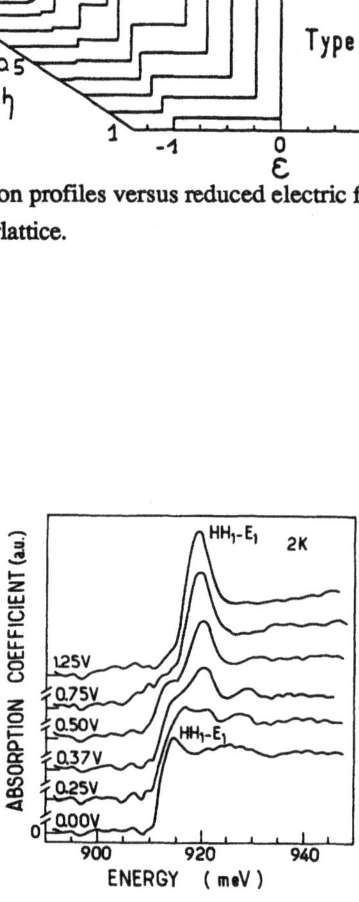

Fig. 7 : Electro-optical absorption spectrum at 2 K for different applied voltages as a function of the incident photon energy in the case of the GaInAs-AlGaInAs superlattice described in the text.

Vcm^{-1}) shifts clearly towards high energies when F is increased. At the same time, satellite structures above and below this exciton peak appear, develop and then fade away at larger fields. These structures correspond to the predicted "oblique transitions" at $E_g^{QW} \pm peFd$, and they are less pronounced at large electric fields, evidencing the increasing field induced localization. At a bias of 1.25 V ($F \sim 20$ $kVcm^{-1}$), the effective absorption edge is blue-shifted by about 5 meV, and the exciton line is clearly reaching saturation in intensity, while the extinction of the oblique transition is not quite perfect yet. The observed blue shift, if we ignore linewidth effects, results from the competition of the blue shift due to the Wannier-Stark quantization, and of the red shift corresponding to the increase of the exciton binding energy E_b. The latter effect can be estimated[12] reliably from earlier works, $E_b^{SL} \approx 2.5$ meV, and $E_b^{QW} \approx 8$ meV. It is thus deduced an experimental value of the bandgap blue shift of 10 meV about, which compares[6] favorably with the calculated value which is 9 meV. It is also noteworthy that the excitonic character of the absorption edge is strikingly reinforced at large electric field as a result of the change from a 3D to a 2D situation.

Another example of these new effects is provided by a study of the low temperature electro-reflectance (ER) in a GaAs-$Al_xGa_{1-x}As$ superlattice[5]. The sample was grown by MBE on an n^+-GaAs substrate. Nominally, this sample is a 160 period 33 Å-33 Å GaAs-$Al_{0.3}Ga_{0.47}As$ SL grown on top of a 8000 Å thick buffer layer of n^+-GaAs. the sample parameters and quality were checked by X-ray double diffraction, luminescence and photo-luminescence excitation (PLE) spectroscopy. Subsequently, a semi-transparent electrode of Cu_2S of large area was evaporated on the sample surface for the ER measurements. The low temperature (2K) PLE spectrum, shown in Fig. 8 indicates clearly the high quality of this sample. In particular, at higher energy than the heavy-hole HH_1-E_1 exciton at 1.658 eV, the onset of the HH_1-E_1 continuum or high order exciton states is well resolved, giving a heavy-hole exciton binding energy of 5 meV, in agreement with theoretical estimates[12]. The same structure can also be observed for the light-hole exciton lying at 1.677 eV. Furthermore, the broader structure observed near 1.708 eV is likely to correspond to the Van-Hove singularity of the joint density of states at $E_g^{SL}+\Delta E_1+\Delta HH_1$. In fact, the observed energies can be fitted to the theoretical ones, calculated within the three-band envelope function model, using parameters close enough to the target values. The calculated band structure parameters are summarized in Table I.

The low temperature and low voltage electroreflectance (ER) spectrum of this

Table 1 : Calculated band structure parameters of the GaAs-AlGaAs superlattice used here. All the energies are in meV. The origin is th bottom of the GaAs conduction band. E_1, HH_1, LH_1 correspond to the ground conduction, heavy-hole and light-hole subbands; ΔE_1, ΔHH_1, and ΔLH_1 give the widths of these subbands in the z direction; d_1 (d_2) is the GaAs (AlGaAs) layer thickness.

d_1(Å) GaAs	d_2(Å) AlGaAs	E_1	ΔE_1	HH_1	ΔHH_1	LH_1	ΔLH_1
31	35	107	45	- 1556	1	- 1575	29

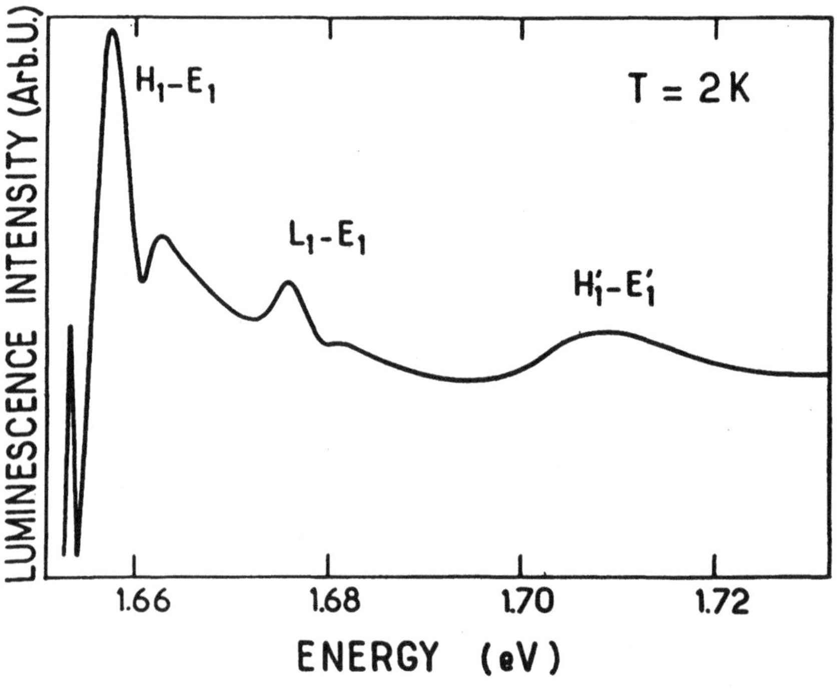

Fig. 8 : Low temperature (2 K) photoluminescence excitation spectrum at zero electric field F of the GaAs-AlGaAs superlattice described in the text.

superlattice is shown in Fig. 9a. In this experiment[5], the relative change in reflectance $(R(U) - R(U_0))/R(U_0)$, is measured, $U_0 = + 1$ V being a constant bias chosen as to compensate the effect of the Schottky barrier and to restaure the flat band conditions in the SL. The spectrum exhibits rich and sharp structures near E_g^{SL}, and also weaker and broader oscillations below and above, which are likely to be due to the Franz-Keldysh oscillations associated with the small depleted part of the GaAs buffer layer. As usual with derivated spectra, it is not easy to determine an accurate transition energy within the observed structure. However, in the present case, the comparison with the PLE spectrum shown in Fig. 8 allows an unambiguous identification, and, clearly, the sharp maxima A at 1.660 eV and B at 1.675 eV can be attributed to the E_1-HH_1 and E_1-LH_1 excitons, respectively, while the less pronounced maxima C at 1.715 eV and D at 1.76 eV involve most probably the zone boundary transitions E'_1 -HH'_1 and E'_1-LH'_1. The behavior of the ER spectra when the reversed modulation voltage U is increased is shown in Fig. 9b. The main line A and B clearly shift at first towards high energies, while their amplitude increases. Then at higher fields, both the blue shift and the amplitudes tend to saturate, and a hint of red shift is finally observed. This latter effect corresponds to the combination of the usual intra-well effect known at the "quantum confined Stark effect" due to the field induced deformation of the quantum well potential[2], and of the enhancement of the exciton binding energy caused by the localization. The maximum values of the blue shifts for the E_1-HH_1 and E_1-LH_1 band to band transitions are ≈ 25 meV and ≈ 40 meV for the E_1-HH_1 and E_1-LH_1 excitons, respectively, which is quite comparable[5] to the half sum of the related conduction and valence subband widths listed in Table I. Besides, other structures, which are likely[5] to correspond to the expected oblique transitions, can be seen in the spectra of Fig. 9b, in particular below the E_1-HH_1 exciton.

4. APPLICATIONS

It is obvious that one can think of the Wannier-Stark quantization to make electro-optical modulators but, to be of practical interest, such devices should be able to work at room temperature. As a native band structure property, Wannier-Stark effects do not depend on temperature, except via short lifetime broadening effects. They have in fact been observed at room temperature in GaAs-AlGaAs[13,14] as well as in InGaAs-AlInGaAs[15] SL's, even though somewhat blurred by thermal absorption tails. As can be seen in Fig. 6, there are two different regions of electro-modulation in the SL absorption. Below the bandgap, oblique transitions can lead to a new type of red shift modulation

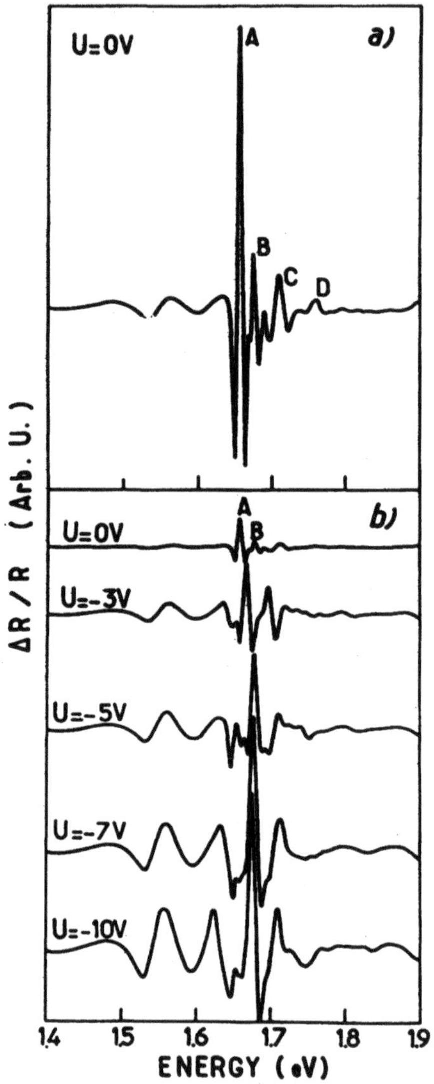

Fig. 9 : Low temperature (2 K) and low voltage electroreflectance spectrum of the GaAs-AlGaAs superlattice used here (a); evolution of the electroreflectance spectrum for increasing applied voltages (b).

challenging the 3D Franz-Keldysh effect or the 2D Quantum Confined Stark Effect and, at the bandgap, there is a region which can give rise to a blue shift modulation. A major feature is the fact that the large field regime (i.e. $\eta \geq 1$) is actually obtained with a rather small electric field, usually well below 100 kV/cm, which is a key for low dissipation in high frequency operation. High contrast blue shift modulation is essential for Self Electro-Optic Effect Devices[16] (SEED), and a SL optical bistability has been demonstrated[14] soon after the first reports of Wannier-Stark effects. Furthermore, the contrast of the blue shift modulator can be improved considerably by using a reflectivity configuration and a carefully designed asymmetric Fabry-Perot cavity[17], which led recently to very good SEED performances[18]. Besides, the oblique transitions have relatively small oscillator strengths, but appear at very low field in the region below the band-gap where the SL is transparent, and they are thus well suited to transmission modulation in a guided-wave configuration, with light propagating in the plane of the layers. A GaInAs-AlInAs SL optical waveguide modulator operating at 1.55 μm has recently been demonstrated[19]. This modulator has a large extinction ratio at very low driving voltage (< 1V), and is expected to have very good high frequency performances. Finally, the use of a symmetric Fabry-Perot cavity to increase artificially the effect of a small electro-absorption would certainly lead to modulation by the oblique transitions in a vertical configuration, resulting perhaps in the monolithic integration of a vertical cavity emitter[20] with an efficient modulator.

5. CONCLUSION

The Wannier-Stark quantization, which has been proposed and observed quite recently, has provided interesting results from the point of view of basic physics, and is likely to lead to new effects in the near future. Besides, it has already proven to be quite promising in the field of electro-optical modulators for long-wavelength optical communication systems, and it is certainly potentially important for electro-optical devices.

6. ACKNOWLEDGEMENTS

The author would like to thank G. Bastard, J. Bleuse, R. Ferreira, B. Soucail and P. Voisin from the Laboratoire de Physique de la Matière Condensée de l'ENS in Paris, M. Allovon, E. Bigan and M. Quillec from CNET in Bagneux, A. Regreny, from CNET in Lannion, and C. Alibert, C. Bouche and S. Gaillard from EM2 in Montpellier for their participation to the work described here.

7. REFERENCES

1. L. Esaki and R. Tsu, IBM J. Res. Develop. 14, 61 (1970).
2. G. Bastard, E.E. Mendez, L.L. Chang and L. Esaki, Phys. Rev. B 28, 3241 (1983).
3. J. Bleuse, G. Bastard and P. Voisin, Phys. Rev. Lett. 60, 220 (1988).
4. E.E. Mendez, F. Agullo-Rueda and J.M. Hong, Phys. Rev. Lett. 60, 1946 (1988).
5. P. Voisin, J. Bleuse, C. Bouche, S. Gaillard, C. Alibert and A. Regreny, Phys. Rev. Lett. 61, 1639 (1988).
6. J. Bleuse, P. Voisin, M. Allovon and M. Quillec, Appl. Phys. Lett. 53, 2632 (1988).
7. G. Wannier, Rev. Mod. Phys. 34, 645 (1962).
8. P. Voisin, G. Bastard and M. Voos, Phys. Rev. B 29, 935 (1984).
9. J. Bleuse, G. Bastard and P. Voisin, unpublished.
10. G. Bastard, C. Delalande, Y. Guldner and P. Voisin, Advances in Electronics and Electron Physics, edited by P.W. Hawkes, Vol. 72, 1, Academic Press, New York, 1988.
11. B. Soucail, N. Dupuis, R. Ferreira, P. Voisin, A.P. Roth, D. Morris, K. Gibb and C. Lacelle, Phys. Rev. B 41, 8568 (1990).
12. A. Chomette, B. Lambert, B. Deveaud, A. Regreny and G. Bastard, Europhysics Lett. 4, 461 (1988).
13. R.H. Yan, R.J. Simes, H. Ribot, L.A. Coldren and A.C. Gossard, Appl. Phys. Lett. 54, 1549 (1990).
14. I. Bar-Joseph, K.W. Goossen, J.M. Kuo, R.F. Kopf, D.A.B. Miller and D.S. Chemla, Appl. Phys. Lett. 55, 340 (1990).
15. B. Soucail, P. Voisin and M. Allovon, unpublished.
16. D.A.B. Miller, D.S. Chemla, T.C. Damen, T.H. Wood, C.A. Burrus, A.C. Gossard and W. Wiegmann, IEEE J. Quantum Electronics QE 21, 1462 (1985).
17. K.K. Law, R.H. Yan, J.L. Merz and L.A. Coldren, to appear in Appl. Phys. Lett. (1990).
18. K.K. Law, R.H. Yan, J.L. Merz and L.A. Coldren, Device Research Conference, Santa Barbara, USA (1990).

19. E. Bigan, M. Allovon, M. Carré and P. Voisin, Appl. Phys. Lett. 57, 327 (1990).
20. F. Koyama, S. Kinoshita and K. Iga, Appl. Phys. Lett. 55, 221 (1989).

SOME UNIQUE FEATURES IN THE PHYSICS OF QUANTUM CONFINEMENT

Raphael Tsu

University of North Carolina at Charlotte
Charlotte, NC 28223, USA

The application of traditional quantum mechanical methods to quantum confinement requires new approaches because states in a quantum well are generally represented by resonant states with relatively short life time. Approximations of these states by stationary states constitute a poor description. Moreover, in 3-dimensional quantum confinement,(3DQC), Pauli's exclusion principle for a system with partial confinement gives rise to situations quite different from the familiar atomic spectra.

1. INTRODUCTION

Since the introduction of superlattice[1)] and quantum wells[2)], intense activities have been devoted to the study of electronic and optical properties of these man-made quantum structures. Due to the finite mean free path, wavefunctions in the confined region lose their phase coherence via inelastic scatterings. In resonant tunneling, the requirement of significant couplings to the outside world, in this case, contacts, the confinement is usually moderate. Specifically, electrons are confined for only several periods of quantum oscillations, the states are far from being eigenstates. Instead of specifying the details of scattering, a complex potential is introduced. In one-dimensionally confined structures(1DQC) as in the usual quantum wells, electron-electron interaction is not important because the average distances between electrons are quite large. However, in 3DQC, the so-called quantum box, Pauli's exclusion principle must be taken into account because electron-electron interaction is vital in the understanding of situations where the occupation of these states leads to large potential shifts. Unlike atoms where electrons are bound to nuclei, confinement is effected by potential barriers, resulting in unusual spectra. For example, the one electron hydrogen-like states are drastically altered when a second electron enters the confinement region, resulting in a net charge.

2. DAMPING VIA A CLASS OF COMPLEX POTENTIALS

A complex potential, known as the optical-model potential[3], was introduced into the Hamiltonian as a highly abbreviated description of a much more complicated physical system. Specifically, the quantum mechanical analog of the Langevin equation may be represented by $H_1 = \pm(\hbar^2/2m)d/dx$, with $q=(2\ell)^{-1}$, and ℓ being the mean free path. It was shown[4] that the peak of the density of states is shifted up in energy together with line-broadening. (Damping in a LC oscillator also introduces broadening and up-shift in frequency.) We shall rehash[5] the consequence of introducing $H_1 = -i\hbar/\tau$. There are two possible solutions. The first has complex energy and real momentum, applicable to time damping, such as cyclotron resonance and atomic transitions. The second has complex momentum and real energy, which is what we shall discuss in more detail here. Substituting a plane wave e^{ikx} with $k = k_r+ik_i$, we arrive at

$$k_r^2 = k_o^2 + k_i^2 \ , \text{ and } \ k_i = Q^2/2k_r \ , \ k_o^2 = 2mE/h^2 \tag{1}$$

with Q defined by $Q^2 = 2m/\hbar\tau$. The density of states

$$n(E) = \text{Im } G_\pm(x=x') = k_r/k_o^2 \ , \tag{2}$$

has a maximum at $k_o^2 = 0.75\ Q^2$. Since the coherence length ℓ is related to k_i by $k_i = (2\ell)^{-1}$, the relationship $k_o^2 > 0.75Q^2$ gives rise to $k_r\ell > 1$. Figure 1 gives the dispersion relation E vs. k_r and n(E). The shaded region defined by $k_o^2 < 0.75Q^2$, represents localized states. Note that without damping, n(E) peaks at E = 0. With damping, n(E) peaks at $k_o^2 = 0.75Q^2$. The shift from E = 0 is precisely the self energy shift. The solution with complex k has been applied to resonant tunneling through a double barrier structure[5]. Results are reproduced in Fig.2. Note that the first peak is shifted up, but the second peak is rather insensitive to damping because confinement is already much reduced for states at higher energies comparable to the barrier height. In amorphous materials, where damping is high, broadening and shift are considerable.

If confinement is complete, elastic scatterings merely transform the eigenstates. However, with partial confinement, elastic scatterings can augment inelastic scatterings, therefore phase coherence will be affected even by elastic scatterings.

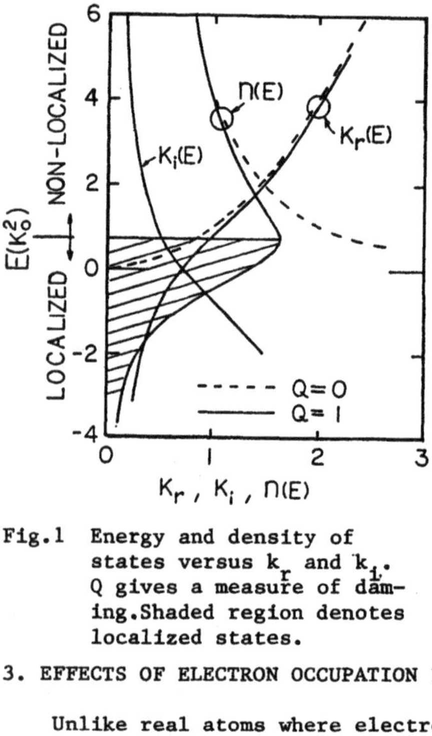

Fig.1 Energy and density of states versus k_r and k_i. Q gives a measure of damping. Shaded region denotes localized states.

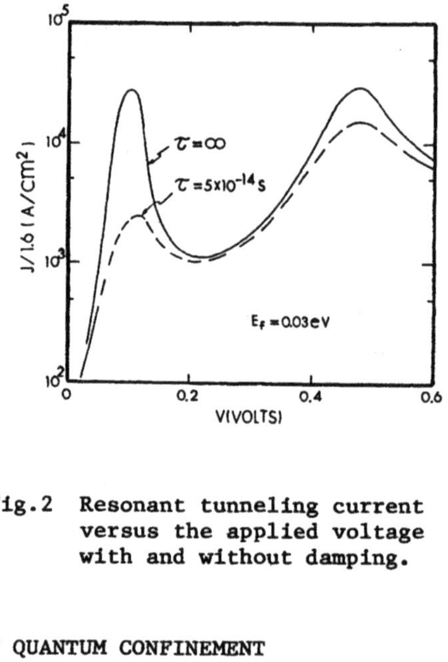

Fig.2 Resonant tunneling current versus the applied voltage with and without damping.

3. EFFECTS OF ELECTRON OCCUPATION IN QUANTUM CONFINEMENT

Unlike real atoms where electrons are bound to the nuclei,where electron charges are balanced by nuclear charges. In quantum confinement, whenever an electron enters the region, potential is raised described by Poisson's equation. In 1DQC, charges spread in a 2-D sheet, a self-consistent calculation is usually not necessary. In 3DQC, for example a sphere of 100Å diameter, each time an electron enters the region, a potential $\phi = e/c$ develops, with c being an effective capacitance. Recent[6] in resonant tunneling via microcrystalline Si embedded in an a-SiO_2 matrix, it was found that ϕ may reach a couple of volts. Thus additional electrons will be blocked not only by Pauli's exclusion principle, but more importantly, by the large increase in potential due to the first electron. Figure 3. gives some calculated one-electron energies for a cube and a sphere of comparable dimension of 75 Å.[7)] A second electron added to the box will occupy, together with the first electron, the ground state of a helium-like state.

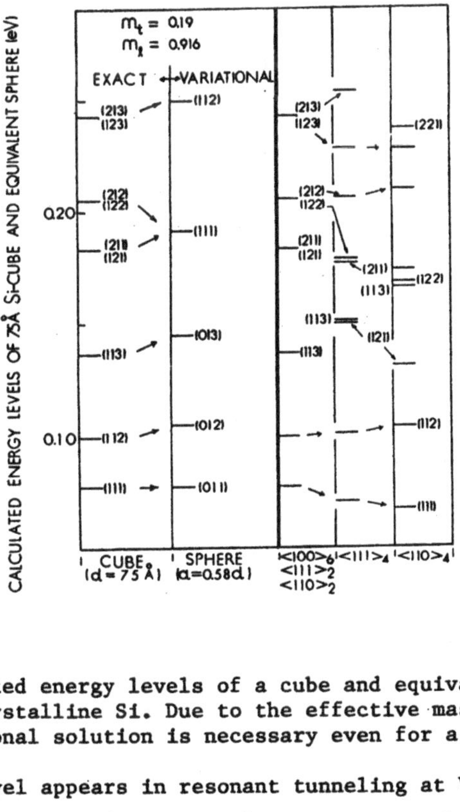

Fig.3 Calculated energy levels of a cube and equivalent sphere microcrystalline Si. Due to the effective mass tensor, a variational solution is necessary even for a sphere.

If the first level appears in resonant tunneling at $V = E_1 + e/c$, where E_1 is the ground state of the one electron state, the second level will appear at a voltage $V = E_2 + 2e/c$ with $E_2 = 2E_1 + E_{ee}$. The up-shift E_{ee} represents the energy due to Coulomb repulsion between the two electrons. The spectra for three electrons should resemble those for an Li atom. The excited states, as in He-atoms, exhibit a singlet state lying above the triplet states. Figure 4 shows the conductance in resonant tunneling through a single layer of microcrystalline silicon sandwiched between two a-SiO_2 barriers. At high gate voltage, the 3D delta function-like structure at low voltage is transformed into a 1D step-like structure which reflects the density of states for IDQC in a quantum well structure.

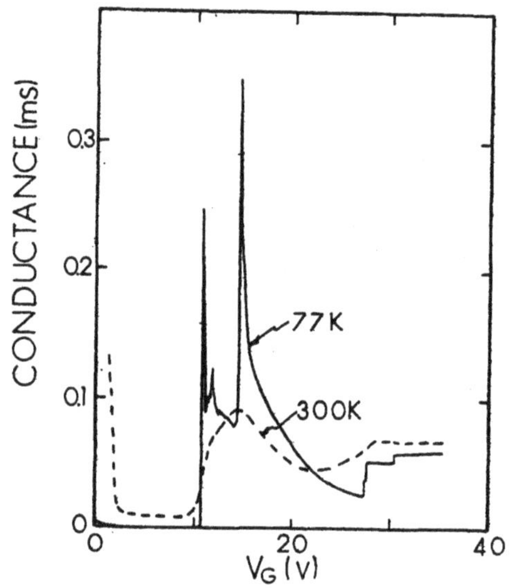

Fig.4 Conductance versus the applied voltage for a single layer of microcrystalline Si sandwiched between two a-SiO_2 barriers. The vertical bars on top give the calculated energies for a cube(C), and sphere(S). Note that unlike resonant tunneling in 1DQC, the voltage corresponding to resonant states is mainly dependent on potential rise from charging.

There is an important difference between the atomic spectra and quantum box spectra. In atomic spectra, Pauli's exclusion principle applies to eigenstates. If the confinement is significant, spectra of a quantum box resemble real atoms. However, in a moderate confinement, the wavefunction of a given state outside of the box is just as significant as that inside of the box. Electron-electron interaction may involve one electron inside with another electron outside such that the antisymmetric wavefunction used in calculating the Coulombic interaction should be affected by the boundary conditions one assigns to contacts. Thus the self-consistent treatment of the problem is much more complicated as the potential function may be quite complex. As far as tunneling is concerned, only a situation having nearly complete confinement is tractable. On the other hand, optical transitions are simpler to treat- no more or less difficult than atomic spectra.

ACKNOWLEDGMENTS

This work is supported in part by the Army Research Office Contract DAAL03-90-G-0067. I would like to thank R.F.Greene for many helpful discussions. And finally, I would like to express my deep pleasure in contributing this work paying special tribute to Prof. Rogerio C.C.Leite on his sixtieth birthday.

REFERRENCES

(1) Esaki,L.,and Tsu,R.,"Superlattice and Negative Differential Conductance in Semiconductors",IBM J.Res. and Dev.,14,61(1970).

(2) Tsu,R. and Esaki,L.,"Tunneling in a Finite Superlattice",Appl. Phys. Lett. 22,562(1973).

(3) For a brief description regarding normalization, see L.I.Schiff, "Quantum Mechanics",(McGraw Hill,NY 1968),p.129.

(4) Tsu,R.,"Effects of Mean Free Path on the Quantum Well Structure of Amorphous Materials",in Tetrahedrally-Bonded Amorphous Semicond. Ed. by D.Adler and H.Fritzsche(Plenum Press,NY 1985).

(5) Tsu,R.,Zypman,F.,"New Insights in the Physics of Resonant Tunneling Surface Science 228,418(1990).

(6) Tsu,R., Ye,Q.Y. and Nicollian,E.H.,"Resonant Tunneling in Microcrystalline Quantum Box Diode", Proc. SPIE, Aachen,1990,SPIE 1361, 232,(1991).

(7) Tsu,R. "Some New Insights in the Physics of Quantum Well Devices", Proc. SPIE,Aachen,1990,SPIE 1361,313(1991).

MULTIPASS AMPLIFICATION OF SHORT LIGHT PULSES

C. HIRLIMANN

Laboratoire de Physique des Solides, associé au CNRS, U.A. 154, Tour 13-3
Université Pierre et Marie Curie
4 Place Jussieu, 75252 Paris Cedex 05, France

The 1980's have seen the development of studies on dynamical processes in the femtosecond regime. These advances have been made possible by the underlying progress in the development of laser sources delivering light pulses substantially shorter than 100 fs [1]. One aspect of the new techniques has been the multipass amplification of laser pulses [2].

At first, light pulses amplification was done by means of dye cells [3], in which a strong population inversion is created in a volume, a few centimeters in length and a few hundred micrometers in diameter. Stimulated emission amplifies a light pulse traveling through the gain volume. This gain region is created by transversally pumping the dye cell with an external laser beam (very often a frequency doubled Nd:YAG laser) and its duration, which depends on both the pumping duration and the radiative recombination time of the dye, is typically on the order of 10 nanoseconds. When amplifying one picosecond light pulses using a 3 cm long cell, the gain duration and the effective amplification duration are one against the other like 1 is to 50 (assuming an index of refraction equal to 2 for the gain medium). The yield of such devices is poor, even if it gets slightly better when amplifying subpicosecond pulses, for in that situation, saturation of the gain is easier to achieve which improves the use of the inverted population. Anyway it is difficult to produce gains larger than a few cm^{-1} in dye cells.

The brite idea of the multipass amplification, which allows to reach the gain range of 100000 cm-1, has been simultaneously worked out by several authors [2]. A strong improvement of the amplification yield is achieved by passing the light pulse to be amplified several times in the same gain volume, making a better use of the stored energy. In the following, we will concentrate on the three mirrors preamplifier we developed [4-6] whose optical diagram is shown on figure 1.

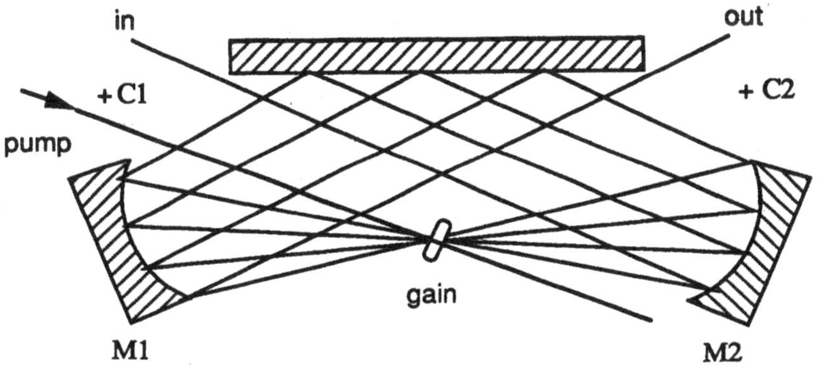

Figure 1 : Optical scheme of a multipass femtosecond amplifier. Mirror M1 (300 mm curvature) focusses the beams to the same point, while mirror M2 recollimates them. The flat mirror shifts the beam from one round trip to the next until the last beam escapes. A gain dye jet is set at the common focal point, whose excitation is provided by a Nd:YAG laser.

Two spherical mirrors (centers C1 and C2) are used in a 1:1 reflecting telescope geometry to focus and recollimate the light beam. Orthongonally to the symmetry axis of the figure, a plane mirror is used to fold the beam back onto the focussing mirror M1 and to insure the shift of the beam which makes multipassing possible. The number of passes is fixed by the position of the flat mirror and can therefore be easily varied. The length of this mirror has to be long enough to minimize the angle of incidence of the beam on the concave mirrors and therefore insure minimization of spherical aberrations. A gain jet is situated at the common focal point of the various passes. We use sulforhodamine 101 as a dye and it is pumped by a frequency doubled Nd:YAG laser having a repetition rate of 20 Hz. The energy of the YAG pulses is 10 mJ at 532 nm and their duration is 8 ns. The light pulses to be amplified are produced by a group velocity dipersion balanced CPM [1], their duration is 80fs at 620 nm and their energy is 20 pJ. The separation between mirror M1 and mirror M2 is 30 cm and a six passes scheme achieves the maximum total gain.

A paradoxical aspect of such a design is that the length of gain volume cannot exceed 1 mm due to the use of a jet, but this is largely overruled by a better use of the stored energy.

The following rate equations can be used to study the dynamics of the amplification in this system :

$$\frac{dN(t)}{dt} = \alpha I_p(t) - \beta\, N(t)\, I_{ASE}(t) - g \sum_n N(t)\, I_a(t)\, \delta(t - t_o - n\Delta) \qquad (1)$$

$$\frac{d\, I_{ASE}(t)}{dt} \propto I_{ASE}(t - \Delta)\, N(t) \qquad (2)$$

$$I_a(t_n) = \sum_1^n I_a(t_{n-1})\, N(t_n) \qquad (3)$$

N(t) describes the time evolution of the avalaible gain in the amplifying volume. The gain source is the pump pulse $I_p(t)$ which transfers energy to the dye molecules with yield α; neglecting the radiative recombinations, the gain decreases for two reasons : i) because of the amplification of the spontaneous emission (ASE) described by the product N(t) $I_{ASE}(t)$ in equation (1). ii) because of the amplification of the femtosecond pulses themselves, last term of equation (1). This last term is a discrete sum, over the number of passes n of the pulse, of finite amplification bursts N(t) $I_a(t)$. The duration of the amplification process is equal to the time of flight of the pulses through the jet and is therefore negligible when compared to the 2 ns total time of flight Δ between the mirrors and this justifies the use of the distibution function $\delta(t-t_0-n\Delta)$ in the last term of equation (1). Equation (2) describes the time evolution of the amplification of the spontaneous emission and takes into account the fact that there is a delay Δ between the emission of the first spontaneous photons and their amplification. Equation (3) accounts for the amplification of the short pulse having the intensity $I(t_0)$ before entering the amplifier.

This set of coupled non-linear equations includes the basic competition between the amplification of the short pulse and the unwanted amplification of the spontaneous emission collected by the mirrors. Figure 2 shows a simulation of the gain dynamics when using in equations (1) to (3) physical data measured in our system. Every time the femtosecond pulse enters the gain volume a strong amplification takes place which strongly depletes the inverted population. It can be observed that after the second pass the gain is completely turned-off by the amplification. This figure clearly illustrates the superiority of multipass schemes in which a larger part of the avalaible gain is used. This model, even if over simplified, also accounts for the saturation of the total amplification which is observed either when one raises the intensity of the pumping pulses or when more passes are added [5]. Competition between ASE and usefull amplification to the advantage of the former is responsible for both the effects. Therefore any improvement of a multipass amplifier has mainly to do with fighting the growth of the ASE.

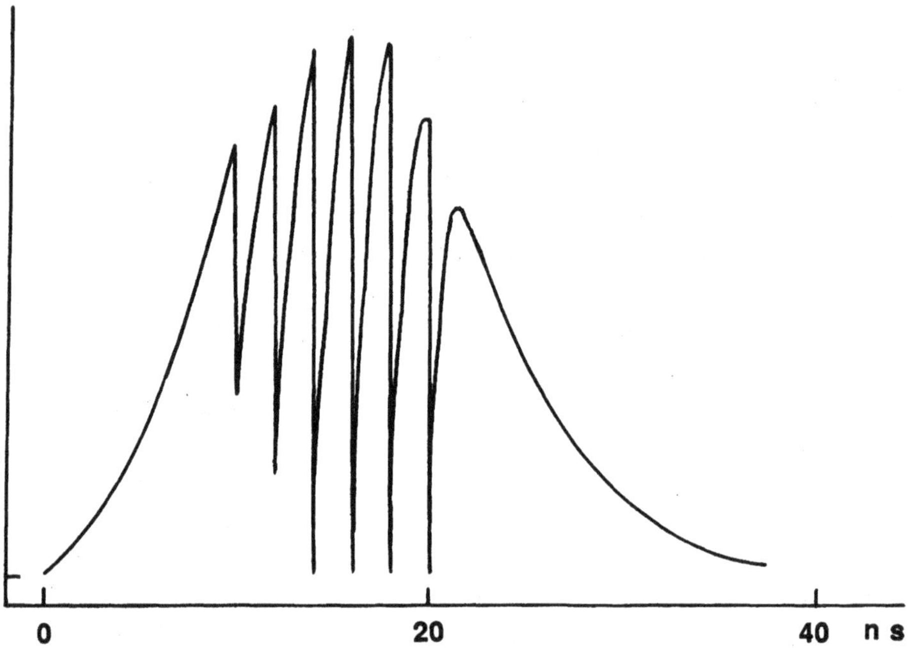

Figure 2 : Dynamics of the gain, in the device described in fig. 1, as obtained from the simple model described in the main text. The gain is depleted every time the pulse to be amplified passes through it, which allows for a better use of the stored energy.

Proposal

In the multipass amplifier described in reference (7), a saturable absorber jet is set between the fourth and the fifth passes, strongly reducing the ASE and enhancing the gain available for the short pulse during the last two passes. The symmetry of our preamplifier allows to replace the flat mirror by a couple of spherical mirrors in a way shown on figure 3. At the second focus of the new design a saturable absorber jet could be placed in order to fight the development of the ASE ; however the optical density of the saturable absorber should be set low enough for it to be able to open after the first amplification pass of the femtosecond pulse and it would therefore be of very low efficiency in discriminating against ASE during the last passes.

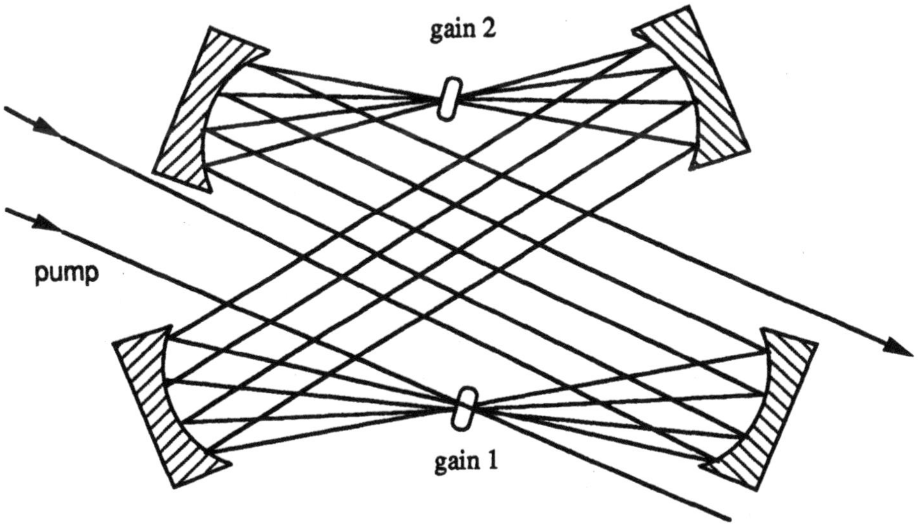

Figure 3 : Optical scheme of a two foci multipass amplifier, which could be used for "compound" amplification. The large focal telescope is used to produce the main gain by external pumping. The second telescope is used to tightly focus the ASE on an auxiliary gain jet and hence produce a second internally pumped amplification region. Variations on the optical scheme described in reference (6) could be used for the same purpose.

Before going to the proposal, let us discuss the following observation. After going through the six passes preamplifier of figure 1 the light goes through a dense malachite green saturable absorber jet to dynamically discriminate between the 20 Hz amplified femtosecond pulses and the ASE plus the remaining unamplified short pulses. This is done by using a 1:1, 3 cm focal length, refractive telescope, ensuring strong focussing on the jet.

It appears that at the output of the malachite green jet, the femtosecond pulse is amplified roughly two times. The origin of this amplification process is probably due to the existance of gain in the dye molecules pumped by the long lasting ASE pulse. In other words the saturable absorber is also a gain medium. A similar observation has been made in the device described in reference (2b p. 392) where gallium arsenide was used as a saturable absorber at a wavelength of 805 nm. This amplification effect is made possible because ASE is generated at energies higher than the energy of the amplified pulse and can therefore match the absorption band of the dye used as a saturable absorber.

These observations lead to the proposal that instead of simply eliminating ASE in multipass amplifiers, it would be better to use it as a pump for another amplifying dye and therefore improve the total yield of the system as well as the contrast ratio of the amplified pulses. This idea could easily be implemented using the optical scheme of figure 3 or any variation on this theme [6]. The main gain jet (the one externally pumped) is placed at one of the foci, while another gain jet, pumped by the ASE, is placed at the second focus. Further studies are necessary to optimize at least the nature and concentration of the auxiliary gain dye. Such a device would be a "compound" amplifier, by analogy with compound steam engines in which the steam leaving the main cylinder is sent, for further expansion, to another low pressure cylinder. Multistage, multipass amplifiers could be imagined, provided one can find a series of dyes all having their gain curve at the same wavelength and an absorption band well suited to absorb the ASE of the previous stage.

references

1. Valdmanis J.A., Fork R.L. and Gordon J.P., Opt. Lett. **10**, 131 (1985).

2. Two very good reviews on the generation and amplification of

femtosecond light pulses can be found in :

a) Simon J.D., Rev. Sci. Instrum. **60**, 3597 (1989).

b) Knox W.H., IEEE J. of Quant. Electron. **24**, 388 (1988).

3. Fork R.L., Shank C.V. and Yen R.T., Appl. Phys. Lett. **41**, 223 (1982).
4. Hirlimann C., Seddiki O., Morhange J-F., Mounet R. and Goddi A. Opt. Commun. **59**, 52 (1986).
5. Hirlimann C., La Revue de Physique Appliquée (F), **22**, 1673 (1987).
6. Fork R.L., Avramopoulos H., Fragnito H.L., Becker P.C., Sherer K.L. and Hirlimann C., Opt. Lett. **14**, 1068 (1989).
7. Knox W.H., M.C. Downer W.H., Fork R.L. and C.V. Shank R.L., Opt. Lett. **9**, 552 (1984).
8. Knox W.H., J. Opt. Soc. Am. **B4**, 1771 (1987).

ON PHOTOINDUCED TRANSIENT PHENOMENA IN METAL SURFACES EXPOSED TO HIGH FLUENCE SHORT LASER PULSES

C. A. S. Lima, M. B. S. Lima and M. A. Algatti
Lab. for Laser Processing of Materials – Dept. of Quantum Electronics – Instituto de Física "Gleb Wataghin"-UNICAMP
Campinas, 13081 - S. Paulo - Brasil

I have had the privilege of knowing Rogério for a long while. Though I was already well acquainted with his outstanding career abroad, as a Physicist, I first met him only in the early 70's when I was finishing my Ph.D. work at U. California, Berkeley. He had returned to Brazil and was at that time actively engaged in contacting promising young Brazilian Physicists abroad and inviting them to come to Campinas, and join a selected group of prominent scientists from several countries that he had invited there, in building up the staff for the Instituto de Física Gleb Wataghin (IFGW) at the Universidade Estadual de Campinas-UNICAMP. The foundations were then being laid for a course of actions which were soon to place the IFGW among the outstanding institutions of international reputation in Physics. We all owe that to Rogério. Brazilian science owes much to him for that. Joining the staff at Campinas gave me the opportunity to see him in action at his best, teaching, educating, exchanging with scholars from all over the world, doing Physics and promoting it to the benefit of his fellow researchers at IFGW, which he steadily and wisely directed for so many years. I had then the opportunity to know him better, both as a Physicist and as a friend. I am very happy to offer, also in the name of my colleagues (M. B. S. L. and M. A. A.), this contribution which was written as a token of our friendship with and admiration for him. (C. A. S. L.)

1 INTRODUCTION

More than two decades have elapsed since the first reports dealing with the experimental observation of nonlinear photoelectric emission from metal as induced by a laser beam, both at the lowest order of nonlinearity[1] and at higher orders[2,3]. These early attempts were prompted by the development of sufficiently powerful lasers, making possible the excitation of the many-photon photoelectric effect. Yet, many such investigations were limited in character since the measurements were frequently disturbed by the contributions to the emission current coming from mechanisms other than the pure many-photon photoeffect. Though measurement techniques were continuously improving, quantitative experiments always faced the difficulties posed by the low values of the nonlinear photocurrent, which demanded increasingly higher power laser pulses. Thus, one could not avoid the laser heating of the photocathode which resulted in a thermoionic contribution to the current, masking the details of the true photoelectric emission. It was not until the past few years that technical developments resulting in ultrashort laser pulses were advanced. In fact, the techniques of mode-locking[4] and pulse compression[5] have made it possible to successively achieve nano, pico and sub-picosecond laser pulses in the visible and IR spectral regions, for a variety of laser types, including CO_2 and Nd-doped YAG or glass lasers. This offered a completely new realm of experimental possibilities to investigate in detail laser induced transient phenomena by probing them under the adequate time scale. Among the many such phenomena[6–9], we have chosen to deal here with some theoretical aspects of multiphoton photoeffect, referred to in the foregoing, and of transient phenomena associated with the generation of nonequilibrium ultrahot electrons in a relatively cold metal lattice, by coupling high fluence ultrafast laser pulses to the metal surface.

The generation of an electron temperature much greater than the lattice temperature has received much attention in recent years in view of the advances in the techniques for production of pico and sub-picosecond laser pulses. Indeed, this has opened up the possibility to probe transient phenomena that take place when the pulse duration is less than the characteristic times for energy exchange between the electron and ion subsystems in metals. The experiments seem to be consistent with the theoretically advanced possibility of significantly large non-isothermality amounting, in some cases, to temperature differences of several hundred degrees between the two subsystems. In Section 3 we discuss the main features of the anomalous thermal emission observed when metal targets are bombarded with ultrashort infrared laser pulses. Numerical computations have been carried out using an energy-balance model based on a 1-d heat flow equation, where the electronic and lattice temperature are coupled together, allowing us to predict the evolution of both electronic and ionic temperatures. The large non-isothermality predicted is discussed and its existence tied to appearance of that anomalous (black-body) emission.

As an almost complementary effect to the forementioned anomalous light emission, the nonlinear pure photoelectric emission from the laser irradiated metal surface shares with the laser assisted thermionic photoemission the build up of the observed nonlinear photocurrent. We describe it in Section 2 as the outcome of a dynamic tunneling of electrons through a laser modified (reduced) metal-vacuum barrier. A non-perturbative approach is formulated based on exploring an unitary transformation (specifically a space translation – ST) to incorporate the effects of the intense laser field upon the surface permeability to the impinging electrons, assumed to form a Sommerfeld's free electron gas within the metal. The WKB approximation is then used to compute the electrons tunneling probability through the modified barrier. The integrated charge and average photocurrent are then calculated therefrom. Computations were made for the nonlinear photoelectric emission from gold surfaces bombarded with 2 *ns*, 0.2 to 1.0 $GWatt/cm^2$, CO_2 laser pulses. The results are shown to be in excellent agreement with experimental results available in the literature[10,11]. They are consistent with a much lower order of nonlinearity for the photoeffect ($n \sim 2$ instead of an extremely unprobable $n = 40$, n being the effective number of photons absorbed) and with a dynamic field induced lowering of the potential barrier height to an effective value of $\phi' \sim 0.2\ eV$.

2 ON A SIMPLE DERIVATION FOR THE LASER INDUCED NONLINEAR PHOTOCURRENT

We start from the Schroedinger equation for an electron under a potential $V(\vec{r})$ and additionally subject to the action of a laser field described by the vector potential $A(\vec{r},t)$:

$$\left[\frac{1}{2m}\left(\frac{\hbar}{i}\nabla + \frac{e}{c}\vec{A}(t)\right)^2 + V(\vec{r})\right]\Psi(\vec{r},t) = i\hbar\frac{\partial\Psi(\vec{r},t)}{\partial t} \qquad (1)$$

where the dipole approximation has been assumed. Now, we apply to it a unitary transformation in the form of a space translation[12,13], which we have also used in several other contexts[14–18], namely:

$$\Omega = e^{i\vec{\delta}(t)\cdot\vec{p}/\hbar}\, e^{-\eta(t)/\hbar}$$

where

$$\vec{\delta}(t) = \frac{-e}{mc}\int^t \vec{A}(t')\,dt'\ ;\ \ \eta(t) = \frac{-e^2}{2mc^2}\int^t \vec{A}^2(t')\,dt'$$

and

$$\Psi(\vec{r},t) = \Omega\phi(\vec{r},t)$$

The resulting equation is

$$\frac{1}{2m}\left[\frac{\hbar}{i}\nabla\right]^2 \phi(\vec{r},t) + V(\vec{r} - \vec{\delta}(t))\phi(\vec{r},t) = i\hbar\frac{\partial\phi(\vec{r},t)}{\partial t} \tag{2}$$

To represent the conduction electrons confined to the metal we work within the Sommerfeld model and take the metal-vacuum barrier as a generalization of Cole's[18] potential in the form

$$V(x) = \begin{cases} \dfrac{2V_0}{1+e^{x/a_0}} & x \geq 0 \\ 0 & x < 0 \end{cases} \tag{3}$$

where V_0 is a constant dependent on the metal chosen and a_0 is the pertinent lattice parameter. This potential is depicted in figure 1.

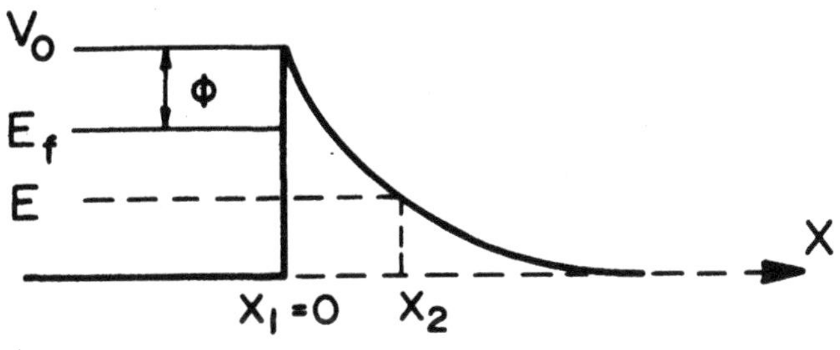

Figure 1: **The metal-vacuum potential barrier**

Accordingly, we have for the ST-transformed potential, in its 1-d version,

$$V(x - \delta(t)) = \frac{2V_0}{1 + e^{(x/a_0) - \lambda\cos\omega t}} \tag{4}$$

where

$$\lambda = \frac{a}{a_0}; \quad a = \frac{e\mathcal{E}}{m\omega^2}; \quad \vec{\mathcal{E}}(t) = \frac{-1}{c}\frac{\partial\vec{A}}{\partial t}; \quad \vec{A} = \vec{A}_0\sin\omega t$$

Now, within the Sommerfeld model the tunneling attempt frequency for normal incidence is given by

$$\omega_t = \frac{1}{2a_0}\left(\frac{2E}{m}\right)^{1/2}$$

which in our case (CO_2 laser, gold sample) turns out to be larger than the laser frequency ω. We can then use the time-average of $V(x - \delta(t))$ which, to second order in λ, turns out to be

$$\langle V(x-\delta(t))\rangle = \frac{2V_0}{1+\left[1+\left(\frac{a}{2a_0}\right)^2\right]e^{x/a_0}} \tag{5}$$

Now resorting to the usual WKB approximation to calculate the tunneling probability we get

$$P(E) = \exp\left\{\frac{-2\sqrt{2m}}{\hbar}\int_{x_1}^{x_2}\left[\frac{2V_0}{1+\left[1+\left(\frac{a}{2a_0}\right)^2\right]e^{x/a_0}} - E\right]^{1/2} dx\right\} \tag{6}$$

which integrates out to

$$P(E,\beta) = \left[\frac{\beta(1+\alpha)}{2+(1-\alpha)\beta+2\sqrt{1+(1-\alpha)\beta-\alpha\beta^2}}\right]^{\gamma} e^{\gamma\sqrt{\alpha}\pi/2}\times \tag{7}$$

$$\exp\left\{\gamma\sqrt{\alpha}\arcsin\left[\frac{-2\alpha\beta+1-\alpha}{1+\alpha}\right]\right\} \tag{8}$$

for $1 \le \beta \le 1/\alpha$, and where the following adimensional quantities have been introduced:

$$\alpha = \frac{E}{2V_0 - E};\quad \gamma = \frac{2a_0}{\hbar}\sqrt{\frac{2mE}{\alpha}};\quad z = \beta e^{x/a_0};\quad \beta = 1+\left(\frac{a}{2a_0}\right)^2$$

In figure 2 we give $P(E,\beta)$ as a function of β (which is linear in the laser intensity as can be seen from its definition above). It can be easily seen that $P(E \to E_F, \beta \to 1/\alpha) \to 1$.

Next, we obtain for the charge emitted over one laser pulse duration τ:

$$Q(\tau) = \int_{\theta=0}^{\pi/2}\int_{t=0}^{\tau}\int_{E=0}^{E_F} e\, g(E)\mathcal{F}(E)P(E,\beta)\omega_t(E)\; ds\; a_0\cos\theta\; d\theta\; dt\; dE \tag{9}$$

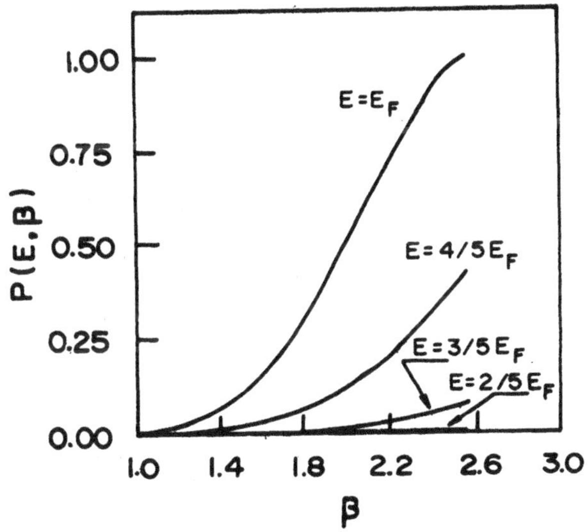

Figure 2: **Tunneling probability for gold. Here E is the electron energy.**

where

$$\omega_t = \left(\frac{2E}{m}\right)^{1/2} \frac{\cos\theta}{2a_0}$$

$$g(E)\mathcal{F}(E) = \frac{(2m)^{3/2}E^{1/2}}{2\hbar^3\pi^2}$$

Then, from (9), we get

$$Q(\tau) = \frac{me\tau E_F^2\; ds}{4\pi\hbar^3}\int_0^1 E'\, P(E',\beta)\; dE' \tag{10}$$

which was solved numerically (VAX-11/785 computer). The results for gold ($E_F = 5.5\ eV$ and $\phi = 4.3\ eV$) are given in figure 3.

We can see from figure 3 that $n = d\log Q/d\log I$ has a value $n = 2.23$ which compares quite favorably with the experimental result $n = 2.2 \pm 0.1$, obtained for this intensity range, under the same irradiation conditions as taken in our calculations. Also, the observed trend of change in n with the laser intensity agrees with experiment.

We conclude that despite the underlying simplicity of our model the relevant features of the nonlinear photoelectric effect in gold were clearly reproduced.

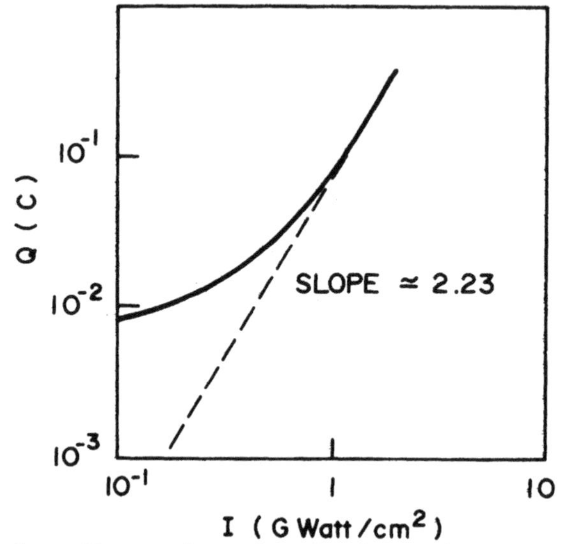

Figure 3: **Total nonlinear photoelectric charge from a gold sample irradiated with a CO_2 laser pulse.**

3 ELECTRON HEATING BY ULTRASHORT LASER PULSES: TRANSIENT ELECTRON-LATTICE THERMAL UNBALANCE

When acted upon by an intense short laser pulse a metal surface will first absorb photon energy via conduction electrons, which are heated up but promptly thermalize collisionally among themselves[19]. This is followed by energy transfer to the lattice by electron-phonon interactions. In a regime where the pulse duration is large as compared to the characteristic electron-phonon relaxation time, thermal equilibrium between electrons and the lattice will set in. On the opposite regime, i. e. intense ultra-short laser pulses, the electronic temperature will substantially differ from that of the lattice, rendering the metal a two-temperature system[20,21]. This situation makes the metal surface the seat of several interesting new phenomena[22–28]. We shall be concerned below only with the one relating to the observation of the anomalous thermal light emission by the ultra-fast optically heated surface. The overall framing of the phenomenon comprises:

- Theoretical considerations indicate that a high degree of non-isothermality between electronic and ionic subsystems in metals *is possible !*

- Thermodynamic non-equilibrium is largely determined by the characteristic time for heat exchange between the forementioned subsystems (estimated for metals to run from 0.1 to 10 ps).

- Normal thermal radiation from heated metals (up to the regime of strong evaporation) in the visible region are ascribable to radiation from bodies in thermal equilibrium (normal grey body).

- To call for the existence of substantive non-isothermality in a metal sample means to recognize the existence of an unusual situation (of both practical and scientific importance) where the electron plasma emits in the visible region of the spectrum like a hot grey body but the lattice is *"cold"* (i. e. has a temperature at which the peak emission would be in the IR or FIR regions).

- To expose (both theoretically and experimentally) this anomalous situation would require separating it from other masking effects. To this end one can, for instance, make use of ultrashort infrared laser pulses. We shall be concerned here with the theoretical consequences of such a procedure.

- The accumulated results from our and other people's work point out in a direction so as to allow us to assert that the phenomenon of anomalous thermal radiation from metals has been theoretically substantiated and experimentally demonstrated.

In short, it can be demonstrated, e.g. for gold and nickel samples, that when metals are heated under certain conditions, previously unknown anomalies become present in the visible glow. In fact, depending on the value of the heating power (laser intensity for instance) and the duration of heating, there appears in the visible region of the spectrum a light emission whose intensity is much in excess to what one would reasonably expect for normal radiation from a metal sample in thermal equilibrium. The numerical calculations needed to expose it are to be made in accordance to a given theoretical ansatz to be specified so as to take into account the relevant optical, thermal and electronic properties of the metal sample, as well as the characteristics of the irradiation source. We have developed such a program and will give below a summary of our procedures and the results obtained. We took, as the heating source, a Neodimium:YAG laser with focussed output of up to tens of $GWatts/cm^2$ in the form of a long train of equally spaced pulses each with a typical width of a few picoseconds.

We started from the pair of coupled equations for the electronic T'_e and ionic T'_i temperatures, describing the thermal coupling between the laser and the metal surface:

$$AT'_e \frac{\partial T'_e}{\partial t'} = B \frac{\partial}{\partial z'} \left[\frac{T'_e}{T'_i} \frac{\partial T'_e}{\partial z'} \right] - (T'_e - T'_i) + C \exp\left(-\alpha \delta z'\right) f(t') \qquad (11)$$

$$D \frac{\partial T'_i}{\partial t'} = (T'_e - T'_i) \qquad (12)$$

In these equations:
$z' = z/\delta$; $t' = t/\tau_{laser}$; $T'_{e,i} = T_{e,i}/\Theta_D$ and

$$A \equiv \frac{\gamma \Theta_D}{g \tau_{laser}}; \quad B \equiv \frac{\chi_0}{g \delta^2} \quad ; \quad C \equiv \frac{(1-R)\alpha I_0}{g \Theta_D}; \quad D \equiv \frac{c_v}{g \tau_{laser}}$$

where

$$g = \frac{\pi^2 n_e m_e v_s^2}{6 \tau_{e-f} T_e} \quad ; \quad \gamma = \frac{c_e}{T_e}$$

τ_{e-f} = electron-phonon relaxation time
χ_0 = equilibrium ($T_e = T_i$) thermal conductivity
δ = optical skin depth
Θ_D = Debye temperature for the metal
τ_{laser} = Laser pulse duration (FWHM)
α = Metal absorptivity
R = Metal reflectivity
I_0 = Laser pulse intensity

We have solved these equations numerically for both the electrons T_e and ions T_i temperatures, using as input the parameters values as given in Table 1, for each metal surface of interest.

The equations (11) and (12) were solved under the boundary condition that the heat flux across the surface is equal to the power irradiated on it i.e.:

$$\left. \frac{\partial T'_{e,i}}{\partial z'} \right]_{z'=0} = E \frac{T'_i}{T'_e} \left[T'^4_{e,i}(z'=0,t') - T'^4_{e,i}(z'=0,t' \to \infty) \right] \qquad (13)$$

In equation (13) $E \equiv \sigma \varepsilon \delta \Theta_D^3 / \chi_0$ is an adimensional parameter, whose values were calculated for Au, Ag, Ni and W and are given in Table 2.

Table 1: Physical parameters used in the calculations

Metal	$n_e \times 10^{22}(cm^{-3})$	$\rho(\mu\Omega cm)$	$v_s(m/s)$	$\Theta_D(K)$
Au	5,931	2,20	3240	165
Ag	5,856	1,61	3650	225
Ni	18,54	7,04	6040	450
W	37,93	5,33	5410	400

Metal	$\chi_0(Wm^{-1}K^{-1})$	$\gamma(mJmol^{-1}K^{-1})$	$g \times 10^{17}(Wm^{-3}K^{-1})$	δ (Å)
Au	318	0,690	1,1646	119
Ag	428	0,650	1,0544	121
Ni	91	7,1	126,6	165
W	170	1,3	321,7	232

Metal	$\alpha \times 10^7(m^{-1})$	n	k	$c_v(calmol^{-1}K^{-1})$	$\tau_{e-f} \times 10^{-14}(s)$
Au	8,3815	0,272	7,07	6,06	2,7158
Ag	8,2867	0,226	6,99	6,07	3,7586
Ni	6,046	2,85	5,10	6,23	0,2714
W	4,315	3,00	3,64	5,81	0,1753

Metal	$A \times \tau_{laser}$	B	$C/(1-R)I_0$	$D \times \tau_{laser}$
Au	$9,6287 \times 10^{-2}$	19,2822	$4,3618 \times 10^{-2}$	21,4436
Ag	0,1349	27,7248	$3,493 \times 10^{-2}$	23,4239
Ni	$3,8859 \times 10^{-2}$	$2,6402 \times 10^{-2}$	$1,0613 \times 10^{-4}$	0,317
W	$1,6969 \times 10^{-3}$	$9,818 \times 10^{-3}$	$3,3535 \times 10^{-5}$	$7,9325 \times 10^{-2}$

Table 2: **Calculated values of the adimensional coefficient E used in the expression for the boundary condition given by Eq. (13). Here ε is the metal emissivity and σ is the Stefan-Boltzmann constant. The other constants are as previously defined.**

Metal	$E \equiv \sigma\varepsilon\delta\Theta_D^3/\chi_0$
Au	4.77×10^{-13}
Ag	9.13×10^{-13}
Ni	4.6839×10^{-11}
W	2.48×10^{-11}

Our calculations were carried out by an IBM-3090 computer, using an appropriately developed numerical computation package to handle the coupled system of differential equations, as given above. The details can be found elsewhere[28]. A sample of the resulting calculated temperature profiles are presented in Figures 4 to 9 corresponding to several specific experimental situations. An inspection of these profiles makes evident the importance of the light pulse regime and beam

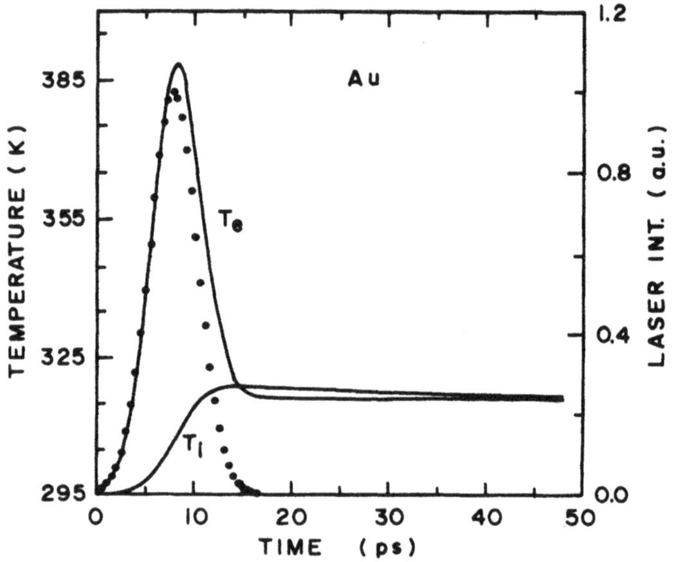

Figure 4: **Electrons (T_e) and lattice (T_i) calculated temperature profiles for gold samples under 500 $MWatt/cm^2$ irradiance, 6 ps long Nd:YAG laser pulses. The • curve is the laser pulse.**

fluence upon the attainment of substantial thermal non-equilibrium between electrons and lattice. It is also evident that in the ultra-short pulse regime the electron temperature follows the laser intensity without delay, implying that the hot electron radiation pulse has a duration close to that of the laser pulse, i. e. the hot electron emission is instantaneous. Of course, for larger times this emission is completely taken over by the normal equilibrium radiation as the electronic and ionic temperatures approach each other asymptotically. For the conditions taken in our calculations, we are entitled to describe the spectrum of the hot electrons radiation by the Planck formula with characteristic temperature T_e[28]. Overall, the characteristics exhibited by the electronic temperature profile while the ultra-fast

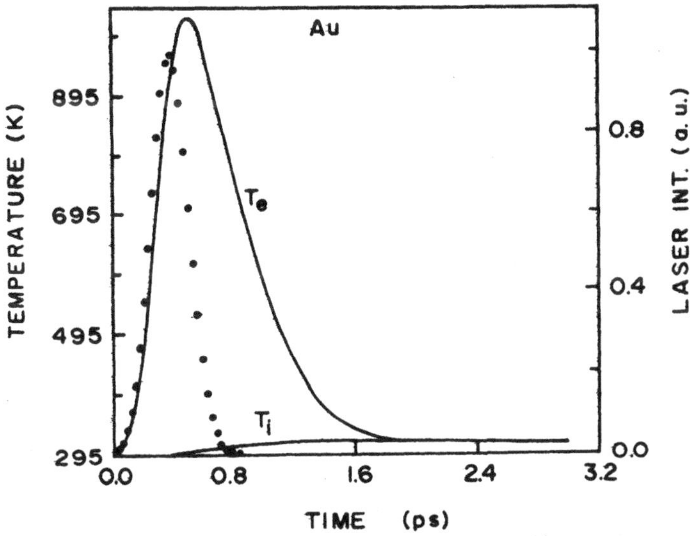

Figure 5: Same as in figure 4 but for 10 $GWatt/cm^2$ irradiance and 300 fs pulses. (Notice that the laser fluence has been kept the same).

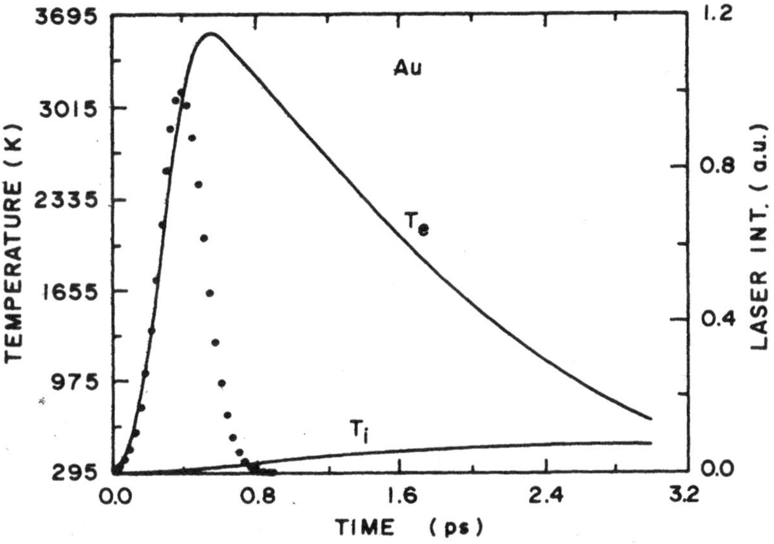

Figure 6: Same as in figure 4 but for 100 $GWatt/cm^2$ irradiance and 300 fs pulses. (Notice that this corresponds to a fluence 10 times higher than in figures 4 and 5).

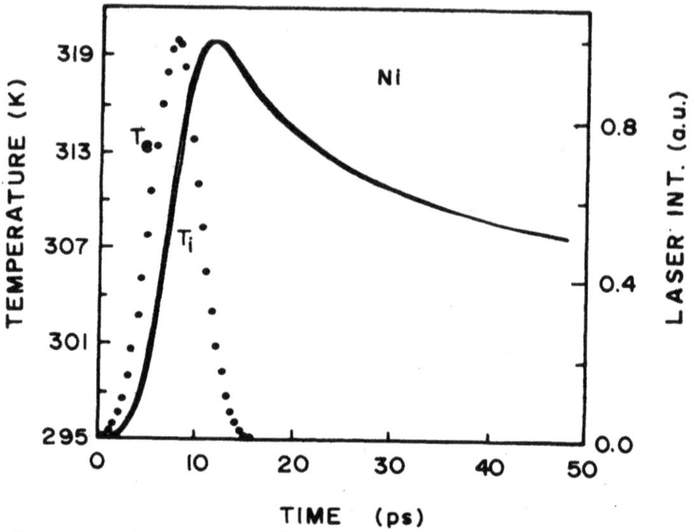

Figure 7: Same as in figure 4 but for nickel samples. Notice that the electronic and lattice temperatures remained the same, as opposed to the corresponding case for gold.

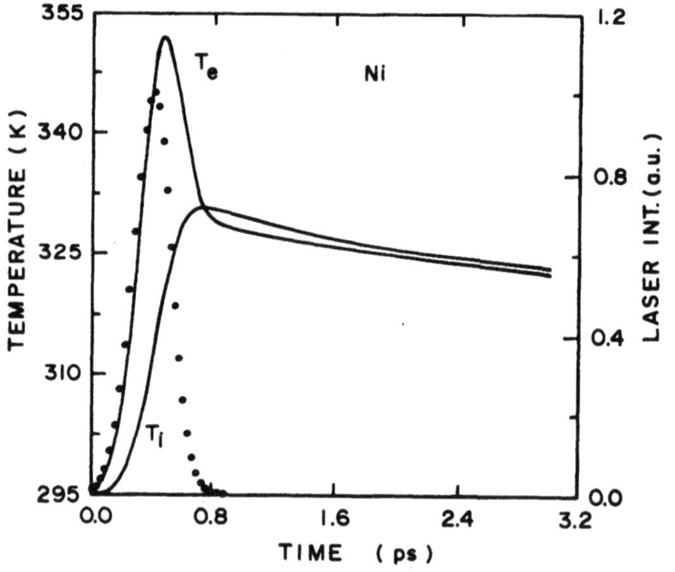

Figure 8: Same as in figure 7 but for 10 $GWatt/cm^2$ irradiance and 300 fs pulses. Again the $T_e - T_i$ temperature difference is much smaller than in the corresponding case for gold, shown in figure 5.

pulse is acting upon the metallic surface, make it entirely plausible to consider it as a non-inertial thermal emitter at a characteristic temperature which can be taken as the peak electronic temperature. Recent experiments seem to give good support to this view[22,24-28].

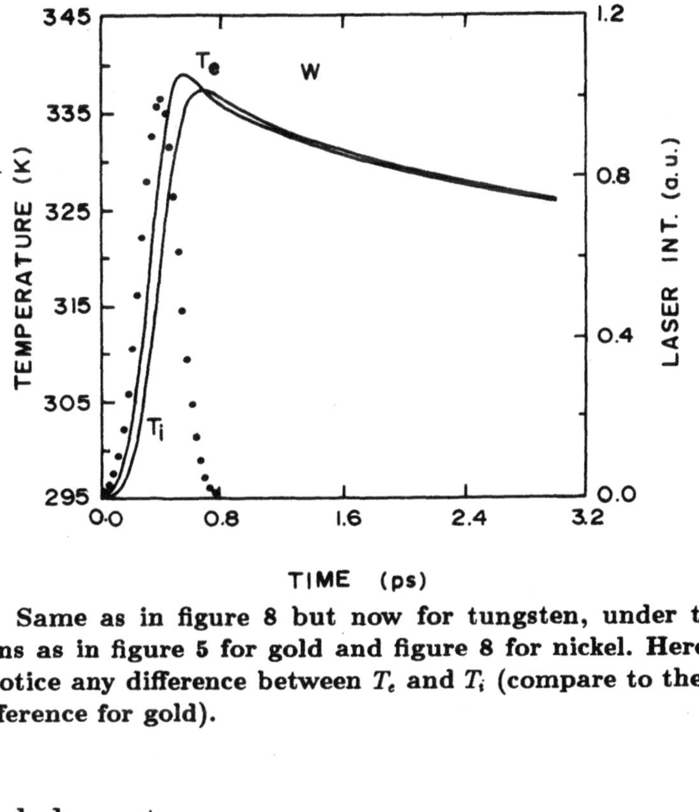

Figure 9: **Same as in figure 8 but now for tungsten, under the same conditions as in figure 5 for gold and figure 8 for nickel. Here we can barely notice any difference between T_e and T_i (compare to the $\sim 600\,K$ peak difference for gold).**

Acknowledgments

We gratefully acknowledge financial support from Conselho Nacional de Desenvolvimento Científico e Tecnológico (CNPq), Brazil, and Fundação de Amparo à Pesquisa do Estado de São Paulo (FAPESP).

References

[1] M. C. Teich, J. M. Schroeer and G. J. Wolga; Phys. Rev. Lett. 13, 611, (1964).

[2] Gy. Farkas, Zs. Náray and P. Varga; Phys. Lett. 24A, 134, (1967).

[3] E. M. Logothetis and P. L. Hartman; Phys. Rev. 187, 460, (1969).

[4] A. J. DeMaria, W. H. Glenn Jr., M. J. Brienza and M. E. Mack; Proc. IEEE, 57, 2, (1969).

[5] *"Ultrashort Light Pulses"* in Topics in Applied Physics vol. 18, edited by S. L. Shapiro, 2nd edition, Springer Verlag Berlin Heidelberg (1984).

[6] J. G. Fujimoto, J. M. Liu, E. P. Ippen and N. Bloembergen, Phys. Rev. Lett. 53, 1837, (1984).

[7] H. E. Elsayed-Ali, T. B. Norris, M. A. Pessot and G. A. Mourou, Phys. Rev. Lett., 58, 1212, (1987).

[8] R.W. Schoenlein, W. Z. Lin, J. G. Fujimoto and G. L. Eesley, Phys. Rev. Lett., 58, 1680, (1987).

[9] D. S. Brorson, J. G. Fujimoto and E. P. Ippen, Phys. Rev. Lett., 59, 1962, (1987).

[10] Gy. Farkas, S. L. Chin, P. Galarneau and F. Yergeau; Opt. Comm. 48, 275, (1983).

[11] Gy. Farkas and S. L. Chin; Appl. Phys. B, 37, 141, (1985).

[12] W. C. Henneberger; Phys. Rev. Lett. 21, 838, (1968).

[13] C. A. S. Lima and L. C. M. Miranda; Phys. Rev. A, 23, 3335, (1981).

[14] C. A. S. Lima and L. C. M. Miranda *"Unitary Transformation Methods in Intense Fields Atomic Physics"* in *"Essays in Theoretical Physics"* in honour of Dirk Ter Haar, edited by W. E. Parry, Pergamon Press, UK, (1984).

[15] C. A. S. Lima and L. C. M. Miranda; Phys. Lett., 86A, 367, (1981).

[16] T. C. Landgraff, J. R. Leite, N. S. Almeida, L. C. M. Miranda and C. A. S. Lima; Phys. Lett, 92A, 131, (1982).

[17] C. A. S. Lima and L. C. M. Miranda; Sol. Stat. Comm., 41, 465, (1982).

[18] L. C. M. Miranda, R. M. O. Galvão and C. A. S. Lima; Phys. Rev. B, 28, 5313, (1983).

[19] M. B. S. Lima, C. A. S. Lima and L. C. M. Miranda, Phys. Rev. A, 19, 1976, (1979).

[20] M. I. Kaganov, M. I. Lifshitz and L. V. Tanatarov, Sov. Phys. JETP, 4, 173, (1957).

[21] S. I. Anisimov, B. L. Kapeliovich and T. L. Perel'man, Sov. Phys. JETP, 39, 375, (1974).

[22] M. B. Agranat, A. A. Benditskii, G. M. Gandel'man, P. S. Kondratenko, B. I. Makshantsev, G. I. Rukman and B. M. Stepanov, Sov. Phys. JETP, 52, 27, (1980).

[23] R. Yen, J. Liu and N. Bloembergen, Opt. Comm. 35, 277, (1980).

[24] M. B. Agranat, S. I. Anisimov, S. I. Ashitikov, B. I. Makshantsev, and I. B. Ovchinnikova, Sov. Phys. Sol. Stat. 29, 1875, (1987).

[25] M. B. Agranat, S. I. Anisimov, and B. I. Makshantsev, Sov. Phys. Sol. Stat. 29, 1966, (1987).

[26] M. B. Agranat, S. I. Anisimov and B. I. Makshantsev, Appl. Phys. B, 47, 209, (1988).

[27] M. A. Algatti, Gy. Farkas, Cs. Tóth, KFKI preprint-66, edited by Pál Király, Hungarian Academy of Sciences, Central Research Institute for Physics, Budapest, Hungary, (1987).

[28] M. A. Algatti, "*Photo-induced and Photo-stimulated Processes in Metal Surfaces Subjected to High Intensity Laser Irradiation*" Ph. D. Dissertation - IFGW - UNICAMP (1990) - in Portuguese.

AMORPHIZATION OF ION-IMPLANTED LAYERS IN SILICON USING PHOTOACOUSTIC DETECTION

H. VARGAS
Instituto de Física
Universidade Estadual de Campinas
13.081 - Campinas, SP, Brazil

ABSTRACT

In this paper in honor of Rogério Cerqueira Leite's 60th birthday the influence of ion implantation in the thermal properties of silicon wafers at room temperature is investigated using the photoacoustic technique. It is suggested that the observed decrease of the values of both thermal diffusivity and conductivity, as the implantation dose increases, is due to the amorphization of the implanted layer.

In the last few years, the photoacoustic and related photothermal techniques[1,2] have been successfully applied to the characterization of optical and thermal properties of semiconductors. In particular , in the case of ion-implanted semiconductors, the photothermal reflectance technique[3-5] has proved to be an important tool for monitoring the dose and uniformity of the implantation process. In this paper, we report on the dependence of the photoacoustically measured thermal diffusivity on the implanted dose of Zn^+ ions in silicion wafers. The importance of the thermal diffusivity as a physical parameter to be monitored is that its knowledge is essential in designing optoelectronic devices, since power dissipation is an important mechanism in limiting the device performance.

The samples used were 388μm thick (100)-oriented, p-type Si wafers cut in the shape of squares having $14x14mm^2$ of area. The experimental batch consisted of four wafers implanted through the natural oxide layer, at room temperature, with Zn^+ ions with doses between 10^{13} and 10^{16} ions/cm^2, at an energy of 150KeV. The thermal diffusivity was measured using the phase-lag method described in Ref. 6. This method consists of measuring the relative phase lag $\Delta\phi = \phi_f - \phi_r$, at a single modulation frequency, between the rear surface illumination as depicted in Fig. 1. Using the thermal diffusion model of Rosencwaig and Gersho[7], the phase lag $\Delta\phi$ is given by

$$\tan(\Delta\phi) = \tanh(\ell a_s)\tan(\ell a_s) \qquad (1)$$

where ℓ is the sample thickness, and $a_s=(\eta f/\alpha)^{1/2}$ is the thermal diffusion coefficient of a sample with thermal diffusivity α at a modulation frequency f. The light source used was a 180mW Ar-ion laser, which, after being mechanically chopped, is divided by a beam splitter and the resulting beams are directed to opposite sides of the photoacoustic cell (c.f., Fig. 1). The output signal amplitude and phase from the microphone was recorded using a lock-in amplifier.

The thermal conductivity, k, was measured using the temperature rise method under continuous white light illumination. The samples had both surfaces sprayed with black paint. In this way, we ensured not only a good light abosorbing surface but also the same heat transfer coefficient for each surface. The samples were adiabatically suspended in a Dewar which was subsequently vacuum sealed. Under these conditions the main heat loss mechanism is the loss by radiation. The Dewar had an entrance glass window through which the continuous white light beam was focused onto one of the sample surfaces. On the opposite surface we had attached a thermocouple to the sample using thermal paste; in this way, the temperature evolution of the back surface could be monitored as a function of time. It can be shown that the long-term time evolution (i.e., for times greater than the heat diffusion time $\simeq \ell^2/\alpha$) of the back surface temperature rise is given by

$$\Delta T = \frac{\alpha \tau I_o}{\ell k} \left| 1 - e^{-t/\tau} \right| \qquad (2)$$

where I_o is the intensity of the incident light beam, and $\tau = \ell k/2\alpha H$ is the rising time. Here, $H = 4\sigma T_o^3$, where σ is the Stefan-Boltzmann constant, and T_o is the ambient temperature, H is the radiation heat transfer coefficient. The thermal conductivity was obtained using the best-fitted values of τ and the previously obtained values of α.

Figures 2 and 3 show the result for the thermal diffusivity and thermal conductivity, respectively, as a function of the Zn^+ dose, over the range of 10^{13}-10^{16} ions/cm^2. At low doses, the measured values of the thermal diffusivity and conductivity were 0.87cm^2/s and 1.47±0.03W/cmK, respectively. These values are in good agreement with the ones typical of undamaged samples, as reported in the literature[2]. As the dose increases, both parameters decrease such that, at a dose of 10^{16} ions/cm^2, the thermal diffusivity exhibits a 16% decrease whereas the thermal conductivity is reduced by 24%. We attribute this sensitivity of the thermal properties to doping levels to the higher degree of lattice damage caused by the implantation. In fact, it has been known for some time[8-11] that the ion-implanted layer undergoes a crystalline-to-amorphous transition as a function of the dose. For given implantion conditions of ion mass, energy, and temperature, the critical dose Φ_c is the experimental quantiy which characterizes the implanted layer amorphization process. According to Christel et al[11], if the implanted ions displace 10% of the lattice ions, the sample can be considered amorphous. If fewer host atoms are displaced, one talks of an inhomogeneous layer composed of amorphous clusters within a crystalline matrix. Thus, the implanted layer's amorphization increases with the dose up to a critical dose; further implantation will not likely cause any additional structural change, but will slightly increase the thickness of the damaged layer. In Fig. 4 we present the variation of the critical dose for amorphization as a function of the atomic number, Z, of the ions implanted in Si, at room temperature with a energy of 150KeV. It follows from Fig. 4 that the critical dose for Zn^+ ion implantation is on the order of $3.5x10^{14}$ions/cm^2. On the other hand, Fig. 2 and 3 tell us that above a dose of roughly $4x10^{14}$ ions/cm^2 there is a saturation tendency of both α and k with increasing dose. Comparing this value of implantation dose with the critical dose for Zn^+ we are led to the conclusion that the observed behavior of α and k above the $4x10^{14}$ions/cm^2 dose may be attributed to the crystalline-to-amorphous transition. Physically, one would expect that the main effect of the amorphization process is to reduce both thermal diffusivity and conductivity, as compared with those values of the pure crystalline phase. However, the observed degradation of the thermal properties as a function of ion implantation reflects the average effect over two regions, the amorphous layer and the crystalline substrate, and not the actual degradation in the implanted region only. To photoacoustically single out the implanted region (estimated to be roughly 1000A thick) we should monitor the PA signal at modulation frequencies greater than, roughly, 16MHz. Finally, the fact that the slopes of the curves of α and k as a function of the implantation are non-vanishing at high doses may be reflecting the expected slight increase of the amorphous layer thickness once the amorphization process has been set in.

To conclude, we have shown in this paper the sensitivity of the thermal properties of silicon to implantation doses. These results are particularly important for the photothermal diagnostics techniques used in the semiconductor industry. To achieve higher device yields, in MOS and bipolar technologies, it is indispensable to characterize and control the various amorphization and recrystallization process. The thermal diffusivity is the essential parameter when performing photothermal depth profile and uniformity studies.

REFERENCES

1. Vargas, H. and Miranda, L.C.M., Phys.Repts., 161, 43 (1988).
2. Mandelis, A., "Photoacoustic and Thermal Wave Phenomena in Semiconductors" (North-Holland, New York,1987).
3. Smith, W.L. and Rosencwaig A. and Willenborg, D., Appl.Phys. Lett., 47, 582 (1985)
4. Wurm, S. and Alpern, P. aad Savignac D. and Kakoschke, R., Appl. Phys.A, 47, 147 (1988).
5. Christofides, C. and Vitkin, I.A. and Mandelis A., J.Appl.Phys., 67, 2815 (1990)
6. Pessoa Jr., O. and Cesar, C.L. and Patel, N.A. and Vargas, H. and Ghizoni, C.C. and Mîranda, L.C.M., J. Appl.Phys, 59, 1316 (1986).
7. Rosencwaig, A and Gersho, A., J.Appl.Phys. 47, 64 (1975).
8. Grove, A.S., "Physics of Semiconductor Devices" (Wile, New York, 1972).
9. Crowder, B.L. and Morehead, F.F., Appl.Phys.Lett., 14, 313 (1969)
10. Morehead, F.F. and Crowder, B.L., Radiat. Eff., 6, 27 (1970)
11. Christel, L.A. and Gibbons J.F. and Sigmon, T.W., J.Appl.Phys. 52, 7143 (1981).

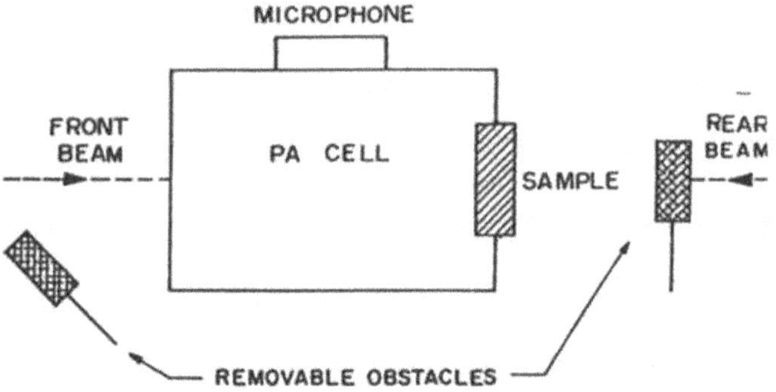

Figure 1: Schematic arrangement for the two-beam photoacoustic measurement of thermal diffusivity.

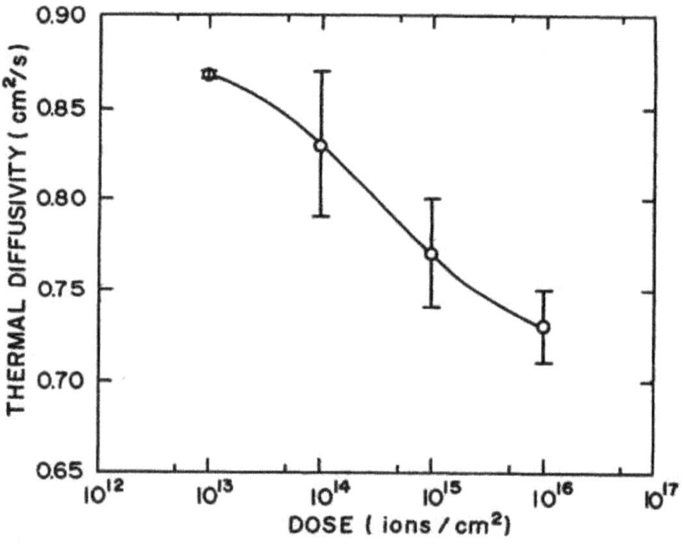

Figure 2: Thermal diffusivity as a function of dose for Zn^+-implanted silicon samples.

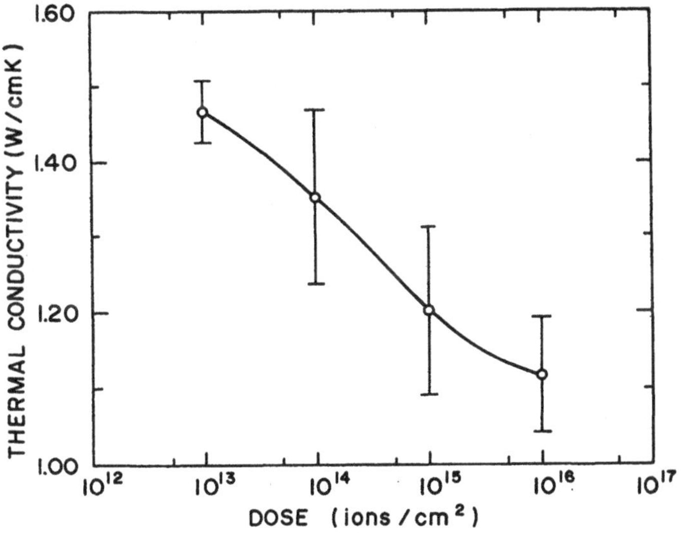

Figure 3: Thermal conductivity as a function of dose for Zn^+-implanted silicon samples.

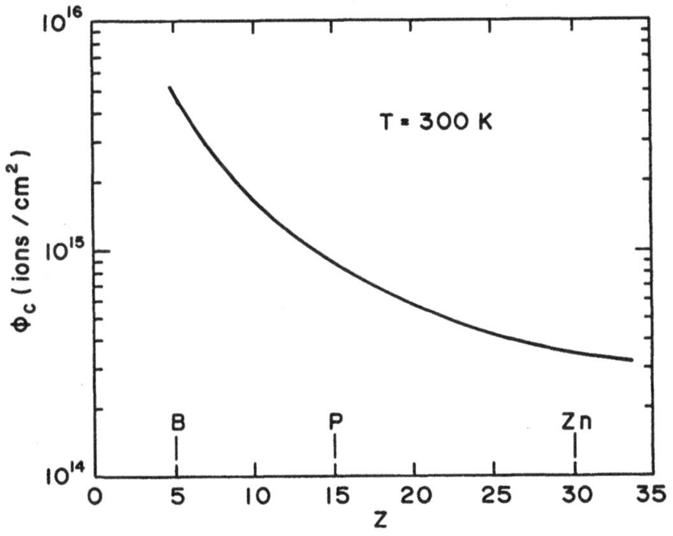

Figure 4: Critical amorphization dose Φ_c as a function of atomic number Z at room temperature.

LASER FIELD EFFECTS ON THE TRANSPORT PHENOMENA: CROSS SECTION TREATMENT, ENERGY LOSS AND STOPPING POWER

Hector R. Torres-Silva and Paulo H. Sakanaka
Instituto de Física, UNICAMP, C.P. 6165,
13081 Campinas, SP, Brazil
December 31, 1990

FOREWORD

I met Rogério, for the first time, at the Instituto Tecnológico da Aeronáutica in 1963, when he convinced me to take the academic career, instead of industry, by showing the beauty and the excitement of scientific research. Completing my M.Sc. Degree at I.T.A., in 1965, I pursued my studies at Columbia University in New York City. There, my friendship with Rogério and Rute (who were living in New Jersey) grew deeply. In 1972 Rogério invited me to come to UNICAMP to establish a Plasma Physics Group. Finally, in 1974, I decided to move to UNICAMP and, then, Rogério gave all the support needed to install a good Plasma Physics Laboratory. His support was essential to the successes achieved by this Group which has graduated 9 students in Ph. D. degree and 25 in M.Sc degree. Thank you very much, Rogério, and happy 60th birthday (P.H.S.).

Abstract

The differential cross-section for inelastic scattering of electrons in the presence of an intense laser field, reported by the authors [1], was confirmed by Ehlotzky (1988) [2], Krstic and Milosevic (1989) [3], and by the authors [4] , when the Kibble parameter is less than unity. However, some difficulties arise when the energy balance and heating by multiphoton process are considered. Bivona et al. (1985), (1986) [5,6] have shown that the absorption coefficient, α, can be negative if $v_0 < v_i$, where v_0 is the amplitude of the oscillatory velocity and v_i is the particle incoming velocity. In this paper we present the proper way to calculate the absorption coefficient from the cross-section treatment and we are able to show that $\alpha \geq 0$, for all values of v_i . On the other hand, the energy loss method has been applied to a large variety of transport problems in optics, solid-state and fusion research [7,8]. In these papers, however, the transport equations were linearized, so there are no multiphoton interactions. Bivona et al. (1982) [9] have shown that, for a one-component plasma, the strong field effects would only be of academic interest. Contrary to this claim, and on the basis of the center of mass approach [1], we obtain a generalization of the energy loss rate which is in accordance with the recent results of Arista et al. (1989) [10].

1 Introduction

The investigation of the absorption of intense electromagnetic radiation due to electron-ion collisions in a fully ionized plasma, is still incomplete. The main difficulty for calculations at high radiation intensities is that a large number of photons must be included. For high intensities of the laser beam, i.e., in the range of 10^{16} to $10^{18}\,W/cm^2$, relativistic effects for potential scattering must be considered [4].

The differential cross section for inelastic scattering in the presence of an intense laser field, obtained by us [11] was recently confirmed [2,3], when the Kibble parameter ($E_k = v_0/c$) is less than unity. The main result is that the Kroll-Watson expression is not valid anymore:

$$\left(\frac{d\sigma}{d\Omega}\right)_L = (1+\varepsilon_k^2)^2 \left(\frac{d\sigma}{d\Omega}\right)_R = \sum_{n=-\infty}^{\infty} \left(\frac{d\sigma}{d\Omega}\right)^n \tag{1}$$

where $\left(\frac{d\sigma}{d\Omega}\right)_R$ is the Rutherford differential cross section.

On the other hand, great difficulties appear when it is necessary to calculate the global absorption coefficient, $\bar{\alpha}$, because equation (1) is independent of the photon number n and the distribution function of electrons.

Similar difficulties appear, in the calculation of the energy balance and heating by multiphoton processes, when a monoenergetic beam of electrons is scattered by a static potential in the presence of a strong laser field. Some authors [5,6] have shown that $\bar{\alpha}$ can be negative if $v_0 < v_i$, ($\vec{v}_i$ = particle incoming velocity, $\vec{v}_0$ = amplitude of the oscillatory velocity). In Section 2, we present the proper way to calculate the absorption coefficient $\bar{\alpha}$ from the cross section treatment, and we are able to show that $\bar{\alpha} > 0$, for all values of $\vec{v}_i$.

However, this treatment cannot give the collective contribution of the electrons to the energy losses and heating process, so, the energy loss method must be considered. This method has been applied with surprising success to a large variety of transport problems [7,8]. In the case of the scattering of charged particles, if an ion is moving with constant velocity $\vec{v}_i$ in the sea of electrons, then, the energy loss of the ion is given by a simple functional F of the charge density ZeN_i , of its velocity $\vec{v}_i$ and of the dielectric properties $\varepsilon(\vec{q},\omega)$ of the system. When we reverse the frame of reference, that is, the ion is fixed in space and the electron gas has the velocity $\vec{v}_{CM} = -\vec{v}_i$, then the scattering process leads to a Joule heating, and the resistivity ρ of the system can be calculated.

In connection with plasma physics, the stopping power is of interest for questions of plasma heating and plasma beam experiments and it is given by [12]

$$S = \frac{dW}{dx} = \frac{Z^2e^2}{\pi v_i^2}\int_0^{\infty}\frac{dq}{q}\int_{-qv_i}^{qv_i} d\omega\omega Im\left\{-\frac{1}{\varepsilon(\vec{q},\omega)}\right\} \tag{2}$$

$$\frac{dW}{dt} = v_i \frac{dW}{dx} \tag{3}$$

In this paper we present a study of the effects of a strong laser field on the interaction of fast non relativistic particles with an electron gas. The problem is formulated in terms of the center of mass approach [1], using a spatial unitary transformation and the random phase approximation (RPA). The present formalism yields expressions which generalize equation (2). In section 3 we sketch the derivation of kinetic equations for the drift velocity of the electrons and of the energy loss rate. In this section we also discuss briefly the formulas and results for the case of a strong high-frequency electric field.

2 Quantum Mechanical Treatment of the Cross Section and the Absorption Coefficient

We begin by considering the problem of a non-relativistic particle of mass m and charge e being scattered by a static local potential $V(r)$, in the presence of a strong laser field. Here we consider that the electrons interact with infinitely heavy ions, via Coulomb potential, and also that the first Born approximation is used for the scattering of electrons by ions. The transition probability $T_{nk}(\vec{p}_i \rightarrow \vec{p}_f)$ from an initial state $\vec{p}_i$ to a final state $\vec{p}_f$, under the perturbation $\phi(\vec{k})$ is given by [11,13,14]

$$T_{nk}(\vec{p}_i \rightarrow \vec{p}_f) = \frac{2\pi}{\hbar} \sum_n \mid \phi(\vec{k}) \mid^2 J_n^2(x) \delta(E_f - E_i - n\hbar\omega) \tag{4}$$

where $\phi(\vec{k})$ is the Fourier transform of the potential interaction $\phi(\vec{r}, E_{fi} = p_{fi}^2/2m$ and

$$x = \frac{\vec{k} \cdot \vec{v}_0}{\omega} = \frac{e\vec{E}_0}{m\omega} \left(\frac{\vec{p}_f - \vec{p}_i}{\hbar\omega} \right) .$$

The corresponding cross section for inverse Bremsstrahlung (I.B.) is obtained from equation (4):

$$d\sigma_{nk}^{a,e} = \frac{4Z^2e^4}{v_i}\frac{1}{\hbar^2}\frac{|R(k)|^2}{k^4}J_n^2(x)\delta\left(\vec{k}\cdot\vec{v}_i - n\omega \pm \frac{\hbar k^2}{2m}\right). \tag{5}$$

In equation (5), Z, $\vec{v}_i$ and n are the ionic charge number, the velocity of the incoming particle beam and the number of photons exchanged with the assisting field, with plus sign for absorption and minus sign for emission, respectively. $R(k)$ is the form factor which takes into account the range of the static screened Coulomb potential. In this work, we will be concerned with the calculation of the absorption coefficient $\bar{\alpha}$. We define it as

$$\bar{\alpha} = \sum_n \alpha_n \tag{6}$$

with

$$\alpha_n = \frac{N_eN_iv_i}{I_0}\int d^3k(n\hbar\omega + \hbar\vec{k}\cdot\vec{v}_i)(d\sigma_n^a - d\sigma_n^e) \tag{7}$$

where N_e , N_i and I_0 are the concentration of the incoming beam and the scattering centers and the intensity of the laser field, respectively. The term $(n\hbar\omega + \hbar\vec{k}\cdot\vec{v}_i)$ corresponds to the final energy of the electrons. Thus, $\bar{\alpha}I_0 \equiv \frac{dW}{dt}$ is the total energy decay rate, which is always positive (for all n, v_i , v_0 and I_0). From equation (5), (6) and (7) we obtain

$$\begin{aligned}\bar{\alpha}(v_i, v_0, \hbar) &= \sum_n \frac{8\pi N_iN_eZ^2e^4}{I_0\hbar v_i}\int\frac{dk}{k^3}|R(k)|^2 \\ &\left\{\left[2n\omega + \frac{\hbar k^2}{2m}\right]J_n^2\left[\frac{v_0}{v_i}\left(\frac{n\omega\hbar k^2/2m}{\omega}\right)\right] - \right.\\ &\left.\left[2n\omega - \frac{\hbar k^2}{2m}\right]J_n^2\left[\frac{v_0}{v_i}\left(\frac{n\omega - \hbar k^2/2m}{\omega}\right)\right]\right\}\end{aligned} \tag{8}$$

Finally taking $\hbar \to 0$ and considering that

$$J_n^2\left(\frac{nv_0}{v_i}\right) = J_{-n}^2\left(\frac{-nv_0}{v_i}\right)$$

we obtain

$$\bar{\alpha}(v_i, v_0) = \sum_{n=1}^{\infty} \frac{16\pi Z^2 e^4 N_i N_e}{I_0 m v_i} \int \frac{dk}{k} \mid R(k) \mid^2 J_n^2 \left(\frac{n v_0}{v_i}\right) . \tag{9}$$

When $v_0/v_i << 1$, we consider only $n = 1$, so $\bar{\alpha}_1$ is given by

$$\bar{\alpha}_1 = \frac{1}{I_0} \frac{4\pi Z^2 e^4 N_i N_e}{m v_i} \left(\frac{v_0}{v_i}\right)^2 \ln \Lambda \tag{10}$$

where $\ln \Lambda$ is the generalized Coulomb logarithm. This expression is always positive. Now we can normalize the absorption coefficient in units of its weak-field value $\bar{\alpha}_1$

$$\frac{\bar{\alpha}}{\bar{\alpha}_1} = \frac{4}{(v_0/v_i)^2} \sum_{n=1}^{\infty} J_n^2 \left(\frac{n v_0}{v_i}\right) . \tag{11}$$

When $v_0/v_i >> 1$ we can approximate the Bessel function $J_n^2(n v_0/v_i)$ as $1/(\pi n v_0/v_i)$, so

$$\frac{\bar{\alpha}}{\bar{\alpha}_1} = \frac{1}{\pi} \frac{4}{(v_0/v_i)^2} \sum_{n=1}^{\infty} \frac{1}{n} \approx \frac{1}{\pi} \frac{4}{(v_0/v_i)^3} \ln(n_{max}) \tag{12}$$

where $n_{max} = m v_0^2/\hbar\omega$. In this form we have recovered the Silin's expression [15], which contains a product of two logarithms and $\bar{\alpha} \sim 1/I_0^{3/2}$.

3 Formulation of the Center of Mass Motion

3.1 Hamiltonian

If we consider N electrons scattered by n_i as randomly distributed ions, applied by a spatially uniform vector potential $\vec{A}(t)$, and if we perform a unitary transformation for the ith electron, the Hamiltonian is [1,15,16]

$$H = \sum_i \frac{p_i^2}{2m} + \sum_{i,j} \frac{e^2}{\mid \vec{r}_i - \vec{r}_j \mid} + \sum_{i,\alpha} U(\vec{r}_i - \vec{R}_\alpha - \vec{a}(t)) \tag{13}$$

where $a(t) = (eE_0/m\omega^2)\sin(\omega_0 t)$, U denotes the potential due to the ion at $\vec{R}_\alpha$, which is randomly located (RPA) and it is modified by the laser field. In the

Heisenberg picture we make a separation of the center of mass motion (CM) from the relative motion of electron in the Hamiltonian (13) so,

$$\hat{H} = \hat{H}_{CM} + \hat{H}_e + \hat{H}_{ee} + \hat{H}_{ei}, \quad \hat{H}_{CM} = \frac{\hat{P}^2}{2M} \tag{14}$$

$$\hat{H}_0 + \hat{H}_{ee} = \sum E_k c_k^\dagger c_k + \frac{1}{2}\sum V(\vec{q})(\rho_q^\dagger \rho_q - N) \tag{15}$$

where $\vec{P}$ and $\vec{R}$ are the CM momentum and coordinate operators respectively. N is the total number of electrons and $M = Nm$ is the total mass of electrons. $\hat{H}_e$ is the free-electron Hamiltonian in relative coordinates, $E_k = \hbar^2 k^2/2m$ with $\hbar\vec{k}$ the momentum of electron in state $\vec{k}$. $c_k^\dagger$ and c_k are electron creation and annihilation operators in relative coordinates. $\hat{H}_{ee}$ is the Coulomb interaction in relative coordinates. The electron-ion interaction is described by

$$\hat{H}_{ei} = \sum_{\vec{q},\alpha} U(\vec{q}) e^{i\vec{q}\cdot(\vec{R}-\vec{R}_\alpha - \vec{a}(t))} \hat{\rho}_q + cc \tag{16}$$

where $\hat{\rho}_q = \sum_k c_k^\dagger c_k$ the electron density operator and $U(q) = 4\pi Z e^2/q^2$. The difference between our formulation and the others [17,18,19] is that we use the velocity gauge and a spatial unitary transformation $\exp(i\vec{p}\cdot\vec{a}(t)/\hbar)$ so, the laser field acts only on $\hat{H}_{ei}$, while in the other references, the electric field acts only on the CM through the interaction $-Ne\vec{E}\cdot\vec{R}$ (length gauge).

3.2 Equations for the Drift Velocity and Average Energy Change

Starting from the quantum-mechanical Liouville equation, we can derive the kinetic equation for a set of macroscopic observables $\{O_i(t)\}$. Here we choose $\{O_i(t)\} = \vec{V}(t),\ f_k(t),\ W(t)$. The corresponding operators are $\hat{O}_i = \{\hat{P}/M,\ c_k^\dagger c_k,$ $\hat{H}_{CM}$ which are connected with observables by $O_i(t) = Tr(\hat{O}_i \rho(t))$ where $\hat{\rho}$ is the statistical density matrix. To the lowest order, the rate of change of any function S of $O(t)$ is given by [1]

$$\frac{d}{dt}S(O(t)) = \frac{i}{\hbar} < [H_0(t), S(O(t))] > + < \frac{1}{\hbar}\left[\hat{V}(t), S(O(t))\right] > -$$

$$\left(\frac{1}{\hbar}\right)^2 \lim_{\epsilon\to 0}\int_{-\infty}^{t} dt\, e^{\epsilon(t-t')} < \left[S(O(t)), \hat{V}(t')\right] > \tag{17}$$

with $[\hat{A},\hat{B}]$ the commutator of operators $\hat{A}$ and $\hat{B}$, $\hat{V} = \hat{H}_{ee} + \hat{H}_{ei}$, $\hat{H}_0 = \hat{H}_{CM} + \hat{H}_e$ and $< \cdots >= Tr\,[\hat{\rho}(\cdots)]$, thus the frictional force operator is

$$\hat{F} = -\frac{1}{\hbar}\left[\hat{p},\hat{V}\right] = -i\sum_{q,\alpha} \vec{q}U(\vec{q})e^{i\vec{q}\cdot(\vec{R}-\vec{R}_\alpha-\vec{a}(t))} + cc \tag{18}$$

and the change of CM energy operator is

$$\dot{H}_{CM} = -\frac{i}{\hbar}\left[\hat{H}_{CM},\hat{V}\right] = -i\sum_{q,\alpha} \vec{q}\cdot\dot{\vec{R}}U(\vec{q})e^{i\vec{q}\cdot(\vec{R}-\vec{R}_\alpha-\vec{a}(t))} + cc \tag{19}$$

then by substituting equations (18) and (19) into (17) and taking the traces we obtain

$$\vec{f} = \frac{Nm}{vol}\frac{d}{dt}\vec{V}_{CM} = -i\sum_{q,n}\vec{q}n_i\frac{|U(\vec{q})|^2}{V(q)}J_n^2(\vec{q},\vec{a})Im\left\{\frac{1}{\bar{\varepsilon}(\vec{q},n\omega+\vec{q}\cdot\vec{V}_{CM})}\right\} \tag{20}$$

and

$$\dot{W} = \frac{<\dot{H}_{CM}>}{vol} = \sum_{q,n}\vec{q}\cdot\vec{V}_{CM}n_i\frac{|U(\vec{q})|}{V(q)}J_n^2(\vec{q}\cdot\vec{a})Im\left\{\frac{1}{\bar{\varepsilon}(\vec{q},n\omega+\vec{q}\cdot\vec{V}_{CM})}\right\} \tag{21}$$

Here $\bar{\varepsilon} + \varepsilon_{RPA} = 1 - V(q)\Pi_0(\vec{q},\omega)$, the complex conjugate terms are taken into account because equations (18) and (19) must be Hermitians and their average values must be real.

3.3 Resitive Force, Energy Loss, and Stopping Power

When a constant electric field is applied on the system, the resistive force $\vec{f}$ (20) is a generalization of the non-linear Lei-Ting force-momentum-balance transport equation for isothermal resistivity [17,18]. Also we have an interesting new interpretation of instabilities in semiconductors with drift velocity. In this case we obtain an induced electric field $\vec{E}_{ind} = \vec{f}/e$. If $v_0 \to 0$; $J_n^2(x) \to \delta_{n,0}$, the result of Gerlach [7] is reproduced.

The problem of energy loss rate or stopping power can also be reinterpreted. If the test-ion has the velocity $\vec{v}_i = -\vec{V}_{CM}$ and making the change $\vec{q}, n \to -\vec{q}, -n$, we obtain

$$\frac{\dot{W}}{n_i} = \frac{Z^2e^2}{2n^2}\int d^3q\vec{q}\cdot\vec{v}_i \sum_{n=-\infty}^{\infty} J_n^2(\vec{q}\cdot\vec{a}) Im\left\{-\frac{1}{\bar{\epsilon}(\vec{q}\cdot\vec{v}_i - n\omega)}\right\} \tag{22}$$

which agrees with the expression obtained by Arista et al.[10]. For small momentum transfer and using the plasmon-pole approximation we obtain

$$\frac{\dot{W}}{n_i} = \sum_{n=-\infty}^{\infty} \frac{Z^2e^2}{V_{CM}}\omega_p \int \frac{dq}{q}\left(\frac{n\omega_0 + \omega_p}{\omega_p}\right) J_n^2\left(\frac{v_0}{V_{CM}}\frac{n\omega_0 + \omega_p}{\omega_0}\right) \tag{23}$$

For weak fields, with $n = 0$, 1, we find the expression given by Bivona et al. [9]. A new result appears when $n\omega_0 >> \omega_p$ (ω_p = plasma frequency). Equation (23) transforms into

$$\frac{\dot{W}}{n_i} = -\frac{Z^2e^2}{v_i}\omega_p^2 \sum_n \int \frac{dq}{q} J_n^2\left(\frac{nv_0}{v_i}\right) \tag{24}$$

where $\int \frac{dq}{q} = \ln \frac{q_D v_i}{n\omega_0}$, and the maximum momentum transfer is taken to be of the order of the Debye momentum $q_D = \frac{\omega_p}{v_t}$. Here, we have a maximum of $\dot{W}/n_i$ when $v_o \to v_i$ [14].

Finally, following Lindhart [20] we can find the stopping number L defined as

$$\frac{1}{n_i}\frac{dW}{dx} = -\frac{\omega_p^2 Z^2 e^2}{v_i^2} L \tag{25}$$

where

$$L = \int \sum_n \left(\frac{n\omega + \omega_p}{\omega_p}\right) J_n^2\left(\frac{v_0}{v_i}\frac{n\omega_0 + \omega_p}{\omega_0}\right) \tag{26}$$

and the stopping power is

$$S = \frac{Z^2e^2\omega_p^2}{v_i^2} L\,. \tag{27}$$

Equations (9), (12), (20), (21), (23), (24), (26), and (27)are the main results of this paper, which, due to the multiphoton exchange can only be evaluated numerically [1,14,16].

4 Conclusion

In this paper we have generalized the transport equations in the presence of an intense laser field. The analysis has been carried out for a geometry in which the laser electric field is parallel to the incoming electron (or ion) velocity in a non-degenerated plasma. Hence, the multiphoton interaction effects appear in the function $\sum_n J_n^2(nv_0/v_i)$ if $n\omega_0 >> \omega_p$ (scattering of a monoenergetic electron beam) [21].

Contrary to the results of Bivona et al. [5,6] we find that $\bar{\alpha}$ is always positive (section 2), so this situation is not reminiscent of a well-known process occuring in plasma physics, namely the two stream instability.

From the center of mass approach proposed here, we can show that equation (23) agrees with equation (9) if $R(k) = 1$, $k \equiv q$ and $n\omega_0 >> \omega_p$, so $\bar{\alpha}$ calculated here is within the framework of the kinetic theory.

The resistive force, energy loss and stopping power have also been generalized in a many body system, so the well-known results are reproduced [7,8,17,20,22]when the laser field is weak.

ACKNOWLEDGEMENT

This work was supported by FAPESP – Fundação de Amparo à Pesquisa do Estado de São Paulo and CNPq - Conselho Nacional de Desenvolvimento Tecnológico e Científico.

References

[1] H. Torres-Silva. *Interação Multifotônica em Plasmas Criados por Laser Superintenso.* PhD thesis, Universidade Estadual de Campinas, Campinas, December 1990.

[2] F. Ehlotzky. *Optics Communications*, 66:265, (1988).

[3] P. Krstic and D. Milošević. *Phys. Rev. A*, 39:1783, (1989).

[4] H. Torres-Silva and P.H. Sakanaka. Charged plasma particle collision in the presence of a strong laser field. *Status: Submitted to J. Phys. Soc. Japn.*, (1990).

[5] S. Bivona, R. Zangara, and G. Ferrante. *Il Nuovo Cimento*, 7D:113, (1986).

[6] S. Bivona, R. Zangara, and G. Ferrante. *Phys. Letter A*, 110:375, (1985).

[7] E. Gerlach. *J. Phys. C*, 19:4585, (1986).

[8] J. Fanvacque and D. Vignaud. *Phys. Rev. B*, 31:1041, (1985).

[9] S. Bivona, G. Ferrante, S. Nuzzo, and M. Zarcone. *J. Phys, C*, 15:169, (1982).

[10] N. Arista, R.M.O. Galvao, and L.C.M. Miranda. *Phys. Rev. A*, 40:3808, (1989).

[11] H. Torres-Silva and P.H. Sakanaka. Charged plasma particle collision in the presence of a strong laser field. In *Proc. of Spring College on Plasma Physics*, Trieste, 1985.

[12] Y. L. Klimontovich and V. P. Silin. *Uspekhi Fiz. Nauk*, 70:247, (1960).

[13] J. Seely and E. Harris. *Phys. Rev. A*, 7:1064, (1973).

[14] H. Torres-Silva and P.H. Sakanaka. Multiphoton effects in electron-ion scattering: a limitation of the cross-section treatment. *Status: Submitted to Phys. Rev. A*, (1990).

[15] R. M. O. Galvao and L. C. M. Miranda. *Am. J.Phys*, 51:729, (1983).

[16] H. Torres-Silva and P.H. Sakanaka. Laser field effects on the transport phenomena: energy loss, stopping power, and inverse bremsstrahlung process. *Status: Submitted to J. Phys. Soc. Japn.*, (1990).

[17] X. L. Lei and C. S. Ting. *Phys. Rev. B*, 32:1112, (1985).

[18] C.S. Ting and T. W. Nee. *Phys. Rev. B*, 33:7056, (1986).

[19] G. Y. Hu and R. F. O'Connell. *Phys. Rev. B*, 36:5798, (1987).

[20] J. Lindhart and K.D. Vidensk-Selsk. 28, (1954).

[21] R. Daniele, F. Trombetta, G. Ferrante, C. Cavaliere, and F. Morales. *Phys. Rev. A*, 1156, (1987).

[22] J. L. Fanvacque. *Phys. Rev. B*, 39:1682, (1989).

ON THE BUILT-IN ELECTRIC FIELD AT MODULATION-DOPED GaAs/GaAℓAs HETEROJUNCTIONS AND DELTA-DOPED GaAs

R.P. Camata, L.M.R. Scolfaro, J.M.V. Martins, and **J.R. Leite**
Instituto de Física, Universidade de São Paulo,
Caixa Postal 20516, 01498 São Paulo, SP, Brazil

and

C.A.C. Mendonça and **E.A. Meneses**
Instituto de Física, Universidade Estadual de Campinas,
Caixa Postal 6165, 13081 Campinas, SP, Brazil

and

I.F.L. Dias, J.C. Bezerra, and **A.G. Oliveira**
Departamento de Física, ICEx, UFMG,
Caixa Postal 702, 30000 Belo Horizonte, MG, Brazil

ABSTRACT

The built-in electric field present at a modulation-doped GaAs/GaAℓAs:Si heterojunction and Si delta-doped GaAs layers is investigated by means of Franz-Keldysh oscillations observed from photoreflectance (PR) measurements taken as a function of temperature. The electric field dependence on temperature, extracted from the PR spectra, is discussed through comparison with theoretical self-consistent calculations of the potential and electric field profiles at the space charge region of the heterostructures.

1. Introduction

The behavior of electric fields inside semiconductor heterostructures has been considered of great physical interest due to its fundamental importance in modern devices technology. The techniques of electromodulation spectroscopy, particularly the photoreflectance (PR) has revealed to be a valuable tool in providing surface or interface electric fields and quantum effect informations in two-dimensional (2D) confined systems. In a PR measurement, the dielectric function is modulated through modulation of the electric field due to changes in the surface photovoltage induced by the absorption of photons around the band gap energy. Its simplicity and contactless character, added to the derivative nature of a modulation spectroscopy have made PR a promissing **in situ** characterization technique for semiconductor interfaces[1-3]. For a comprehensive review of modulation spectroscopy of semiconductors, see Ref. 4).

The various applications of PR in the characterization of low-dimensional structures include quantum well heterostructures[5-7], strained layer superlattices[8,9], modulation-doped heterojunctions[10-15] and delta-doped layers[16,17]. The modulation mechanisms and line shapes observed in the spectra of quantum wells and superlattices seem to be reasonably well understood[5-9]. As far as modulation-doped and delta-doped structures are concerned, however, there have been some controversies related to the assignment of the observed PR structures to 2D electron gas features[14-17]. In these systems, one observes Franz-Keldysh (FK) oscillations-like structures, from which the magnitude of the built-in electric field can be evaluated. Nevertheless, the origin of the electric field giving rise to these FK oscillations remains still unclear.

In this work we investigate the built-in electric field present in GaAs/GaAℓAs:Si heterojunctions and Si delta-doped GaAs by carrying out self-consistent electronic state calculations taking into account temperature effects. The results are compared to the values of the electric field extracted from recent PR measurements[14-16]. The aim of the work is to analyse the dependence on temperature of the observed FK oscillations and its correlation with the presence of a 2D electron gas in these systems. The theoretical analysis of the results are based on self-consistent calculations of the potential profiles and subband energies at the space charge region, including temperature effects.

2. Samples Description

We considered here two MBE grown samples, a single modulation doped GaAs/$Ga_{1-x}A\ell_xAs$:Si heterojunction and a Si delta

-doped GaAs. The heterojunction (sample 1) consists of a 2.3 µm-thick undoped GaAs layer grown on a semi-insulating GaAs substrate. Then a 186 Å-thick spacer $Ga_{1-x}A\ell_xAs$ (x = 0.3) has been introduced, followed by a 450 Å-thick n(Si) doped GaAℓAs layer ($n \approx 1 \times 10^{17}$ cm^{-3}) and a capping layer of undoped GaAs 65 Å-thick. The 2D electron density in the GaAs channel is $N_s = 5 \times 10^{11}$ cm^{-2} assuming that all donors in the doped GaAℓAs:Si layer supply their electrons to the space charge region.

The delta-doped sample (sample 2) is lightly p-type ($N_a \sim 5 \times 10^{15}$ cm^{-3}) and was grown on a semi-insulating substrate. The growth is interrupted to introduce the silicon donors. The nominal sheet doping N_d is 5.2×10^{12} cm^{-2}. The 2D electron density (N_s) is 4.9×10^{12} cm^{-2} as obtained from Hall and Shubinikov-de Haas (SdH) measurements[18]. From SdH data the number of occupied subbands and their occupations are determined. At least three subbands are occupied, at T = 0 K, in sample 2 investigated here. For sample 1 only the lowest subband is occupied at T = 0 K.

Details about the setup utilized to conduct the PR experiments are reported elsewhere[14,15].

3. Results

3.a) Photoreflectance Spectra

We start by first discussing the results for the heterojunction (sample 1). Shown in Figure 1 are the PR spectra of sample 1 for three different temperatures. The structure labelled I bellow the GaAs band gap energy (E_g) is probably due to transitions involving carbon acceptors[14]. The sharp structure denoted as E_x is ascribed to excitonic transitions in the GaAs buffer layer[19]. The pronounced oscillatory structure seen above the GaAs band edge at high temperatures is characteristic of a FK effect. As is well known for bulk materials, this FK oscillatory behavior arises from regions of the sample where strong electric fields are present[20,21]. The

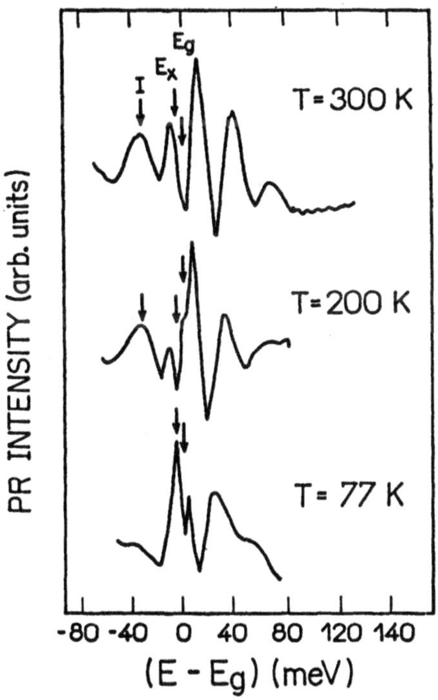

Fig. 1. Photoreflectance spectra of a GaAs/GaAlAs:Si hetero-junction (sample 1) at different temperatures. FK oscillations above the GaAs band gap (E_g) are observed in the spectra at high temperatures. The structures seen below E_g are described in the text.

extrema in the FK oscillations display a periodicity which shows a functional dependence of the form $(E_n - E_g)^{3/2}$, where E_n is the photon energy of the nth extremum[20]. Figure 2 shows plots of $(E_n - E_g)^{3/2}$ as function of index n for three different temperatures. Following, for example, Aspnes and Studna[21], we can extract the magnitude of the built-in electric field from the angular coefficient of the straight lines shown in Fig. 2, by using the following expressions:

$$(h\Omega)^3 = e^2\hbar^2 F^2/2\mu_{\parallel} \tag{1}$$

and

$$n\pi = \frac{4}{3} \times \left[(E_n - E_g)/\hbar\Omega\right]^{3/2} + \varphi \tag{2}$$

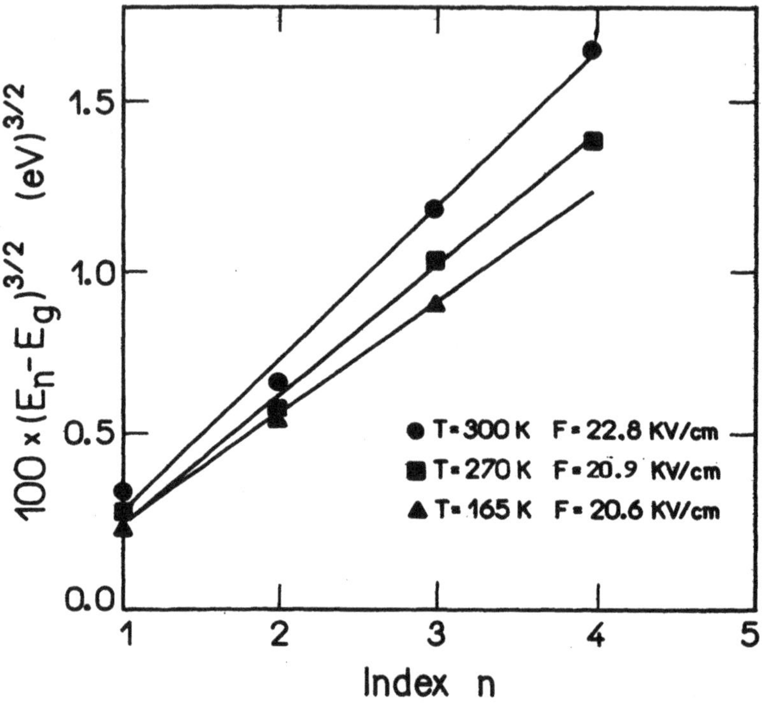

Fig. 2. Plots of $(E_n - E_g)^{3/2}$ as a function of index n for different temperatures for sample 1. The values of the electric field F obtained from the slope of the straight lines are shown.

where $h\Omega$ is the electro-optical energy, F is the magnitude of electric field, $\mu_{\parallel}$ is the reduced interband effective mass and φ is an arbitrary phase factor. For the reduced interband effective mass $\mu_{\parallel}$ we use the value 0.058 m_e, where m_e is the electron mass. The values of F are shown in Fig. 2.

As we can observe in Fig. 1, the decreasing in temperature is followed by a decreasing in the energy spacing between the oscillations. Consequently, a reduction in the values of the built-in electric field is obtained. No FK oscillatory behavior has been observed at 77 K. At this temperature, only the well known Aspnes's third derivative functional form of the PR lines is seen, indicating that a lower electric field regime was reached[20].

Now we discuss the results for the delta-doped sample. In Figure 3 we show plots of $(E_n - E_g)^{3/2}$ versus n, for sample 2, for four different temperatures. The characteristic FK oscillations observed in the PR spectra are shown in the insert of the figure. As also observed for the heterojunction (sample 1), the features seen in the spectra are displaced to a region closer to the band edge as the temperature is lowered inducing a decrease in the value of the built-in electric field.

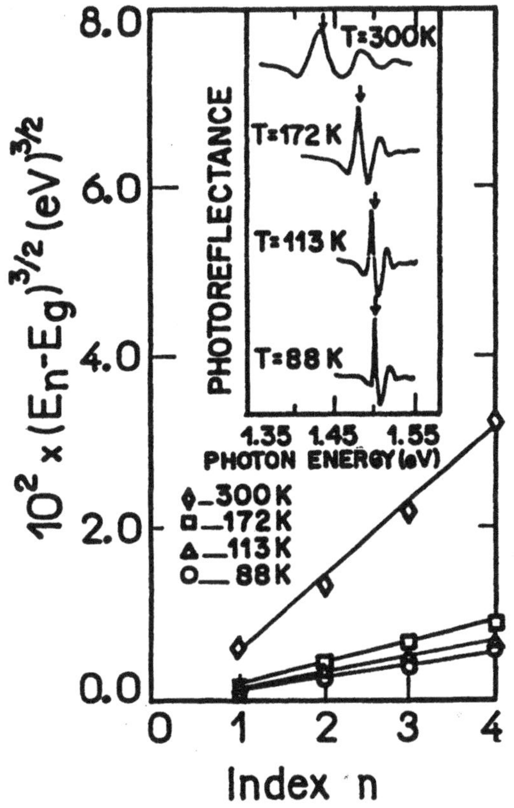

Fig. 3. Plots of $(E_n - E_g)^{3/2}$ as a function of index n for different temperatures for sample 2. Shown in the insert are the corresponding PR spectra. The arrows indicate the GaAs band gap energy.

3.b) Theoretical Calculations

Theoretical self-consistent calculations of subband energy levels and potential profiles of the two samples investigated here were carried out. The calculations are based on a simultaneous solution of the one-dimensional Schroedinger equation for motion perpendicular to the heterojunction, or δ-layer, and Poisson's equation which determines the electrostatic potential[16,22]. Exchange-correlation effects were taken into account within the local density approximation[15]. Temperature effects were included by considering the Fermi-Dirac distribution function for $T \neq 0$[23]. We start with the solution for $T = 0$; each wave function is then populated with a number N_i of electrons which, for a finite value of temperature, is given by

$$N_i = \frac{m^* k_B T}{\pi \hbar^2} \ln\left[1 + \exp\left(\frac{\mu_F - \varepsilon_i}{k_B T} \right) \right] \tag{3}$$

where m^* is the electron effective mass and μ_F is the chemical potential which is determined from the condition

$$N_s = \frac{m^* k_B T}{\pi \hbar^2} \sum_{i=1}^{N} \ln\left[1 + \exp\left(\frac{\mu_F - \varepsilon_i}{k_B T} \right) \right] \tag{4}$$

For the delta-doped system we have assumed that all Si donors are ionized with part of the supplied electrons compensating the acceptors in the depletion-layer region. Also, the possibility of diffusion of the Si donors out of the doped plane was taken into account by introducing a uniform spread of the positive charges[16]. Although the sample has been grown as a δ-doping, some spread of the silicon donors (80 Å) was obtained by matching the calculated occupancies to SdH data. We should mention that we have performed calculations considering a more realistic Gaussian diffusion profile and detected no significant difference between the Gaussian and uniform widths[18].

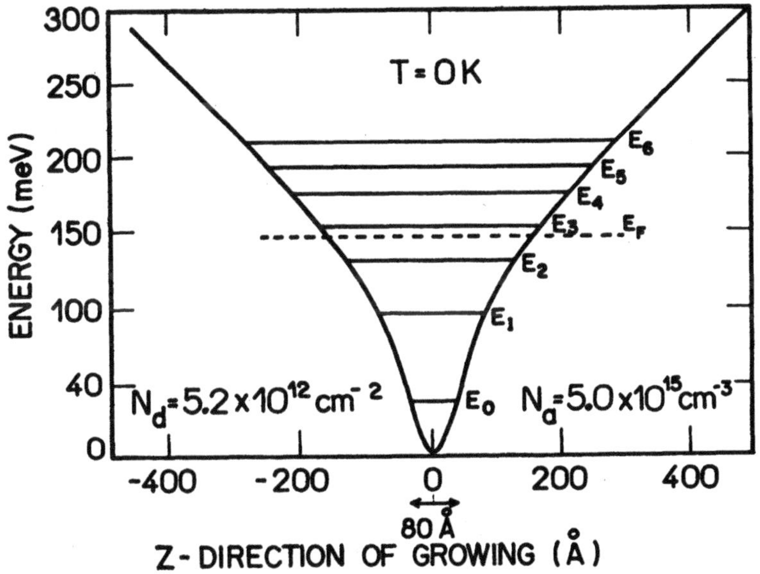

Fig. 4. Conduction-band potential profile for sample 2 at $T = 0\,K$ obtained from the self-consistent calculations. The first seven energy subbands are shown. The Fermi energy, E_F, is indicated by the dashed line. The value obtained for the spread of the ionized donors is also shown.

Figure 4 shows the calculated potential profile and subbands for the delta-doped sample investigated here, at $T = 0\,K$. The first seven subbands have been taken into account in the calculations. In Figure 5 the temperature dependence of the energy subbands and occupancies are shown. The subband energies increase with increasing temperature since higher subbands start to be populated. As a consequence the occupancies of the two lowest subband levels decrease and an increase of the mean electric field at the delta-doped region is observed. A similar behavior has been obtained for the heterojunction[15]. We should mention, however, that as the 2D electron density for the heterojunction is one order of magnitude less than in the case of the delta-doped sample both the energy subbands and occupations for the heterojunction show a less sensitive dependence on the temperature.

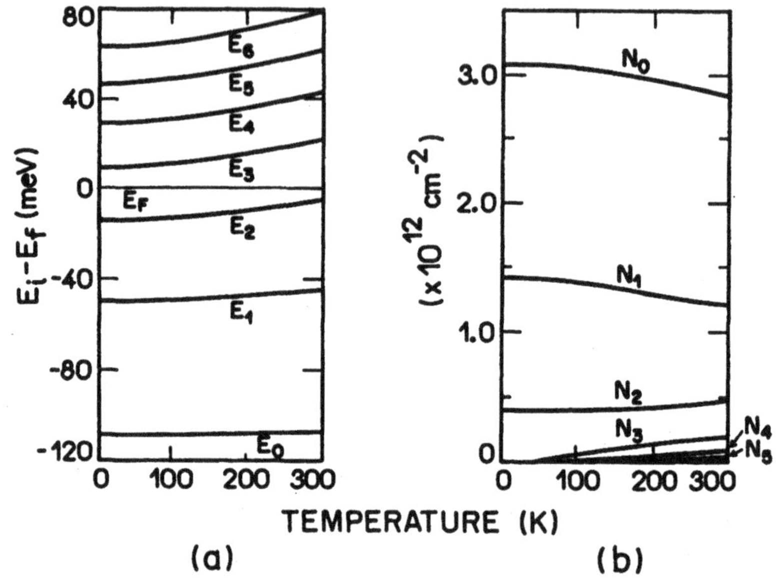

Fig. 5. Temperature dependence of (a) the energy subbands and (b) the subband occupancies for sample 2 as obtained from the self-consistent calculations.

4. Discussion and Conclusions

In Figure 6 we compare the values of electric field extracted from the experiment for sample 1 with those obtained from the calculations by numerically differentiating the electric potential with respect to z (direction of growing). The question which arises now is at what value of z such derivative should be taken? In other words, at what point z should the electric field be evaluated? Taking into consideration the fact that the most probable electronic transitions will occur to the regions at the Fermi level or to the first excited subband just above the Fermi level, we have a guide to compute the built-in electric field. Curve labelled A in Fig. 6 was obtained by computing the values of the built-in electric field at the Fermi level. Curve labelled as B was obtained considering the maximum of the density probability at the second excited subband. Figure 7 compares the measured

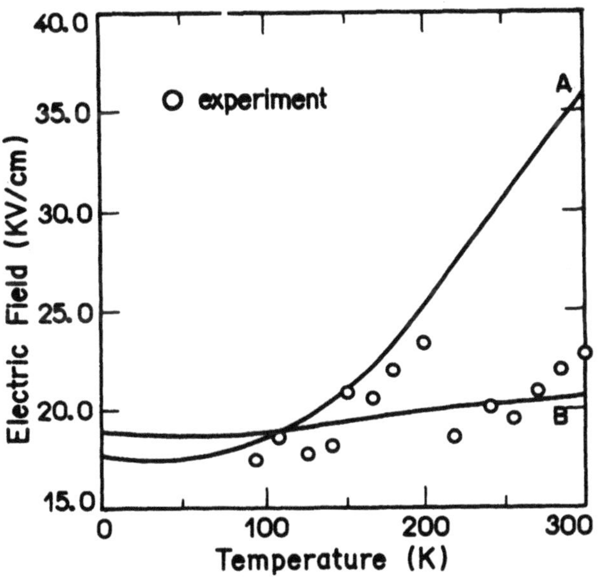

Fig. 6. Temperature dependence of the built-in electric field as obtained from the self-consistent calculations for sample 1. Curve A corresponds to computed values of the electric field at the chemical potential. Curve B corresponds to computed values of the electric field at the maximum of the probability density at the second excited subband. The values of electric field extracted from experiment are indicated.

values of the electric field with the calculated ones for sample 2, also computing the electric field at the Fermi level and at the maximum of the density probability at the first available excited subband (at T = 0 K) for different values of the acceptor background concentration N_a .

The analysis of the results depicted in Fig. 6 for sample 1 shows that curve B is in fairly good agreement with experiment, suggesting that the observed FK oscillations could be originated from the space charge region at the heterojunction. Although the trend in the calculated values of the electric field with temperature for the delta-doped GaAs (sample 2) is very similar to that observed experimentally, we are not at a position to definitively conclude that the

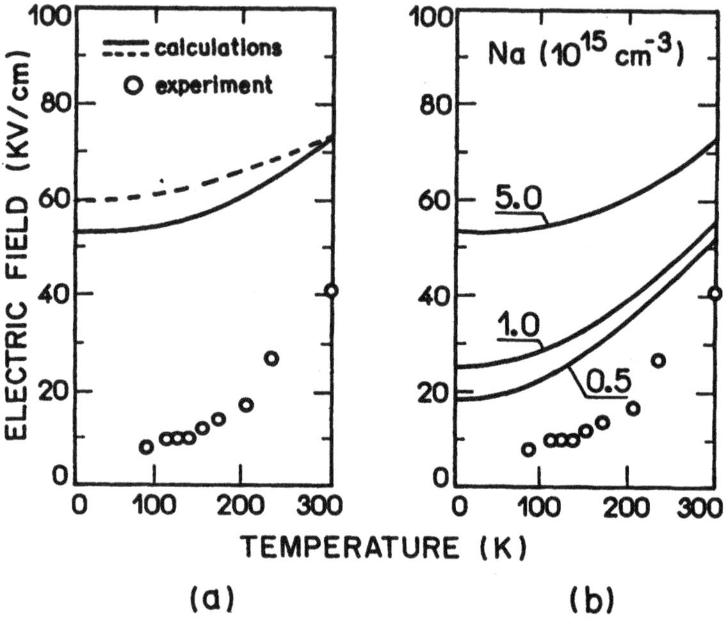

Fig. 7. Temperature dependence of the built-in electric field as obtained from the self-consistent calculations for sample 2. (a) Dashed and solid lines correspond to computed values of the electric field at the maximum of the probability density at the fourth excited subband (E_3) and at the chemical potential, respectively, for $N_a = 5 \times 10^{15}$ cm^{-3}. (b) Calculated values of the electric field (solid lines) at the chemical potential for different acceptor concentrations N_a. Experimental values are indicated by open circles.

measured fields are due only to the confinement region. As can be seen in Fig. 7b, even for lower values of N_a ($< 0.5 \times 10^{15}$ cm^{-3}) it was not possible to obtain the experimental values for the electric field. Despite the fact that FK oscillations from other regions of the sample, whose electric fields were not taken into account in the calculations, could also contribute to the measured built-in electric field, the electric field dependence on temperature and its relative insensitivity to changes in illumination intensity seem to point towards an evidence of FK oscillations coming from the space region at the delta-doping.

References

1. F.H. Pollak and H. Shen, Superlattices and Microstructures **6**, 203 (1989)
2. R.N. Bhattacharya, H. Shen, P. Parayanthal, F.H. Pollak, T. Coutts and H. Aharoni, Phys. Rev. **B 38**, 4044 (1988)
3. H. Shen, F.H. Pollak, J.M. Woodall and R.N. Sacks, J. Vac. Sci. Technol. **B 7**, 804 (1989)
4. M. Cardona, "Modulation Spectroscopy", Solid State Physics, Suppl. II (Academic, New York, 1969)
5. B.V. Shanabrook, O.J. Glembocki and W.T. Beard, Phys. Rev. **B 35**, 2540 (1987)
6. W.M. Theis, G.D. Sanders, C.E. Leak, K.K. Bajaj and H. Morkoç, Phys. Rev. **B 37**, 3042 (1988)
7. De-Sheng Jiang, Yin-Sheng Tang, Jian-Bai Xia and R. Enderlein, Superlattices and Microstructures **6**, 387 (1989)
8. G. Ji, W. Dobbelacre, D. Huang and H. Morkoç, Phys. Rev. **B 39**, 3216 (1989)
9. H. Shen, Z. Hang, Jing Leng, F.H. Pollak, L.L. Chang, W.I. Wang and L. Esaki, Superlattices and Microstructures **5**, 591 (1989)
10. E.S. Snow, O.J. Glembocki and B.V. Shanabrook, Phys. Rev. **B 38**, 12483 (1988)
11. Y.S. Tang, Y.W. Xu, D.S. Jiang, W.H. Zhuang and M.Y. Kong, Superlattices and Microstructures **6**, 391 (1989)
12. X. Yin, F.H. Pollak, L. Pawlowicz, T. O'Neill and M. Hafizi, Appl. Phys. Lett. **56**, 1278 (1990)
13. N. Pan, X.L. Zheng, H. Hendriks and J. Carter, J. Appl. Phys. **68**, 2355 (1990)
14. C.A.C. Mendonça, L.M.R. Scolfaro, J.M.V. Martins, E.A. Meneses and J.R. Leite, Proc. SPIE (1990) - in press
15. J.M.V. Martins, L.M.R. Scolfaro, C.A.C. Mendonça, E.A. Meneses and J.R. Leite, Superlattices and Microstructures (1990) - to appear

16. L.M.R. Scolfaro, C.A.C. Mendonça, E.A. Meneses, J.M.V. Martins and J.R. Leite, Int. J. Quant. Chem. **S24**, 447 (1990)

17. A.A. Bernussi, C.A.C. Mendonça, P. Motisuke, E.A. Meneses, F. Cerdeira, F.H. Pollak, P. Basmaji and I.F.L. Dias, Proc. 20th Int. Conf. Phys. Semic., Greece, Vol. 2, p. 1065 (1990)

18. R. Rodrigues, P.S.S. Guimarães, J.F. Sampaio, R.A. Nogueira, A.T. Oliveira Jr., I.F.L. Dias, J.C. Bezerra, A.G. de Oliveira, A.S. Chaves and L.M.R. Scolfaro, Solid State Commun. - to appear

19. M. Sydor, J. Angelo, J.J. Wilson, W.C. Mitchel and M.Y. Yen, Phys. Rev. **B 40**, 8473 (1989)

20. D.E. Aspnes, Surf. Sci. **37**, 418 (1973)

21. D.E. Aspnes and A.A. Studna, Phys. Rev. **B 7**, 4605 (1973)

22. V.M.S. Gomes, G.M.G. Oliveira, J.R. Leite and A.S. Chaves, Int. J. Quant. Chem. **S32**, 655 (1987)

23. F. Stern and S. das Sarma, Phys. Rev. **B 30**, 840 (1984)

ABSORPTION OF INFRARED RADIATION AND DIELECTRIC FUNCTION BY DONOR-TRIAD MOLECULES IN GERMANIUM

S. Canuto
Departamento de Física - UFPE
50739 Recife, PE, Brazil

A. Ferreira da Silva
Instituto Nacional de Pesquisas Espaciais
INPE/LAS - São José dos Campos, SP, Brazil
and Instituto de Física - UFBA

F. de Brito Mota
Instituto de Física - UFBA
40210 Salvador, BA, Brazil

and

A. Fazzio
Instituto de Física - USP
01498 São Paulo, SP, Brazil

Abstract

We have investigated the absorption spectrum and the real and imaginary parts of the dielectric function of donor-triad molecules in n-doped germanium. The calculations were performed at the "ab initio" second-order many-body perturbation theory level utilizing a large gaussian-type basis set. The results indicate that the observed absorption edges around 4 meV are due to the different electronic transitions of the donor-triad molecules. A very good agreement with experiment is obtained for arsenium doped germanium.

In a lightly n-doped semiconductor, at low temperature, the absorption spectrum exhibits a series of sharp atomic like lines which correspond to the optical transitions between isolated impurity atoms. As the impurity concentration N increases, a continuous absorption appears to extend to lower energy below the ionization level of the isolated atom due to interactions with other donors.[1]

In this work, we show that the appearence of lower energy tails of optical spectra, at a certain range of energy and impurity concentration of arsenium doped germanium (Ge:As), is due to the electronic $H_3 \rightarrow H_3^+$ transitions from their bonding energies. In our approximation, this ionization energy of H_3 corresponds to the energy required to promote an electron from the donor-triad-molecule to the bottom of the host conduction band (HCB). These transitions are also responsible for the anomalies found in the real (ε_1) and imaginary (ε_2) parts of the dielectric function.

The calculations for the donor-triad molecules involve "ab initio" calculations on H_3 and H_3^+. Within the Born-Oppenheimer approximation the total energy of H_3 and H_3^+ are calculated assuming a triangular form in D_{3h} symmetry[2]. First, the Hartree-Fock-Self-Consistent-Field (HF-SCF) equations are solved expanding the atomic orbitals in a basis of cartesian gaussian-type functions. This basis consists of 9s functions contracted to 4s, three sets of uncontracted 3-component p-type functions and one set of uncontracted 5-component d-type functions. This basis is usually referred to as (9s 3p 1d / 4s 3p 1d) and has been suggested by Siegbahn and Liu[3]. After the SCF calculations, the effects of electron correlation are included by second-order Many-Body Perturbation Theory (MBPT) utilizing the usual Moller-Plesset partitioning of the Hamiltonian[4]. This is referred to as MP2 and involves the calculation of one Brandow diagram[5]. This diagram involves only doubly excited configurations and is the dominant contribution to the total electron correlation energy. Analysis of the systematic contribution of the electron correlation effects to the absorption of infrared radiation (AIR) by donor triad molecules is underway up to the complete fourth-order MBPT[6]. In the present contribution the Spin-Unrestricted HF configuration is the reference state for the MP2 calculations on H_3. For H_3^+ the restricted HF configuration is used as reference. For increasing inter-atomic distance R, H_3 dissociates properly into three H atoms. For H_3^+ proper dissociation is obtained only after the inclusion of the electron correlation effects.

The ionization energy, $E_I(R) = E(H_3^+) - E(H_3)$, of the ground-state donor-triad molecules changes with R. For R > where a_o, is the Bohr radius, $E_I(R)$ is less than 1Ry. The dependence of the energy difference $E(R) = 1Ry - E_I(R)$ with R is shown in Fig. 1. In a semiconductor, the only change we have to make is a change of units. The effective Bohr radius a_o is calculated from the experimental ionization energy E_I^* via $a_0^* = 1Ry\ a_0/E_I^*\kappa$, where κ is the dielectric constant of the host material. Thus, the donor-triad with separations $R > a_o^*$ have ionization energies smaller than E_I^* of an isolated donor. This aspect which is of our interest, should influence the absorption spectra, dielectric functions and transport properties. We will use the following atomic units for the system Ge:As; $a_0^* = 37Å$, $\kappa \approx 14$ and $E_I^* = 14.2$ meV[7,8].

The AIR spectrum is given by

$$\alpha\ (E) \approx -\frac{1}{\pi}\ \mathrm{Im} \ll G^+\ (E,\ R) \gg \tag{1}$$

where << >> means an average disorder of the Green function propagator. For the ground state energies we obtain[2]

$$\ll G(E.R) \gg\ = \int P(R)\ [E - E_I(R) + i0^+]^{-1}\ dR \tag{2}$$

then

$$\alpha(E) \approx \int P(R)\delta\ [E - E_I(R)]\ dR, \tag{3}$$

where P(R) is the donor-triad distribution function[2,9]

$$P(R) = 4\pi\ R^2\ N^2\ C_3(R)\ \exp\ [-C_3'(R)N], \tag{4}$$

with

$$C_3(R) = \frac{4}{3}\pi\left[R_0^3 - \frac{3}{4}R_0^2R + \frac{R^3}{16}\right] \tag{5}$$

and

$$C'_3(R) = \frac{4}{3}\pi\left[R_0^3 + \frac{3}{2}R_0^2R - \frac{R^3}{8}\right] \tag{6}$$

where $R_0 = 4a_0$, i.e., a sphere with this radius centered at the donor[9,10].

In Fig. 2 we show the results of the calculated AIR spectra for different impurity concentrations, below the impurity critical concentration, $N_c = 3.5 \times 10^{17}$ cm^{-3}, for the metal-insulator transition of Ge:As[7]. Here we have considered the spin and the many-valley effects which appear in the HCB of germanium in an approximative way. For our isocoric system, Ge:As, the energy was reduced by a factor $\nu^{-1/2}$ where $\nu = 4$ being the number of valleys (minima) of the HCB i.e., Ge[7,11]. The absorption edge of the spectra becomes extended towards zero energy with increasing impurity concentrations. The same behaviour is observed in the measurements of Kobayashi et al[12].

As the dielectric function, $\varepsilon(E) = \varepsilon_1(E) + i\varepsilon_2(E)$, describes the optical response of the host material, we calculate the real (ε_1) and imaginary (ε_2) parts through the Kramers-Kronig relations[13,14] for very low frequency[14], i.e., $E \rightarrow 0$

$$\varepsilon_1(E) = \varepsilon_{Ge} + \frac{2}{\pi}\int_0^\infty \frac{E'\varepsilon_2(E')}{(E')^2 - E^2}\,dE' \tag{7}$$

and

$$\varepsilon_2(E) = -\frac{2}{\pi}\int_0^\infty \frac{E'\varepsilon_1(E')}{(E')^2 - E^2}\,dE' \tag{8}$$

These dielectric function are strongly connected with the AIR obtained from Eq. 3[14].

In Fig. 3 we show ε_1 and ε_2 for Ge:As at N= 5.0 x $10^{16} cm^{-3}$. The anomalies are found around 4meV as in the absorption spectra. Experimental study of ε_1 and ε_2 in-doped semiconductors, as well as a theoretical treatment to include such transitions, are called for to further check the above ideas. Other optical properties are currently under investigation.

References:

1. Thomas, G.A., Capizzi, M., De Rosa, F., Bhatt, R.N. and Rice, T.M., Phys. Rev. B 23, 5472(1981); Gemma, N., Natori, A. and Kamimura, H., J. Phys. Soc. Japan 51, 1207(1982).
2. Ferreira da Silva, A. and Canuto, S., Solid State Commun. 75, 939(1990).
3. Siegbahn, P. Liu, B., J. Chem. Phys. 68, 2457(1978).
4. Moller, C. and Plesset, M.S., Phys. Rev. 46, 618(1934).
5. Bartlett, R.J., Ann. Rev. Phys. Chem. 32, 359(1981); J. Phys. Chem. 93, 1697(1989).
6. Canuto, S., Ferreira da Silva, A. and Fazzio, A (to appear).
7. Ferreira da Silva, A., Phys. Rev. B 37, 4799(1988).
8. Edwards, P.P. and Sienko, M.J., Phys. Rev. B 17, 2575(1978).
9. Nagasaka, K. and Naritu, J., J. Phys. Soc. Japan 35, 787(1973).
10. Toyotomi, S., J. Phys. Soc. Jpn. 38, 175(1975).
11. Fabbri, M. and Ferreira da Silva, A., Phys. Rev. B 29, 5764(1984).
12. Kobayashi, M., Sakaida, Y., Taniguchi, M. and Narita, S., J. Phys. Soc. Japan. 47, 138(1973).
13. Stern, F., Phys. Rev. 133, A1653(1964).
14. de Brito Mota, F. and Ferreira da Silva, A., Int. J. Quantum Chem. S 24, 411(1990).

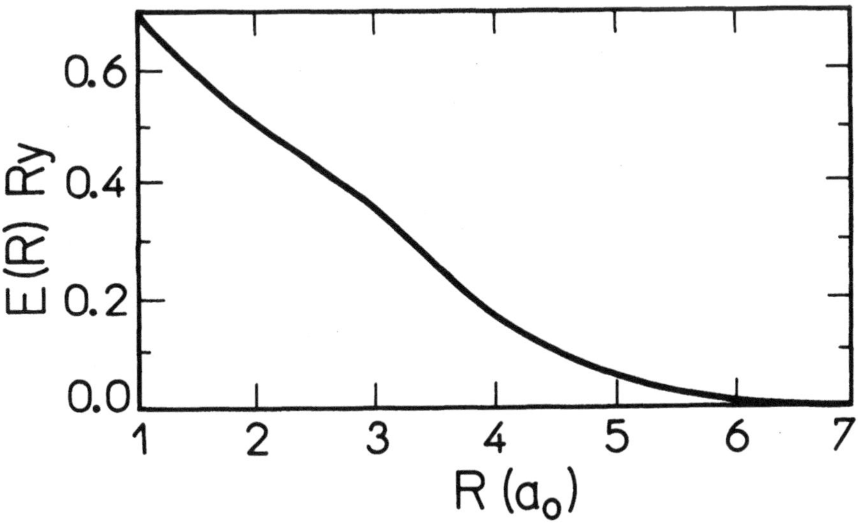

Fig. 1 - Dependence of the ionization energy difference, on the interatomic separation R.

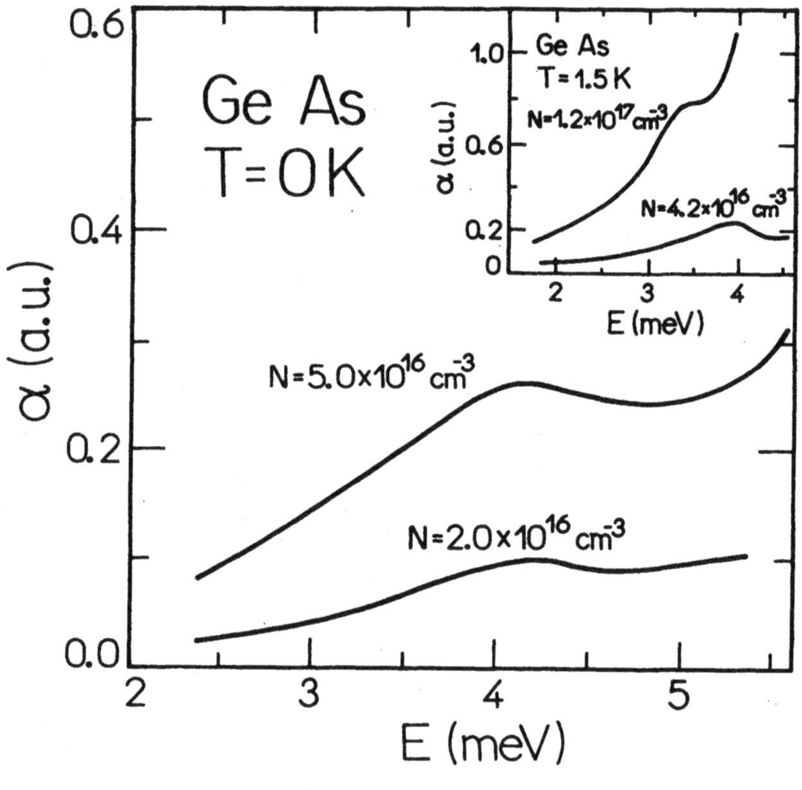

Fig. 2 - Donor impurity concentration N dependence of the infrared absorption spectra α of as-doped Ge at OK due to the donor-triad molecules. Inset shows the experimental data from Kobayashi et al.

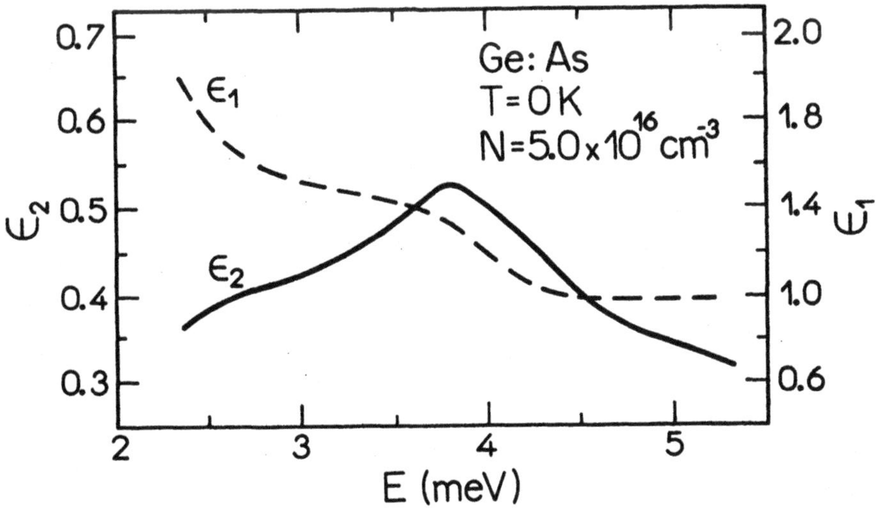

Fig. 3 - Real (ε_1) and imaginary (ε_2) parts of the dielectric funtion in arsenium doped germanium. N = 5.0 x 10^{16} cm^{-3}.

A STUDY OF THE HUBBARD MODEL FOR FINITE U

Mario E.Foglio
Instituto de Física "Gleb Wataghin" - UNICAMP - C.P. 6165
13081 Campinas - S.P. - BRASIL

To Rogerio C.Cerqueira Leite
on his sixtieth birthday.

The Hubbard model with finite U was studied in the lowest non-trivial approximation by a method of Kadanoff and Baym that makes a systematic approximation of the equation of motion. For appropriate values of the parameters, our results coincide with the exact atomic limit and with a direct calculation of a model for $U \to \infty$ in which the doubly occupied states are automatically excluded. A higher order approximation would be necessary to obtain the correct $U = 0$ limit.

1. INTRODUCTION

The importance of electronic correlations in the modern theory of solids has been well known since the early days of quantum mechanics. Molecular orbital theory treats correlations through an average self-consistent field and it is the basis of the successful band theory, which has dominated the study of metals for a long time. The alternative Heitler-London treatment starts from an atomic description, so that correlations are included at the beginning, but the inherent difficulties of the problem have relegated its use to the treatment of insulators and to the magnetic properties of f-electrons. The recent discovery of the Heavy-Fermion systems and of the high-T_c superconductors has renewed the interest in the study of electronic correlations, which are essential in the understanding of these systems. The narrow band problem was studied in a series of papers by Hubbard [1], who considered a simplified Coulomb repulsion between electrons at the same site. Although the Hubbard Hamiltonian seems very simple, rigorous results are only known for limiting cases that do not correspond to most situations of interest, and it seems rather important to develop reliable approximations. Several analytical approaches have been employed to study the Hubbard model, like decoupling of the Green's Functions (GF) equations of motion,[1] variational techniques,[2] and methods based on the auxiliary boson field.[3]

To describe localized ionic states, Hubbard introduced operators $X_{\alpha\beta}(j) = |j,\alpha\rangle\langle j,\beta|$ that transform the local state $|j,\beta\rangle$ at site j into the local state $|j,\alpha\rangle$ at the same site. These operators are particularly useful because rather complicated local correlations at site j can be described implicitly by the generic states $|j,\alpha\rangle$. As these X-operators are not Fermions or Bosons, Wick's theorem is not valid, and the decoupling of the equations of motion was initially employed by Hubbard [1] to study the problem. Besides difficulties with the violation of sum rules, the factorizations employed are rather arbitrary, and one would like a more systematic procedure. A method that seems to fulfill this requirement was introduced[4] by Kadanoff and Baym (KB) and it was employed[5] by Ruckenstein and Schmitt-Rink (RS) to study the Hubbard model in the $U \to \infty$ limit. The same authors (RS) have recently shown that this method is related to a projection technique that was used for the Anderson model.[6]

In the present work we apply the KB method to the Hubbard model with finite U, extending the work of RS for the $U \to \infty$ case. It was necessary to introduce some modifications to the technique because of the presence of the doubly occupied local states, and careful consideration was given to the meaning of "the lowest order non-trivial approximation".

In Section 2. we present the details of the derivation, in Section 3. we consider three limit cases in which we can compare our results with those of exact or independent calculations, and in Section 4. we discuss our results.

2. APPLICATION OF THE KB EXPANSION TO THE HUBBARD MODEL.

2.1 Model Hamiltonian

The Hamiltonian introduced by Hubbard[1] to describe electronic correlations can be written in the form

$$H = H_t + H_u = \sum_{ij\sigma} t_{ij} C^{\dagger}_{i\sigma} C_{j\sigma} + U \sum_{j} C^{\dagger}_{j\sigma} C_{j\sigma} C^{\dagger}_{j-\sigma} C_{j-\sigma} \tag{1}$$

where $C_{i\sigma}$ is the usual destruction operator of an electron in the Wannier state at site i and with z-component of the spin equal to $\sigma/2$, and we assume that $t_{jj}=0$. Employing the Hubbard operators $X_{\alpha\beta}(j)$ we can write

$$C_{j\sigma} = X_{o\sigma}(j) + \sigma X_{-\sigma d} \tag{2}$$

where $|j,0\rangle$ is the local state with no electrons, $|j,\sigma\rangle$ the state with only one electron with spin $\sigma/2$ and $|j,d\rangle$ the doubly occupied state, all at site j . To abbreviate we introduce the operators $Y_{\sigma}(j,1) \equiv X_{o\sigma}(j)$ and $Y_{\sigma}(j,2) \equiv \sigma X_{-\sigma d}(j)$, so that $C_{j\sigma} = Y_{\sigma}(j,1) + Y_{\sigma}(j,2)$.

2.2 Green's Functions

Following KB we shall use GF of the Matsubara type (imaginary time) in the presence of an external potential

$$G_{1\sigma}(ja\tau,j'a'\tau';A) = -\langle(S(A)Y_\sigma(ja\tau)\bar{Y}_\sigma(j'a'\tau'))_+\rangle/\langle S(A)\rangle \quad (3)$$

where $Y_\sigma(ja\tau)=\exp(H\tau)Y_\sigma(ja)\exp(-H\tau)$ and $\bar{Y}_\sigma(ja\tau)=\exp(H\tau)\ Y^\dagger_\sigma(ja)\ \exp(-H\tau)=(Y_\sigma(ja-\tau))^\dagger$. The subindex + in Eq.(3) indicates τ-ordering increasing to the left, and we shall often abbreviate $Y_\sigma(1)=Y_\sigma(j_1a_1\tau_1)$. The external potential has two contributions: $A_\sigma(\tau) = A^c_{-\sigma}(\tau) + A^d_{-\sigma}(\tau)$;

$$A^c_{-\sigma}(\tau) = \sum_{ll'}{}' \sum_{ss'} W_\sigma(ls,l's',\tau)\ \bar{Y}_\sigma(ls\tau)\ Y_\sigma(l's'\tau), \quad (4a)$$

$$A^d_{-\sigma}(\tau) = \sum_{ls}\Big[\ V_\sigma(1,ls\tau)\ Y_\sigma(ls\tau)\ \bar{Y}_\sigma(ls\tau) + V_\sigma(2,ls\tau)\ \bar{Y}_\sigma(ls\tau)\ Y_\sigma(ls\tau)\ \Big] \quad (4b)$$

where $S_\sigma(A)$ is a functional of the external field $A_\sigma(\tau)$:

$$S_\sigma(A) = \exp_+\left\{\int_0^\beta d\tau\ (\ A^c_\sigma(\tau) + A^d_\sigma(\tau)\ )\right\} \quad (5)$$

and the $W_\sigma(ls,l's',\tau)$ and $V_\sigma(a,ls\tau)$ determine its strength. Note that only fields and operators with subindex $-\sigma$ appear in $A_\sigma(\tau)$ (cf. definition in Eq.(4)), and the symbol A indicates that we consider the presence of a non-zero $A_\sigma(\tau)$, and we shall not write it unless necessary for clarity.

2.3 Equations of Motion.

Following KB we obtain

$$\left\{\frac{\partial}{\partial\tau} + \Delta E_\sigma(a_1)\right\}G_{1\sigma}(1,1') + G_{2\sigma}(1,1') = \quad (6)$$

$$= \langle(S(A)[Y_\sigma(1);\ A_\sigma(\tau)]\bar{Y}_\sigma(1'))_+\rangle/S(A) - Q_\sigma(1)\delta(1,1')$$

where

$$Q_\sigma(1;A) = \langle(S(A)\{Y_\sigma(1);\bar{Y}_\sigma(1)\})_+\rangle/\langle S(A)\rangle \quad , \quad (7a)$$

$$\delta(1,2) = \delta_{a_1,a_2}\,\delta_{j_1,j_2}\,\delta(\tau-\tau) \quad , \quad (7b)$$

and

$$G_{2\sigma}(1,1') = \langle (S(A)[H_t;Y_\sigma(1)]\overline{Y}_\sigma(1'))_+\rangle / \langle S(A)\rangle \qquad (7c)$$

The $\Delta E_\sigma(a)$ is defined in Table 1, together with several other functions we shall employ to avoid repetition of equations associated with the two values of a in $Y_\sigma(ja\tau)$. Note that in all our equations, both U and the magnetic field h appear explicitly only through the function $\Delta E_\sigma(a)$. For zero external potential the $\langle X_{ab}(j\tau)\rangle$ are independent of τ and j because of the time and lattice translational invariance, and in Table 1 we have used n_o, n_σ and n_d to indicate the occupation numbers per site of the local states $|j,0\rangle$, $|j,\sigma\rangle$ and $|j,d\rangle$ respectively.

Table 1

a	$\bar{a}$	s(a)	$\tau^+(a)$	$\Delta E_\sigma(a)$	$Q_\sigma(a)$
1	2	+1	$\tau+0$	σh	n_o+n_σ
2	1	-1	$\tau-0$	$\sigma h+U$	$n_{-\sigma}+n_d$

When $A_\sigma(\tau) = 0$ we find

$$G_{2\sigma}(ja\tau,1') = \qquad (8)$$

$$-s(a)\sum_i\Big\{t_{ji}\Big[\sum_s\langle(Y_{-\sigma}(js\tau^+(a))\overline{Y}_{-\sigma}(js\tau)Y_\sigma(ib\tau)\overline{Y}_\sigma(1'))_+\rangle+$$

$$+\langle(\overline{Y}_{-\sigma}(ja\tau^+(a))Y_\sigma(ja\tau)Y_{-\sigma}(ib\tau)\overline{Y}_\sigma(1'))_+\rangle\Big]+$$

$$+t_{ij}\langle(Y_{-\sigma}(j\bar{a}\tau)Y_\sigma(ja\tau^+(a))\overline{Y}_{-\sigma}(ib\tau)\overline{Y}_\sigma(1'))_+\rangle\Big\}$$

When the relations

$$(\overline{Y}_{-\sigma}(ja\tau^+(a))Y_\sigma(ja\tau))_+ = X_{-\sigma\sigma}(j\tau)\ , \qquad (9)$$

$$(Y_{-\sigma}(j\bar{a}\tau)Y_\sigma(ja\tau^+(a)))_+ = \sigma X_{od}(j\tau)$$

are substituted in Eq.(8), one easily recovers the equations of motion of the Fermion GF $-\langle(C(1)\overline{C}(1'))_+\rangle$.

2.4 The Decoupling of the Equations of Motion.

We shall try and apply the method described by KB to decouple Eq.(6) in a systematic way. The GF that appear in $G_{2\sigma}(1,1')$ are expressed as functional derivatives of $G_{1\sigma}$ with respect to the coefficients W and V of the external field $A_\sigma(\tau)$. In the Hubbard model with $U\to\infty$, RS were able to repeat the tricks employed by KB to approximate the functional derivatives, and the same technique was also applied to the Anderson lattice in the same limit.[7] In the case of finite U, we have been able to repeat the same procedure step by step only for the derivation with respect to the fields A^d_σ which are "diagonal" in the sites; for the "cross" fields A^c_σ that contain pairs of different sites, some modifications were necessary.

2.4.1 Functional derivatives related to A^d_σ. Taking the functional derivative of $G_{1\sigma}(ib\tau,1')$ with respect to $V_{-\sigma}(a,js\tau)$ we find (for $A_\sigma(\tau)=0$)

$$\langle (Y_{-\sigma}(js\tau^+(a))\overline{Y}_{-\sigma}(js\tau)Y_\sigma(ib\tau)\overline{Y}_\sigma(1'))_+\rangle = \ldots\ldots \quad (10)$$

$$-s(a)\frac{\delta G_{1\sigma}(ib\tau,1')}{\delta V_{-\sigma}(a,js\tau)} + G_{1-\sigma}(js\tau^+(a),js\tau)G_{1\sigma}(ib\tau,1')$$

To take advantage of this relation, KB found a way to express the functional derivative so that a systematic approximation could be made. To follow their method, we introduce the self-energy $\Sigma_\sigma(ja\tau,2)$:

$$G_{2\sigma}(ja\tau,1') = \sum_{ib} t_{ji}G_{1\sigma}(ib\tau,1') + \int d2\,\Sigma_\sigma(ja\tau,2)G_{1\sigma}(2,1') \quad (11)$$

where $\int d2 = \int_0^\beta d\tau_2 \Sigma_{j_2 a_2}$, and we define the inverse $G^{-1}_{1\sigma}(1.2)$:

$$\int d2\,G_{1\sigma}(1,2).G^{-1}_{1\sigma}(2,1') = \int d2\,G^{-1}_{1\sigma}(1,2).G_{1\sigma}(2,1') = \delta(1,1') \quad (12)$$

Introducing Eq.(11) into Eq.(6) and taking the functional derivative with respect to $V_\sigma(a,js\tau)$, we obtain, after some manipulation, a relation between the functional derivative that appears in Eq.(10) and terms of the type $\delta Q/\delta V$ and $\delta\Sigma/\delta V$. When we neglect these two functional derivatives, we find, after some manipulation, that

$$-\sum_{ib} t_{ji} s(a) \sum_{s} G_{1-\sigma}(js\tau^{+}(a),js\tau) G_{1\sigma}(ib\tau,1') =$$
$$= \sum_{ib} Q_\sigma(a) t_{ij} G_{1\sigma}(ib\tau,1') \qquad (13)$$

which corresponds to the simplest Hartree-Fock (HF) decoupling of the first term in the r.h.s. of the Eq.(8). Note that this is only true for the Σ_s and not for the individual terms.

2.4.2 The functional derivatives related to A^c_σ. To discuss the remaining contributions to Eq.(8), we define

$$x_\sigma(ja,ib,\tau,1') = \langle (\bar{Y}_{-\sigma}(ja\tau^{+}(a)) Y_\sigma(ja\tau) Y_{-\sigma}(ib\tau) \bar{Y}_\sigma(1'))_{+} \rangle$$
$$y_\sigma(ja,ib,\tau,1') = \langle (Y_{-\sigma}(j\bar{a}\tau) Y_\sigma(ja\tau^{+}(a)) \bar{Y}_{-\sigma}(ib\tau) \bar{Y}_\sigma(1'))_{+} \rangle \qquad (14)$$

so we can write

$$x_\sigma(ja,ib,\tau,1') = \qquad (15a)$$
$$= \frac{\delta G_{1\sigma}(ja\tau,1')}{\delta W_{-\sigma}(ja,ib,\tau^{+}(a))} + G_{1\sigma}(ja\tau,1') G_{1-\sigma}(ib\tau,ja\tau)$$

$$y_\sigma(ja,ib,\tau,1') = \qquad (15b)$$
$$= -\frac{\delta G_{1\sigma}(ja\tau^{+}(a),1')}{\delta W_{-\sigma}(ib,j\bar{a},\tau)} - G_{1\sigma}(ja\tau,1') G_{1-\sigma}(j\bar{a}\tau,ib\tau)$$

To calculate the functional derivatives we repeat the KB technique, and when we neglect the $\delta Q/\delta V$ and $\delta\Sigma/\delta V$ we find the relation

$$\frac{\delta G_{1\sigma}(ja\tau,1')}{\delta W_{-\sigma}(ls,l's',\tau_\sigma)} = \frac{s(s)}{Q_\sigma(s)} G_\sigma(ja\tau,ls\tau_s)\, x_\sigma(ls,l's',\tau_s,1') +$$

$$+ \frac{s(\bar{s}')}{Q_\sigma(\bar{s}')} G_\sigma(ja\tau,l'\bar{s}'\tau_s)\, y_\sigma(l'\bar{s}',ls,\tau_s,1') \qquad (16)$$

By substitution in Eq.(15) we obtain

$$x_\sigma(ja,ib,\tau,1') = G_{1\sigma}(ja\tau,1')G_{1-\sigma}(ib\tau,ja\tau) + \qquad (17a)$$
$$+[s(a)/Q_\sigma(a)]G_{1\sigma}(ja\tau,ja\tau^+(a))\,x_\sigma(ja,ib,\tau,1')$$
$$+[s(\bar{b})/Q_\sigma(\bar{b})]G_{1\sigma}(ja\tau,i\bar{b}\tau)\,y_\sigma(i\bar{b},ja,\tau,1')$$

$$y_\sigma(ja,ib,\tau,1') = -G_{1\sigma}(ja\tau,1')G_{1-\sigma}(j\bar{a}\tau,ib\tau) - \qquad (17b)$$
$$-[s(b)/Q_\sigma(b)]G_{1\sigma}(ja\tau,ib\tau)\,x_\sigma(ib,j\bar{a},\tau,1')-$$
$$-[s(a)/Q_\sigma(a)]G_{1\sigma}(ja\tau^+(a),ja\tau)\,y_\sigma(ja,ib,\tau,1')$$

At this point, one alternative would be to solve the system of linear equations for x_σ and y_σ, but this would treat these two quantities as if they were exact, while we have neglected $\delta Q/\delta V$ and $\delta\Sigma/\delta V$ in Eq.(16). When this method is followed, the GF obtained are not consistent with the expression derived[5] by RS for $U\to\infty$. Another alternative would be to use HF decoupling of the x and y at the r.h.s.: this gives a GF that coincides with that of RS in the $U\to\infty$ limit. We shall follow this alternative, leaving the other for a future higher order calculation of Eq.(6). By HF decoupling of the x and y we find

$$x_\sigma(ja,ib,\tau,1') = G_{1-\sigma}(ib\tau,ja\tau)G_{1\sigma}(ja\tau,1') \qquad (18)$$
$$y_\sigma(ja,ib,\tau,1') = -G_{1-\sigma}(j\bar{a}\tau,ib\tau)G_{1\sigma}(ja\tau,1')$$

and substituting into Eq.(17) we obtain

$$x_\sigma(ja,ib,\tau,1') = G_{1-\sigma}(ib\tau,ja\tau)\{B_{x\sigma}(a)G_{1\sigma}(ja\tau,1')+$$
$$+ C_\sigma(ja,i\bar{b})G_\sigma(i\bar{b}\tau,1')\} \qquad (19a)$$

$$y_\sigma(ja,ib,\tau,1') = G_{1-\sigma}(j\bar{a}\tau,ib\tau)\{B_{y\sigma}(a)G_{1\sigma}(ja\tau,1')+$$
$$+ C_\sigma(ja,ib)G_\sigma(ib\tau,1')\} \qquad (19b)$$

where

$$B_{x\sigma}(a) = 1 + s(a)G_\sigma(ja\tau,ja\tau^+(a))/Q_\sigma(a) \qquad (20)$$
$$B_{y\sigma}(a) = -1 + s(a)G_\sigma(ja\tau^+(a),ja\tau)/Q_\sigma(a)$$
$$C_\sigma(ja,ib) = - s(b)G_\sigma(ja\tau,ib\tau)/Q_\sigma(b)$$

2.4.3 The approximate Green's functions. Substituting Eq.(13,19) into Eq.(18) and the result into Eq.(6) we find (for $A_\sigma(\tau) = 0$)

$$\left\{ \frac{\partial}{\partial\tau} + \tilde{E}_\sigma(a) \right\} G_{1\sigma}(ja\tau,1') + \qquad (21)$$
$$+ \Sigma\tilde{t}_{ji}(a,b)G_{1\sigma}(ib\tau,1') = - Q_\sigma(a)\delta(ja\tau,1')$$

where $\tilde{E}_\sigma(a)$ and $\tilde{t}_{ji}(a,b)$ are renormalized energies and hopping terms respectively, that will be explicitly written after solving Eq.(21). For this purpose, we Fourier transform the GF both in space and imaginary time:

$$G_\sigma(a,b,\mathbf{k},i\omega_\nu)=\int_0^\beta d\tau\Sigma_j G_{1\sigma}(ja\tau,ib0)\exp\{-i\mathbf{k}.\mathbf{R}_{ji}+i\omega_\nu\tau\} \qquad (22)$$

where $\omega_\nu=(2\nu+1)\pi/\beta$ (ν integer), $\beta=1/kT$ and $\mathbf{R}_{ji}=\mathbf{R}_j-\mathbf{R}_i$. Employing the standard techniques we find

$$G_\sigma(a,b,\mathbf{q},i\omega_\nu)=Q_\sigma(b)[\delta_{ab}i\omega_\nu+A_\sigma(a,b,\mathbf{q})]/\Delta(i\omega_\nu) \qquad (23)$$

where

$$A_\sigma(\bar{a},\bar{a},\mathbf{q})=(\tilde{E}_\sigma(a)+\varepsilon_{aa}(\mathbf{q})) \quad , \qquad (24)$$
$$A_\sigma(a,b,\mathbf{q})=\varepsilon_{ab}(\mathbf{q}) \quad (\text{for } a\neq b) \quad ,$$

$$\Delta(z)=(z+A_\sigma(1,1,\mathbf{q})).(z+A_\sigma(2,2,\mathbf{q})-A_\sigma(1,2,\mathbf{q}).A_\sigma(2,1,\mathbf{q}) \qquad (25)$$

and

$$\varepsilon_{ab}(\mathbf{q})\doteq\Sigma_j\tilde{t}_{ji}(a,b)\exp(-i\mathbf{q}.\mathbf{R}_{ji}) \quad . \qquad (26)$$

The Eqs.(23-26) give $G_\sigma(a,b,\mathbf{q},i\omega_\nu)$ as a function of the renormalized $\tilde{t}_{ji}(a,b)$ and $\tilde{E}_\sigma(a)$ which should be obtained self-consistently from the GF. To give the explicit expression for $\tilde{E}_\sigma(a)$ and $\tilde{t}_{ji}(a,b)$, we shall consider non-zero hopping $t_{ji}=t_o$ only between the Z_{nn} nearest neighbors of each site. When we consider also the symmetry properties of the lattice, we find

$$G_{1\sigma}(ia\tau,jb\tau)=Q_\sigma(b)\{\delta_{ab}F_{1\sigma}+F_\sigma(a,b)\} \tag{27}$$

where

$$NF_{1\sigma}=\Sigma_q(\varepsilon^o(q)/Z_{nn}t_o).(\omega_1 f_1-\omega_2 f_2)/(\omega_1-\omega_2) \ , \tag{28}$$

$$NF_\sigma(a,b)=\Sigma_q(\varepsilon^o(q)/Z_{nn}t_o).A_\sigma(a,b,q).(f_1-f_2)/(\omega_1-\omega_2) \tag{29}$$

and the $\varepsilon^o(q)$ is the dispersion relation for the band with $U=0$. The ω_j $(j=1,2)$ are the two roots of $\Delta(z)=0$, which are functions of q , and the $f_j\equiv f(\omega_j)$ are the corresponding Fermi functions for given values of μ and β :

$$f(x)=\{\exp[\beta(x-\mu)]+1\}^{-1} \ . \tag{30}$$

2.4.4 The relations of self consistence . To complete the relations that should be solved self-consistently, we write the $\varepsilon_{ab}(q)$ and $\tilde{E}_\sigma(a)$ as functions of $F_{1\sigma}$ and $F_\sigma(a,b)$: with the assumptions made above, the $C_\sigma(ja,ib)$ in Eq.(19) do not depend on i,j (cf. Eq.(20)) and we have

$$\varepsilon_{ab}(q) = \varepsilon^o(q)\{Q_\sigma(a) + C_t^\sigma(a,b)\} \tag{31}$$

$$\tilde{E}_\sigma(a) = \Delta E_\sigma(a) + C_E^\sigma(a)$$

where

$$C_t^\sigma(a,b) = -s(a)\{(Q_{-\sigma}(a)+Q_{-\sigma}(b))\delta_{ab}F_{1-\sigma}+ \tag{32}$$

$$+s(\bar{b})[Q_{-\sigma}(a)F_{-\sigma}(\bar{b},a)+Q_{-\sigma}(b)F_{-\sigma}(\bar{a},b)].[\delta_{ab}F_{1\sigma}+ F_\sigma(b,a)]$$

and

$$C^{\sigma}_{E}(a) = -s(a).Z_{nn}t_{o}.\sum_{b}\Big\{B_{x\sigma}(a)Q_{-\sigma}(a)[\delta_{ab}F_{1-\sigma}+F_{-\sigma}(b,a)]+$$
$$+B_{y\sigma}(a)Q_{-\sigma}(b)[\delta_{\bar{a}b}F_{1-\sigma}+F_{-\sigma}(\bar{a},b)]\Big\} \qquad (33)$$

Employing the property $G^{*}_{AB}(z)=G_{B^{\dagger}A^{\dagger}}(z^{*})$ one obtains another relation that should be satisfied

$$C^{\sigma}_{t}(a,b)Q_{\sigma}(b)=C^{\sigma}_{t}(b,a)Q_{\sigma}(a) \qquad (34)$$

(this is equivalent to $F_{\sigma}(a,b)Q_{\sigma}(b)=F_{\sigma}(b,a)Q_{\sigma}(a)$).

To calculate n_{o} , n_{σ} , $n_{-\sigma}$ and n_{d} we introduce

$$NF_{o\sigma} = \Sigma_{q}(\omega_{1}f_{1}-\omega_{2}f_{2})/(\omega_{1}-\omega_{2}) \quad , \qquad (35)$$

and

$$NF_{o\sigma}(a) = \Sigma_{q}A_{\sigma}(a,a,q)(f_{1}-f_{2})/(\omega_{1}-\omega_{2}) \; . \qquad (36)$$

With the usual GF formulas we then find

$$n_{\sigma}=Q_{\sigma}(1)(F_{o\sigma}+F_{o\sigma}(1)) \quad , \qquad (37a)$$

$$n_{d}=Q_{\sigma}(2)(F_{o\sigma}+F_{o\sigma}(2)) \quad . \qquad (37b)$$

The calculation of n_{o} and $n_{-\sigma}$ from the GF gives values that automatically satisfy

$$n_{o}=Q_{\sigma}(1)-n_{\sigma} \quad , \qquad (37c)$$

$$n_{-\sigma}=Q_{\sigma}(2)-n_{d} \quad . \qquad (37d)$$

because of the form of the $G_{1\sigma}$ in this approximation.

The condition $n_{\sigma}=n_{-\sigma}$ for paramagnetism, together with the completeness relation $Q_{\sigma}(1)+Q_{\sigma}(2)=1$ gives

$$Q_{\sigma}(1) = [1-(F_{o\sigma}+F_{o\sigma}(2))]/[1+F_{o\sigma}(1)-F_{o\sigma}(2)] \qquad (38)$$

To solve the problem, it would then be enough to find self-consistent constants $C^{\sigma}_{t}(a,b)$ and $C^{\sigma}_{E}(a)$ (after fixing the model parameters $\varepsilon^{o}(q)$, $Z_{nn}t_{o}$, μ and $\Delta E_{\sigma}(a)$).

The numerical study of the self-consistent problem for several interesting sets of parameters is under study.

3. LIMIT CASES OF THE APPROXIMATE SOLUTION.

We shall consider our approximate solution for three sets of parameter values in which the solution is independently known.

3.1 The Atomic Case.

This case corresponds to $t_{ji}\to 0$, i.e. $\varepsilon^{o}(q)\to 0$ and therefore $\tilde{E}_{\sigma}(a)=\Delta E_{\sigma}(a)$. By inversion of the spatial Fourier transform, Eq.(23) gives

$$G_{1\sigma}(ja,ib,i\omega_{\nu}) = \delta_{ab}\cdot Q_{\sigma}(a)/(i\omega_{\nu}-\Delta E_{\sigma}(a)) \qquad (39)$$

The quantity $Q_{\sigma}(2)=n_{-\sigma}+n_{d}$ is the average number of $-\sigma$ electrons per site (both in singly and doubly occupied states), and $Q_{\sigma}(1)=1-Q_{\sigma}(2)$. From Eqs.(2,39) we then recover the exact solution in the atomic limit.[1] Substitution of $i\omega_{\nu}$ by the complex frequency z in Eq.(39) gives the GF of Zubarev[8] for each spin component and for each individual atomic state, i.e. for the singly and doubly occupied states

3.2 The Limit of Infinite U.

For finite frequencies the doubly occupied state is empty, and most of the averages and GF are equal to zero. From Eqs.(23-26) one obtains a self-consistent solution with $A(1,1,q)\sim U$ and the remaining $A(a,b,q)\sim 1$. As also $\omega_{1}(q)\sim U$ and $\omega_{2}(q)\sim 1$, the only non-zero GF is

$$G_{1\sigma}(1,1,q,z) = Q_{\sigma}(1)/(z-[\tilde{\varepsilon}(q)+\tilde{E}_{o\sigma}]) \qquad (40)$$

where

$$\tilde{\varepsilon}(q) = Q_{\sigma}(1)\varepsilon^{o}(q) \quad , \qquad (41)$$

and in the absence of magnetic field (i.e. h=0) we find

$$Q_\sigma(1)\tilde{E}_{o\sigma} = -\Sigma_i t_{ij} \langle \bar{Y}_{-\sigma}(j1\tau)Y_{-\sigma}(i1\tau)\rangle \quad . \tag{42}$$

This is the solution obtained by the KB method when one starts directly from the Hamiltonian

$$H = \sum_{ij\sigma} t_{ij} X^{+}_{o\sigma}(i)X_{o\sigma}(j) \tag{43}$$

We should mention, that if one solves Eq.(17) without decoupling x_σ and y_σ at the r.h.s., one obtains $n_o\tilde{E}_o/Q_\sigma(1)$ rather than $Q_\sigma(1)\tilde{E}_o$ at the l.h.s. of Eq.(42). The corresponding ground state energy is then different from that calculated by the KB method when starting directly from the Hamiltonian in Eq.(43). I believe that this difference is because we neglect $\delta\Sigma/\delta W$ in Eq.(16), while the correlations are kept to a higher order in x_σ and y_σ when we do not decouple them with Eq.(18).

3.3 The Limit of U = 0.

In this limit we should obtain a band of uncorrelated electrons with dispersion relation $\varepsilon^o(\mathbf{q})$, but our present treatment does not give this result.

Substituting Eq.(9) into Eq.(14) shows that $x_\sigma(ja,ib,\tau,1')$ and $y_\sigma(ja,ib,\tau,1')$ should not depend on the component a. This property is not satisfied in our approximation because we decouple the x_σ and y_σ with Eq.(18) which leads to our final result through Eq.(19). When the x_σ and y_σ do not depend on a, we find from Eq.(8) together with the local completeness

$$X_{oo}(j)+X_{\sigma\sigma}(j)+X_{-\sigma-\sigma}+X_{dd}(j)=I \tag{44}$$

that $\Sigma_a G_{2\sigma}(ja\tau,1')=\Sigma_{ib} t_{ji} G_{1\sigma}(ib\tau,1')$. In the $U\to 0$ limit, this relation leads to the expected result: a band of uncorrelated electrons with dispersion relation $\varepsilon^{0}(q)$. It seems then clear, that if we had followed the alternative mentioned in Section 3.2 of solving Eq.(17) without decoupling the x_{σ} and y_{σ}, we would have obtained the correct limit for U=0, but as discussed in that section we would also obtain the wrong limit for $U \to \infty$.

4. SUMMARY AND DISCUSSION.

Although the Hubbard Hamiltonian has a very simple form, there are rather few exact results[9] available. Vollhardt[10] has briefly discussed the approximation methods, giving strong emphasis to the variational treatment with the Gutzwiller trial wave function. This wave function is obtained by starting from the uncorrelated wave function and projecting out a fraction of the doubly occupied states; to calculate the average energy it is necessary to make approximations that are essentially of the mean field type. Being a variational method, it is not adequate to calculate finite temperatures and dynamical properties, and other shortcomings of this treatment have also been pointed out.[5] Nevertheless, the treatment is simple and gives the essential physical aspects of the problem, and it would be convenient to have it as a starting point for further improvements. A recent work of Kotliar and Ruckenstein[11] (KR) has shown a way to pursue this program by generalizing the auxiliary Boson treatment of Coleman[3] to the case of finite values of U. Four bosons, one for each possible local state were introduced, and the system was described by a Hamiltonian for the electrons and the four bosonic fields. Unphysical states with too many bosons were eliminated by requiring the validity of several local relations between the electronic and bosonic occupations. The simplest solution is of the mean field type, and is equivalent to the lowest order (the saddle point approximation) of the functional integral treatment of the problem. The simplest Hamiltonian that one can write gives the correct $U\to\infty$ limit, but does not reproduce the uncorrelated bands in the $U=0$ limit. To avoid this problem, KR noticed that different Hamiltonians could describe the system, and although they would all give the same exact solution, different results would be obtained by the use of the same type of approximation. Between the different Hamiltonians, they managed to select one that correctly gives the two limits discussed above, and in the saddle point approximation it gives the same results as the Gutzwiller calculation. This result opened the way to study finite temperature and dynamic effects by employing the functional integral formalism one step further, i.e. considering Gaussian fluctuations: several authors have recently developed this idea.[12,13]

The motivation of the present work was to extend for finite U the RS treatment,that in the $U\to\infty$ limit gives the same results that the mean field approximation of the auxiliary boson treatment do, and that can also treat dynamical effects and finite temperatures in a natural way. As in the work of KR we have found for finite U some ambiguity in the treatment, but in our case the Hamiltonian was uniquely defined and the alternatives occurred in the details of the approximation. The situation was different in the previous treatment of the Hubbard model[5] and of the Anderson model[7] for $U\to\infty$, because the lowest-order nontrivial approximation was unique in both of them. As discussed in Section 2.4.2 we chose the alternative that gave reasonable results for $U\to\infty$ but did not reproduce the correct $U=0$ limit: I believe that this corresponds to treating all the approximations on an equal footing. We have also shown that there is an alternative approximation that gives the correct $U=0$ limit, but at the cost of spoiling the accord in the $U \to \infty$ limit. I conjecture that the agreement at the two limits can be obtained by following this last alternative but going to the next order in the treatment.

There are several aspects that we have not discussed. First of all, we have not obtained solutions to the equations in Section 2.4.4. Although it is only necessary to determine the constants $C_t^\sigma(a,b)$, $C_E^\sigma(a)$ and μ for fixed values of $Z_{nn}t_o$, $\Delta E_\sigma(a)$ and n (the total number of electrons per site), the unknowns appear in the arguments of the integrals $F_{1\sigma}$, $F_\sigma(a,b)$,$F_{o\sigma}$ and $F_{o\sigma}(a,b)$ through ω_1 and ω_2, and we have not been able to obtain simple relations even for zero T and a simple density of states. A numerical study of this problem is under way for $T=0$. The study of finite temperatures does not require any change in the formalism, but other GF should be calculated to study dynamical susceptibilities.

From the magnetic point of view,we have only considered the paramagnetic phase, but no essential changes are necessary to consider ferromagnetic phases. The study of antiferromagnetic phases would require the abandoning of the independence of averages with respect to the site position, but a modified translational invariance would still be available.

5. ACKNOWLEDGEMENTS.

I would like to express my gratitude to the "Instituto de Física Balseiro", S.C. de Bariloche, Argentina, for its hospitality. Most of this work was done while I was visiting that Institution , and I had a very fruitful interaction with the members of the Solid State Theoretical Group. In particular, I would like to acknowledge many stimulating discussions with Blas Alascio, Carlos Balseiro, Armando Aligia and Jorge Sofo.

I am very grateful to Roberto Luzzi for critically reading the manuscript.

This work was partially supported by the Conselho Nacional de Desenvolvimento Científico e Tecnológico, Brasil.

6. REFERENCES.

1. Hubbard, J., Proc. R. Soc. (London). Ser. A276, 238 (1964); 277, 237 (1964); 281, 401 (1965); 285, 542 (1965); 296, 82 and 100 (1966).
2. Gutzwiller, M. C., Phys. Rev. 134, A923 (1964); 137, A1726 (1965); Phys. Rev. Lett., 10, 159 (1963).
3. Coleman, P., Phys. Rev., B29, 3035 (1984).
4. Kadanoff, L. P. and Baym, G., Quantum Statistical Mechanics, (Benjamin, New York, 1962)
5. Ruckenstein, A. E. and Schmitt-Rink, S., Phys. Rev. B38 7188 (1988); Int. J. of Mod. Phys. B 3, 1809 (1989).
6. Hewson, A. C. and Wasserman A. L., J. Phys.: Condensed Matter 1, 403 (1989).
7. Foglio, M. E., Phys. Rev. B 43 (1991) (in press).
8. Zubarev, D. N., Soviet Physics, Usp. 3, 320 (1960).
9. Cyrot, M., Physica 91B, 141 (1977)
10. Vollhardt, D., Rev. Mod. Phys. 56, 99 (1984)
11. Kotliar, G. and Ruckenstein, A. E., Phys. Rev. Lett. 57, 1362 (1986).
12. Rasul, J. W. and Li, T., J. Phys. C: Solid State Phys. 21, 5119 (1988).
13. Lavagna, M., Phys. Rev. B41, 142 (1990).

ON THE SECOND DERIVATIVES OF A LEGENDRE TRANSFORM *

Newton Bernardes

Instituto de Física
Universidade Estadual de São Paulo
Box 20516
01498 São Paulo, SP, Brazil.

Abstract

A matrix method is used to obtain, in a simple way, the several second derivatives of the partial Legendre transform $\phi_{\{k\}}$ of a function of several variables, $f(x_0, x_1, \ldots x_n)$.

0* This paper is dedicated to Rogerio C. C. Leite on the occasion of this 60th birthday.

1. INTRODUCTION

In many physical problems, especially those of the equilibrium theory of Thermodynamics, one encounters the question of relating the several second derivatives of a function f of several variables to the corresponding second derivatives of the Legendre transform of this function (Callen 1960, Lima and Bernardes 1991).

Even though this is a question of elementary Calculus, these relations are not readily available. None the less they may be easily obtained if one uses a shorthand notation, as we show in this paper.

2. THE LEGENDRE TRANSFORMATION

Given a function $f(x_0, x_1, \dots x_n)$ one may define a class of Legendre transforms $\Lambda_{\{x_j\}}\{f\}$ of f by the equation

$$\Lambda_{\{x_j\}}\{f(x_j; x_\lambda)\} = \underset{x_j}{\text{supremum}}\left\{f - \sum_0^k p_j x_j\right\} = \phi_k(p_j; x_\lambda) \tag{1}$$

where $\{x_j\}$ is an abbreviation for the set of variables $x_0, x_1, \dots x_k$, $0 \le j \le k$ and $k < \lambda \le n$.

From the definition (1) of a Legendre transformation, if f is a differentiable function, it is easy to see that the differential of $\phi_{\{k\}}$ is given by

$$d\phi_{\{k\}}(p_j; x_\lambda) = -\sum_0^k x_j^* dp_j + \sum_{k+1}^n p_\lambda dx_\lambda \tag{2}$$

while

$$df(x_j^*; x_\lambda) = \sum_0^k p_j dx_j^* + \sum_{k+1}^n p_\lambda dx_\lambda \tag{3}$$

and thus

$$\frac{\partial \phi_{\{k\}}}{\partial p_j} = -x_j^*\,, \qquad 0 \le j \le k \tag{4}$$

$$\frac{\partial \phi_{\{k\}}}{\partial x_\lambda} = \frac{\partial f}{\partial x_\lambda}(\equiv p_\lambda)\,, \quad k < \lambda \le n\,. \tag{5}$$

Equation (4) gives us those values x_j^* of the variables x_j for which $\left\{f(x_j; x_\lambda) - \sum_0^k p_j x_j\right\}$ is an extremum.

3. THE SECOND DERIVATIVES

If $f(x_j;x_\lambda)$ is twice differentiable we may relate the several second derivatives f^{mn} of the function f and the corresponding derivatives of its Legendre transform $\phi_{\{k\}}(p_j;x_\lambda)$.

The second derivatives of ϕ are of four types, viz.:

$$\frac{\partial^2\phi_{\{k\}}}{\partial p_i\partial p_j}\equiv\phi^{ij}_{\{k\}}\ ;\quad \frac{\partial^2\phi_{\{k\}}}{\partial x_\mu\partial p_j}\equiv\phi^{\mu j}_{\{k\}}\ ;$$

$$\frac{\partial^2\phi_{\{k\}}}{\partial p_i\partial x_\lambda}\equiv\phi^{i\lambda}_{\{k\}}\ ;\quad \frac{\partial^2\phi_{\{k\}}}{\partial x_\mu\partial x_\lambda}\equiv\phi^{\mu\lambda}_{\{k\}}$$

where a latin subscript runs from 0 to k, while a greek subscript runs from $k+1$ to n. From eqs.(4)-(5) we have

$$\phi^{ij}_{\{k\}}=-\frac{\partial x^*_j}{\partial p_i}=-\frac{\partial x^*_i}{\partial p_j} \tag{6}$$

$$\phi^{\mu j}_{\{k\}}=-\frac{\partial x^*_j}{\partial x_\mu}=\frac{\partial p_\mu}{\partial p_j} \tag{7}$$

$$\phi^{i\lambda}_{\{k\}}=\frac{\partial p_\lambda}{\partial p_i}=-\frac{\partial x^*_i}{\partial x_\lambda} \tag{8}$$

$$\phi^{\mu\lambda}_{\{k\}}=\frac{\partial p_\lambda}{\partial x_\mu}=\frac{\partial p_\mu}{\partial x_\lambda} \tag{9}$$

Thus we must write each differential dx^*_i and dp_μ in terms of the differentials dp_j and dx_λ as follows. We have

$$dp_i=\sum_{i=0}^{k}f^{ij}dx^*_j+\sum_{\lambda=k+1}^{n}f^{i\lambda}dx_\lambda \tag{10}$$

$$dp_\mu=\sum_{i=0}^{k}f^{\mu j}dx^*_j+\sum_{\lambda=k+1}^{n}f^{\mu\lambda}dx_\lambda \tag{11}$$

where

$$f^{qr}=(\partial^2 f/\partial x_q\cdot\partial x_r)^*\ ,\quad q,r=0,1,2\ldots n\,.$$

We may consider the coefficients f^{qr} as forming a $n \times n$ matrix $[\boldsymbol{f}^{(2)}]$, evidently symmetric, which may be broken into four block matrices: $f^{ij} \equiv [\boldsymbol{A}]$, $f^{i\lambda} \equiv [\boldsymbol{B}]$, $f^{\mu j} \equiv [\widetilde{\boldsymbol{B}}]$ and $[f^{\mu\lambda}] \equiv \boldsymbol{C}$, $\boldsymbol{A}$ and $\boldsymbol{C}$ being symmetric matrices, while $[\widetilde{\boldsymbol{B}}]$ is the transposed of $\boldsymbol{B}$,

$$[f^{qr}] \equiv [\boldsymbol{f}^{(2)}] \equiv \begin{bmatrix} \boldsymbol{A} & \boldsymbol{B} \\ \widetilde{\boldsymbol{B}} & \boldsymbol{C} \end{bmatrix} \tag{12}$$

We may write for short, $(dx_0^*, dx_1^* \ldots dx_k^*)$ as a vector $d\boldsymbol{x}^*$, $(dx_{k+1} \ldots dx_n)$ as a vector $d\boldsymbol{\xi}$ to form a column vector $[d\boldsymbol{x}^*; d\boldsymbol{\xi}]$, and at the same time $(dp_0 \ldots dp_k)$ as $d\boldsymbol{p}$ and $(dp_{k+1} \ldots dp_n)$ as $d\boldsymbol{\pi}$ to form a column vector $[d\boldsymbol{p}; d\boldsymbol{\pi}]$, in order to maintain the distinction between latin and greek indices. In view of eq. (12) we may write eqs. (10)-(11) in matrix form

$$d\boldsymbol{p} = \boldsymbol{A}d\boldsymbol{x}^* + \boldsymbol{B}d\boldsymbol{\xi} \tag{13}$$

$$d\boldsymbol{\pi} = \widetilde{\boldsymbol{B}}d\boldsymbol{x}^* + \boldsymbol{C}d\boldsymbol{\xi} \tag{14}$$

or

$$[d\boldsymbol{p}; d\boldsymbol{\pi}] = \begin{bmatrix} \boldsymbol{A} & \boldsymbol{B} \\ \widetilde{\boldsymbol{B}} & \boldsymbol{C} \end{bmatrix} \cdot \begin{bmatrix} d\boldsymbol{x}^* \\ ; \\ d\boldsymbol{\xi} \end{bmatrix} \tag{15}$$

or yet

$$\frac{\partial \boldsymbol{p}}{\partial \boldsymbol{x}^*} = \boldsymbol{A}\,, \qquad \frac{\partial \boldsymbol{p}}{\partial \boldsymbol{\xi}} = \boldsymbol{B} \tag{16)(17}$$

$$\frac{\partial \boldsymbol{\pi}}{\partial \boldsymbol{x}^*} = \widetilde{\boldsymbol{B}}\,, \qquad \frac{\partial \boldsymbol{\pi}}{\partial \boldsymbol{\xi}} = \boldsymbol{C}\,. \tag{18)(19}$$

We wish to solve eqs. (13)-(14) to obtain $d\boldsymbol{x}^*$ and $d\boldsymbol{\pi}$ in terms of $d\boldsymbol{p}$ and $d\boldsymbol{\xi}$. If we multiply eq. (13) by the matrix $\overline{\boldsymbol{A}}$, the inverse of matrix $\boldsymbol{A}$, we obtain

$$d\boldsymbol{x}^* = \overline{\boldsymbol{A}}d\boldsymbol{p} - \overline{\boldsymbol{A}}\,\boldsymbol{B}\,d\boldsymbol{\xi} \tag{20}$$

which gives a partial answer to our problem. Then we substitute eq. (20) into (14) to obtain

$$d\boldsymbol{\pi} = \widetilde{\boldsymbol{B}}\overline{\boldsymbol{A}}d\boldsymbol{p} + (\boldsymbol{C} - \widetilde{\boldsymbol{B}}\overline{\boldsymbol{A}}\boldsymbol{B})d\boldsymbol{\xi} \,. \tag{21}$$

Eqs. (20)-(21) give us the several second derivatives in eqs. (6)-(9), and they may be written in matrix form

$$[d\boldsymbol{x}^*; d\boldsymbol{\pi}] = \begin{bmatrix} \overline{\boldsymbol{A}} & -\overline{\boldsymbol{A}}\boldsymbol{B} \\ \\ \widetilde{\boldsymbol{B}}\overline{\boldsymbol{A}} & \boldsymbol{C} - \widetilde{\boldsymbol{B}}\overline{\boldsymbol{A}}\boldsymbol{B} \end{bmatrix} \cdot \begin{bmatrix} d\boldsymbol{p} \\ ; \\ d\boldsymbol{\xi} \end{bmatrix} \tag{22}$$

or, in the manner of eqs. (16)-(19)

$$\frac{\partial \boldsymbol{x}^*}{\partial \boldsymbol{p}} = \overline{\boldsymbol{A}} \,, \qquad \frac{\partial \boldsymbol{x}^*}{\partial \boldsymbol{\xi}} = -\overline{\boldsymbol{A}}\boldsymbol{B} \tag{23)(24}$$

$$\frac{\partial \boldsymbol{\pi}}{\partial \boldsymbol{p}} = \widetilde{\boldsymbol{B}}\overline{\boldsymbol{A}} \,, \qquad \frac{\partial \boldsymbol{\pi}}{\partial \boldsymbol{\xi}} = \boldsymbol{C} - \widetilde{\boldsymbol{B}}\overline{\boldsymbol{A}}\boldsymbol{B} \,. \tag{25)(26}$$

From eqs. (23)-(26) and (6)-(9) we may write the matrix $[\phi^{qr}] \equiv [\phi^{(2)}_{\{k\}}]$ of the second derivatives of the Legendre transform of f as the symmetric matrix

$$[\phi^{qr}] \equiv [\boldsymbol{\phi}^{(2)}_{\{k\}}] = \begin{bmatrix} -\boldsymbol{A} & \overline{\boldsymbol{A}}\boldsymbol{B} \\ \\ \widetilde{\boldsymbol{B}}\overline{\boldsymbol{A}} & \boldsymbol{C} - \widetilde{\boldsymbol{B}}\overline{\boldsymbol{A}}\boldsymbol{B} \end{bmatrix} \tag{27}$$

where the matrices $\boldsymbol{A}, \boldsymbol{B}, \boldsymbol{C}$ are defined in terms of the matrix $[\boldsymbol{f}^{(2)}]$ of the original function f by eq. (12).

4. CONCLUSION

The various second derivatives of a partial Legendre transform $\phi_{\{k\}}$ of a function f of several variables are easily obtained, in matrix form $[\boldsymbol{\phi}^{(2)}_{\{k\}}]$, eq. (27), in terms of the matrix $[\boldsymbol{f}^{(2)}]$, eq. (12), formed by the second derivatives of the original function. These relations are useful in the theory of Thermodynamics, as discussed elsewhere (Lima and Bernardes 1991).

Acknowledgements: The author appreciates many discussions with Niels Fontes Lima.

References

[1] Lima, N.F. and Bernardes, N. 1991, to published.

[2] Callen, H.B. 1960, *Thermodynamics*, J. Wiley, New York, Appendix G, pg. 361.

Shades of a New Color Printer

by
Kerry L. Shaklee
Principal Research Fellow
Metrum Information Storage
A Business of Alliant Techsystems Inc.
Littleton, Colorado 80122, USA

ABSTRACT

In this world of the "Information Age" the bandwidth of the communications channel becomes a critical factor in determining the rate at which information can be exchanged. In the case of interpersonal information exchange, the channels (speech, written word, pictures etc.) also limit the communications rate. In recent years the compute power and storage capacity of personal computers and low end workstations have crossed the threshold which now permits wide spread acquisition and manipulation of digital images and graphical data. While digital imagery has further increased our interpersonal communications capacity, the exploitation of the added bandwidth afforded to us by color has been slow in color. One reason for this lag in color utilization has been the absence of image quality color printers. Over the past five years I have been performing advanced research and development on a line of image quality digital color printers. These printers, which use CRT illumination to expose a variety of photographic papers, stand to set new price/performance standards in the color imaging marketplace. This paper will examine the characteristics of these printers and explore some of the underlying science and folklore involved in communicating with color images. Unfortunately, this publication has no provision for color figures to be used to demonstrate the salient features. I will have to try to do it with words (a very low bandwidth approach) so the reader will have to use his/her imagination and trust me.

The preparation of this tribute to Rogerio Leite on the occasion of his sixtieth birthday has been an interesting experience. In thinking about Rogerio, it became clear that while his personal accomplishments speak for themselves, the effects of his influence and impact on others is not so obvious and therefore more difficult to identify. I believe that the work that I will describe here is, in some part, an element of Roger's legacy.

What we are is a combination of the sum total of our life's experiences and the genetic cocktail bequeathed us by our parents and their ancestors.

My Brazilian experience taught me many lessons. Rogerio, as the mastermind of the UNICAMP phenomenon and by inviting me to participate in it, was the prime figure behind that experience. It is difficult, probably impossible, to specifically identify the traits of who I am today (the same goes for all the others who have interacted directly or indirectly with Rogerio) that have been derived from interactions with Roger. I will not attempt any such differentiation. Rather, let it be said that there is a piece of Rogerio in some machines and techniques which today make beautiful color images.

Before launching into the story of my conversion from the role of a scientist studying the physical properties of light to that of a technologist exploiting the perceptual effects of the same electromagnetic radiation, I would like to dedicate a few lines in section I to the memory two special friends with whom I shared a piece of my Brazilian experience, two people who also crossed paths with Rogerio. These people are John Rogers who passed away in 1985 and Billy Michael Kale who had a mercifully short bout with leukemia before passing away in 1990.

During the past five years I have guided and/or executed research and development efforts leading to the implementation of two image-quality color printers. By effectively increasing the communications bandwidth (person to person, machine to person, person to self) these machines stand to influence how and how well we live our lives in the technological society of the twenty first century.

In the bulk of this tome I will attempt to describe the printers and their technology, some applications of color printing, to present some of the basic concepts of image capture, manipulation and reproduction and to wax eloquently about some things I have learned about business, creativity and the enjoyment of life. The subject and field are large and growing. There is no way that I can do it justice in the time and space available here. I do hope, however, that I can convey some of the excitement and some of the opportunity that I see for machines that can electronically capture, manipulate and print beautiful color images. The nearly explosive growth of available computer horsepower and data storage capacity has made electronic imaging available to professionals in a wide range of disciplines. As these people learn how to exploit these new and powerful tools, they will bring more and more of electronically processed images into the lives of us all.

Since my Brazilian experience, my life has taken a different math than most of the authors and readers of this book. Having hung up my research spurs, I have spent the balance of my career at the interface between science/technology and business. Therefore, the experiences that I have had and the insights I have to share are probably quite different from those of my colleagues who have continued to ply their research interests.

For those who have not kept track of me over the last fourteen years or so, a brief synopsis of my activities is included in the box. During this time I have also become interested in the cognitive and creative processes - how our minds work, where ideas come from and how do we stimulate them to happen. According to my current model, our thought processes are multidimensional. As is discussed in detail in section II, it is difficult to squeeze or project a multidimensional concept onto a communications vehicle (such as this book) of limited dimensionality. Therefore, rather than trying to present a linear and logical story, I have chosen to present my thoughts as a series of short cameos, each pertaining to the central theme of interpersonal communications, which individually stand alone and can be read in any (almost) order. I hope that the projection of my thoughts onto the hyperspace of the reader's mind might lead to some new insights or revelations.

INDEX

Pictures and images are communications vehicles. In many cases, they convey large amounts of information. The issues involved in the generation, transmission, reception and processing of visual information are equally as important and interesting as is the technology behind the communication process. Therefore, the early sections deal with the processes involved in the generation and communication of thoughts and ideas. The last sections describe the results of my recent work on color printers which serve as a mechanism for increasing the bandwidth for the communication of thoughts, concepts and ideas. If this effort causes one of the readers to think about how we communicate, to have an insight, to maybe contribute something to the field or to buy and use one of my printers to generate and communicate with beautiful images then my efforts here will have been justified.

Kerry Shaklee's Travels Since the Brazilian Adventure

1978 - Vice President at Vikonics, a manufacturer of electronic security systems for nuclear power stations

1979-1981 - President of Tranquonix Associates, a high tech consulting operation - my own business

1981-1982 Director of Engineering at Barnes Engineering - Jim Moore, Mike Kale & Georgy Csnak were there too

1982 - 1984 - Technical Director at Display Interface Corp. a startup making computer peripherals

1984-1985 - Consulting with Mike Kale - made a CO_2 laser based machine to produce fiberoptic directional couplers

1985 - Present - Honeywell (now Metrum) as a staff scientist inventing & developing new color printer technologies

-----I A tribute to two departed friends -----

John Rogers

John Rogers was a bright physicist who went to Brazil for an adventure and ultimately chose to stay, marry, raise a family and to contribute extensively to the Brazilian technical family. While John easily could have had a job at any one of a number of top laboratories in the United States, he felt that he could have a bigger and more effective impact in Brazil.

John was a born and dedicated mentor. He strongly influenced many of Brazil's best young scientists, a number of whom are also contributors to this testimonial publication. As a mentoree, I was probably not one of John's greater achievements since I did not do a very good job of learning to deal with the culture and Brazilian way of life. However, his teachings have made a significant contribution to my later accomplishments. John was a true friend and good ones are hard to come by.

Mike Kale

Mike Kale was doing his research in Campinas in the mid 1970s as part of some kind of exchange with Rice University. Although Mike and I did not have very much contact in Brazil, we found ourselves working closely and well together on several subsequent occasions. Several times, in a dark laboratory with laser beams bouncing all around, while waiting for something to happen, we found ourselves discussing the Brazilian experience. It was clear to us that some of those hard southern hemisphere lessons had significantly enhanced our ability to compete and produce in the North American technological arena. Tenacity, resourcefulness and the understanding of how to deal with an unpredictable environment have all helped us compete with those who have not had the privilege of such lessons.

Mike and I (together and separately) have had achievements and accomplishments which are part of the legacy of Rogerio Leite. In many cases the legacy follows because Rogerio played such a significant role in creating an environment within which the rest of us could learn and grow. From all this Mike too became a good and cherished friend. I miss his regular phone calls which always began "Hello Kerry".

----II Communications Theory According to Shaklee----

Sophisticated and effective methods of communication are one of the ways humans are distinguished from the other members of the animal kingdom. While we stand out from the rest of the biological universe because we can more effectively communicate information and ideas, we still find that the available communications channels are restricted in bandwidth and that this limits the rapidity with which we can exchange information and places restrictions on the quality of that information. The human mind is, by virtue of the tangled neural network, an entity of many dimensions (degrees of freedom) most of which can not be simultaneously squeezed onto a communications channel of limited dimensions. Additionally, as is the case with more conventional communications channel concepts, noise tends to modify or distort the information. Finally, the receiver also plays an important role in the information transmittal process. The perceived data is filtered by the receivers own perceptions of the universe and his mind takes an active (perhaps aggressive) role in filling in the missing parts.

This multidimensional, bandwidth-limited concept of human communications is illustrated in Fig. 1. Let us suppose that Mr. Smiley, on the left, wishes to communicate some concept, a pencil for example, to Mr. Frown on the right. Mr. Smiley's concept is some projection onto the multidimensional space spanned by the tangled neural network which constitutes his brain. As we proceed through this complex process, figures 2 and 3 tell a more detailed story of how ideas flow between people.

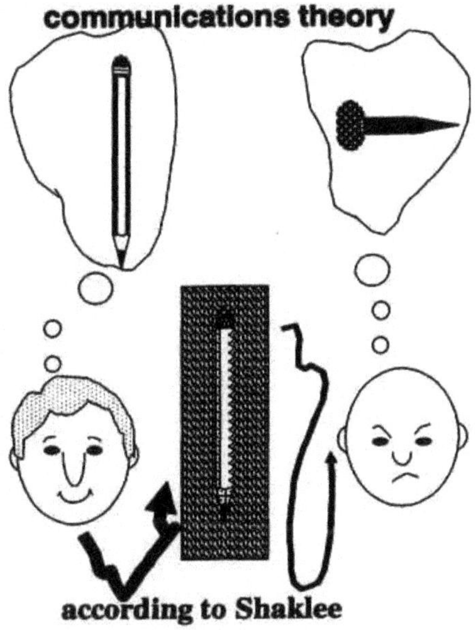

Figure 1. An illustration of the effects of dimensional compression, channel noise and receiver mapping on the interpersonal communication of a concept.

The typical communications channels which Mr. Smiley might choose to use (spoken word, written word, sketch, picture etc.) are characterized by a limited number of degrees of freedom (dimensions). Therefore, in order to attempt to communicate his idea, Mr. Smiley must project it onto the space appropriate to the chosen channel. This is analogous to how we represent 3- dimensional objects by projecting them onto a 2-dimensional surface.

The projection of an object onto a subspace generally results in the loss of information. In the case of the 3-dimensional objects, numerous views (rotations) and cuts are often necessary in order to adequately communicate the important features. Similarly, in the case of an idea, a number of transformations (analogous to rotations, cuts, zooms etc.) may be necessary to place the important features (as Mr. Smiley sees it) onto the channel.

Once an idea is placed on the channel, it is subject to the noise of the channel. This may be due to the the imprecise nature of the language (for spoken or written communications) or to the inaccuracies of a sketch or drawing or to the lighting or equipment defects in the case of a photograph. All this means that the information arriving at Mr. Frown's sensors (the ones appropriate to the channel chosen by Mr. Smiley) represents, at best, a subset of Mr. Smiley's idea. More likely the channel noise has also modified the information which Mr. Smiley placed on the channel.

Fig. 2 The idea communication process illustrating multiple exchanges. Note that the two perceptions of the object are similar but not identical at the end of ths exercise. Turn to Fig. 4 to learn the outcome of the process.

While projection and channel noise may have already done a number on Mr. Smiley's idea, the real fun begins when Mr. Frown attempts to process the received information. This process involves the mapping of the channel dimensions onto the hyperspace of Mr. Frown's mind. Such a mapping can also be thought of a filtering process whereby Mr. Frown's mind attempts to relate this incoming information to things he already knows or has experienced. Additionally, the received information is totally inadequate for mapping onto all the available dimensions in Mr. Frown's mind. Therefore, his mind will tend to fill in the blanks with details with which it feels most comfortable. Since these details are based on the sum total of Mr. Frown's life experience (which is undoubtedly quite different from Mr. Smiley's) the final idea which may (quite likely will) be much different from the one which Mr. Smiley had hoped to send.

Once Mr. Smiley and Mr. Frown think that they have a common concept a clock begins ticking. Ideas have a relatively short half life and the half life of a shared idea is even shorter. Following the communications exchange, Mr. Smiley and Mr. Frown go their separate ways and have new individual life's experiences added to their present ones. At some level, they continue to process the idea of their experiences (De Bono - see section IV - refers to this as mapping). The processing modifies the idea or concept, and probably does so differently for each person, and probably without either of them being aware of the nature or extent of the modifications. When they subsequently meet and attempt to continue the dialogue they most likely will find that they are not in such close agreement as they had earlier believed.

Fig. 4 The consequences of idea half life, on the part of both parties, at the time of reduction to practice.

This rather wordy nontechnical treatment of an interpersonal communications theory makes it sound like it is a miracle that humans manage to exchange any information at all. Maybe that is really the case. When I have time, in retirement maybe, I hope to attempt to apply some of the tools and concepts (entropy, bandwidth, signal to noise etc.) of conventional communications theory to the problem. Such a treatment, coupled with our ever better understanding of how the mind works, may lead us to improved communications and sharing of ideas and concepts.

Meanwhile, there are some things we can do to better the situation:

1) Transmit multiple projections of an idea or concept. Do not merely repeat the same projection multiple times,

2) Have the receiver transmit back his version of the concept. This requires receiving, mapping, projecting and transmitting on the part of the receiver,

3) Use as many and as wide-a-bandwidth channels as possible. Words, by themselves, are of extremely limited bandwidth and dimensionality. Pictures, color, models, examples, sound, odors, tastes, gestures etc. all add bandwidth and dimension.

4) Seek to explore common experiences a vehicle for facilitating receiver mapping.

5) Keep the dialogue alive. Don't let the half life get between the sender and the receiver.

Numerous times, in my business life, I have had some emphatically tell me "I told you (or sent you a memo or a letter etc.) six days (weeks or months) ago telling you exactly what I wanted (meant). Why are you bothering me now?" All five points above got missed. There was a single projection (letter, memo etc.). Memos, letters provide little opportunity to transmit back the processed information. The channel was of very limited bandwidth and dimensionality and probably did not seek to explore common experience. Finally, the statement assumes that nothing happened, in the meantime, which might have altered the sender's or the receiver's perception of the issues.

----- III The Bandwidth Progression -----

The color printers described in this article represent the logical progression of communications products (lines on paper) from one organization going back more than forty five years. This is illustrated in Fig. 5.

Mentrum Information Storage got its start in the 1940s as Heiland Research Corporation. Starting with pen recorders, the company progressed to galvanometer recorders with direct print photo paper and then to oscillographic recorders. The oscillographic recorder led to 2-dimensional image recording and now the dimensions of color have been added.

While the company has changed hands twice, the trend has been in the direction of increasing the amount of information which can be recorded on various forms of human readable media (paper). **I equate this growth of stored (recorded, printed)information density with bandwidth, not so much in the sense of electrical bandwidth but rather in terms of the amount and rate of information communicated to the observer by the paper and the marks thereon.**

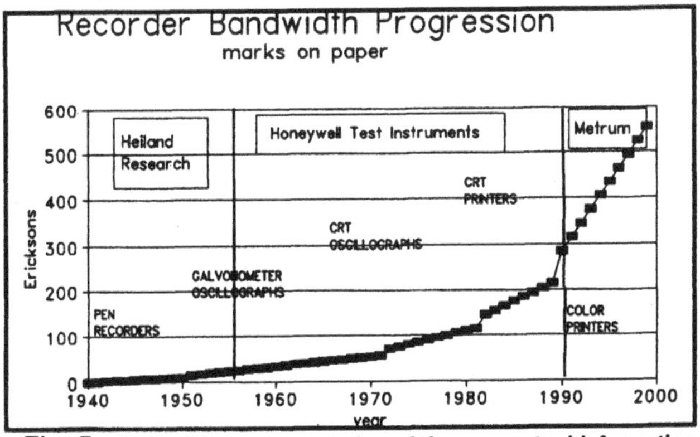

Fig. 5. An intuitive representation of the perceptual information bandwidth of Honeywell (Metrum) data hard copy recorders for the past fifty years. The units (Ericksons) are a new measure that I have coined for use in this nonquantitative environment.

The recent advent of color recording (printing) has again increased the bandwidth. In terms of added dimensions (from monochrome to three primary colors) the bandwidth gain would be expected to be three fold. In practice, it seems that color adds something extra (more than three times the perceived visual information) to an image or a chart on a piece of paper. While artists have long understood how to effectively communicate using color, the technical community is only now (and slowly) coming to grips with how to progress beyond a physical description of color to effectively manage and use the perceptual aspects of combining light of various wavelengths. Composition of shapes and shades adds dimensions when using color that are not there in monochrome applications.

It is widely held that the absence of affordable image quality color printers has held back the development of systems and software to capture, generate, manipulate and print striking electronic color prints. I believe that the products that we have developed will help to move the field along, to encourage new applications, new software and the development of systems to take advantage these new capabilities.

During his technical career, Rogerio Leite was a student of the physical properties of light. Now he has also unwittingly left a part of his legacy in the arena of the perceptual utilization of the interaction of photons with matter.

----- IV Curse You Aristotle! -----

Some time after I had forsaken the hallowed halls of research for the more practical world (so they say) of product development I had a rather abrupt and painful encounter with the organizational tools of project management. If I didn't learn how to plan (things like PERT), schedule and budget I was never going to manage to get anything done. After the crisis was past I found myself wondering why I hadn't been taught these tools during the course of my education. Surely, if I had only known these skills of planning and organization early in my career my research and productivity would have been on much higher plains. Who knows what I could have been by today?

When I had aged even more and had become (hopefully) somewhat wiser I came to realize that such indoctrination during my (more?) impressionable years might well have put a stop to my research career before it even got started. Traditional analysis and logic techniques, of which project management tools are a subset, are rule-based, rooted in the centuries' old teachings of Aristotle. You apply the rules and you get the right answer. What could be better? Well, today I wonder if Aristotle didn't set the world (or at least the western part) back by several hundred years.

Edward De Bono, a British educator and spokesman for the creative process, has coined the term "Lateral Thinking". By contrast, his contention is that "Linear Thinking", the Aristolean logic discussed in the previous paragraphs, will always give the right answer (mathematics teaches that if you apply the rules you get, by definition, the right answer) but not necessarily the best answer. The rules that we apply in analyzing a problem may not be appropriate, could be overly (or underly) restrictive or they could be downright wrong!

Lateral Thinking, as defined by De Bono, takes advantage of the multidimensional processing power of the (less logical) right brain to work the problem. His contention is that the right brain, through the complex processes of the tangled neural network, seeks to find applicable sets of rules based on our total accumulated experience, with a little chaos thrown in for good measure. In this manner, he seeks to legitimize such processes as intuition, insight, creativity, innovation and invention. Rather than viewing these concepts as manifestations of magic, instability etc., he sees them as the proper functioning of our rather complex brain going about its business of finding relationships among our accumulated life's experiences.

De Bono asserts that we should learn to trust our lateral thinking processes. This, in itself, requires training and experience and is an up hill battle in the face of our educational and cultural systems which promote linear thinking almost exclusively.

Processes such as intuition, insight etc. typically cause discomfort (especially to the external observer) because they appear to break the rules with which the observer is comfortable. Further, the thought processes leading to the answer very often are not obvious, a situation makes it difficult for the possessor of the insight to convince himself, let alone anyone else, of the validity of the conclusions reached.

Perhaps even more perplexing than the question of where insights come from (at least for those of us who live by them) is the dilemma of how to make them happen when we need (or want) them. While there are no "innovation" switches which we can turn on, we now know that there are tricks or techniques which can sometimes be used to stimulate the process. Brainstorming, relaxation, diversion (performing some task which cause the right brain to exercise), mind mapping, metaphors and LARC (Left and Right Creativity, a process described by Williams and Stockmyer) are all techniques which I use to stimulate the problem solving process. In practice, I have found that the more I exercise the lateral thinking process the easier it is to energize it on demand.

Another way of looking at the issue is in terms of determinism. In the realm of linear thinking the world is deterministic. Event A begets event B - we somehow know what the determining rule is. In the lateral thinking environment Event A begets event B or event C or event D - we don't (consciously) know what the rule is but our intuition or insight or whatever is working on it. I believe that the universe is still deterministic. Some of us only figure out the rules after the fact.

Tom Peters recently took another approach to the problem in a syndicated column entitled "Smiling Through the Illusion of COntrol:. He states: "I believe that we are not, by and large, in control of our destinies; we are (individual, firm, nation) victims or beneficiaries of timing and luck". He goes on to say; "..., I think that as leaders if we own up to the role of timing and luck, we can greatly improve our chances of success". Not much determinism there for old Tom. He seems also to be advocating figuring out the rules after the fact. The bulk of the article deals with how we might develop strategies aimed at the seduction of lady luck. "Implement many strains of seed in many varieties of soil. To wit, implementing radical decentralization, providing enormous entrepreneurial incentives and encouraging obstreperous champions to run rampant, overseen by a minuscule central staff." In other words, he is suggesting that the business world employ the same techniques that we have used in research for many years.

-----The Seersucker Theorem -----

Another approach to this issue is embodied in the Seersucker Theorem which I believe should be attributed to Al Whetstone. Al was a communist during his graduate student days at Penn in the early fifties. In those days accelerator technology and safety concerns will still in their infancy. The students routinely lined up the linear accelerator by looking into the electron beam and maximizing the fluorescence of the retina. This technique earned many of them cataracts in later life, Al included.

> **SEERSUCKER THEOREM**
>
> **In spite of the wealth of evidence proving that SEERS do not exist there exists a class of SUCKERS who insist on seeking the services of the SEERS**
>
> Al Whetstone 1983

As a child of the depression, Al grew up in an orphanage. However he went on to found seven companies of which one, Summagraphics, became quite successful. Al never managed to justify his capitalist activities in the light of his earlier communist activism. Some of the conflict he tried to hide by drinking. At other times, he instigated very complex head games with those who worked for him. Often he would attack the bastions of the very capitalist structure which supported his companies.

Al took Summagraphics from nothing to thirty million dollars over the course of several years. Finally, the investors and the board of directors had had it with him and forced him out. His response was to form another company, Display Interface Corporation, to compete with Summagraphics in the data tablet and mouse business. He recruited me to be his technical director, to manage the research, development and implementation of the products. In reality, Al wanted a sparring partner who could give him a good tussle in the head game department. Life was very interesting for a while, trying to keep up with Al's trickery and still try to bring a couple of gadgets to fruition.

When Al had finally worn me down he turned his attention to his financial backers. An interesting example was his introduction of the Seersucker Theorem. It all happened at a Venture Capitalist seminar in Hartford in the fall of 1983. The seminar, which was sponsored by the local organization of Venture Capitalists, was intended to introduce potential entrepreneurs to the wonders of venture financing. As part of the program, several venture funded businessmen were asked to participate in a panel discussion and share their experiences.

Al, having recently gone through several rounds with his Venture Capitalists' analysts was fit to be tied. In order to do due diligence and to determine how rich they were going to get from a subsequent round of financing, they were requesting more and more detailed projections on the revenue and profits. Since this activity centered around a mouse (computer pointing device) which we had barely invented and certainly did not yet have perfected and since the market, at that time, had not yet crystallized, Al could see no sense in this busywork.

When it came Al's turn to talk, he announced that he would present this theorem which he had recently discovered. The crowd quieted, waiting to hear words of wisdom which would certainly help them all become much richer. The reader can rest assured that not a single Venture Capitalist smiled even once as the theorem was read and the sad consequence was that Display Interface Corporation was closed by the investors less than a year later.

The sad part of this rather humorous episode is that Al hit the nail square on the head. Having left the sheltered halls of research for the real world of business (so they say), I am frequently astonished at the amount of precise (to the nearest dollar if not the penny) results that are generated from imprecise and uncertain inputs. Tom Peters, in a recent column, contended that good fortune in business is really a matter of chance and luck. The successful ones are those who recognize and take advantage of opportunities wherever and whenever they occur.

----- V Digital Representation of an Image -----

An image is typically represented by dividing it into a two dimensional array of pixels (picture elements). Associated with each pixel is a value indicative of the respective image content. For a monochrome image the value will typically correspond to the average light intensity over the area of the pixel. For a color image the value might be a triplet corresponding to the average intensities of the red, green and blue components respectively. A more detailed discussion of pixel values will be given later.

The choice of the dimensions of a pixel has a significant impact on the amount of information captured (in a scanned image) or printed. Two concepts, resolution and addressibility, are used to describe the potential information content of a digital image. Resolution is determined by the size of the sensor element (scanner) or printing element (printer). Resolution is typically represented as pixels per unit distance (eg. pixels or dots per inch - dpi) and is generally indicative of the smallest structure which can be recognized in a digital image. Typical resolutions available for page scanners are of the order of 300 dpi, with higher resolutions just now appearing. Typical resolutions for printers are also of the order of 300 dpi.

Addressiblility, on the other hand, is a measure of the ability to control the center position of the sensing element (scanner) or printing element (printer) within an image. It seems reasonable that the addressiblility (expressed in dpi for example) should be greater than or equal to the resolution. Resolution may be degraded with respect to addressibility due to the physical size of the sensor or printing element or because several neighboring elements are necessary to represent a pixel value for the neighborhood. This latter condition is typical of how a laser printer may represent a shaded image. In most laser printer images, the resolution is typically an integral fraction (1/3 or 1/4) of the addressibility of the printer.

Binary printers (a laser printer or a thermal wax transfer printer) generate shades using a process knows as dithering. In this process a super pixel composed of several neighboring pixels (typically a square region of 3X3 or 4X4 pixels) is generated where the ratio of the on pixels to the off pixels is used to approximate a shade value over the entire super pixel. A 3X3 super pixel can represent ten different shade values (counting the all zero state) while a 4X4 super pixel can represent 17 different shade values. Obviously, the process of dithering reduces the effective resolution of the image with respect to the addressibility. A 3X3 dither pattern will reduce the resolution by a factor of three. Therefore, a 300 dpi laser printer will typically print shaded images at a resolution less than 100 dpi.

This discussion points up the value of continuous tone scanners and printers which can detect or produce a large number of shades at each dot. Such continuous tone devices can achieve significantly higher resolutions and therefore more pleasing and more informative images than can binary printers.

---- Representation of the value of a pixel ----

There are two common classifications of images: monochrome and color. For monochrome images each pixel is typically represented by a number corresponding to none of n possible states. Historically, n has been a number ranging from 1 to 256 with values n =4, 8, 16 being quite common. Currently, most new imaging applications are beginning to standardize on 250 (8-bits) as a convenient representation of a continuous tone image. For the purposes of our discussion, a continuous tone representation occurs whenever neighboring states (different light intensity levels reaching the eye) are spaced sufficiently close together that the human eye can not distinguish between them. In most applications, 32 to 64 appropriately chosen states have been found sufficient to meet this criterion. In a few cases involving representations that are not keyed to the eye's response function, more states (up to 1024) may be required to display continuous tone over the entire range.

The representation of a color image is significantly more complicated. Extensive research in color perception has shown that the human eye contains three color receptors (nominally red, green and blue). Further, it has been demonstrated that the perceived response to most spectral colors can be reproduced (to a high degree of reliability) through judiciously chosen combinations of red, green and blue light intensities. This has led to the development of three-color methods for generating color display (CRT, TV) and color printing (half tone, photography, color hard copy).

A common digital method of representing pixel values is by means of the respective red, green and blue intensities. This may be accomplished by means of a packed pixel representation where the triplet is stored in contiguous memory locations or by means of a planar method where the respective color values are stored in separate red, green and blue planes.

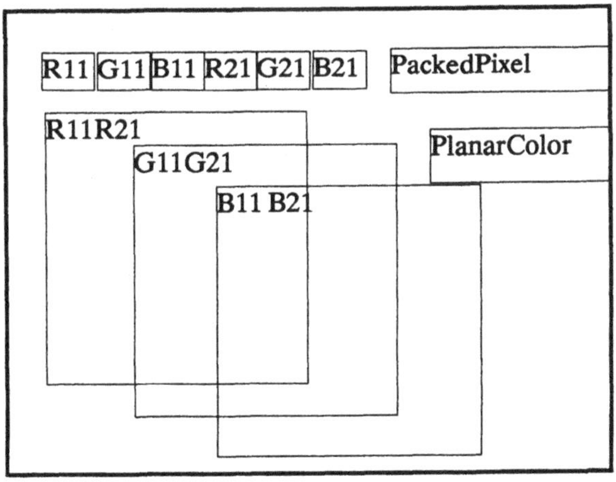

Pixel colors can also be represented in other coordinate systems. A coordinate system used by color television is referred to as H, S, V (hue, saturation and value). Several color systems attempt to quantify colors in terms of the response of the eye. The objective of representations such as the CIE chart is to have equal steps on the chart correspond to equal changes as perceived by a human observer.

For printed images, the color system is a subtractive one based on Cyan, Magenta and Yellow as the primaries instead of the Red, Green, Blue of the additive systems such as television. The CMY system is the complement of the RGB system in that C = 1 - R etc.. For many printing applications, the gray value of a pixel is printed using black ink. This approach, known as undercolor removal, involves computing the gray content of a pixel (the minimum of CMY) printing it separately and then subtracting the gray amount from the CMY values.

The photographic printing schemes used with the Metrum printers use the CMY approach. Provision has been made in the exposure system to account for color impurities contained in the photographic dyes. This makes it possible to obtain more accurate color representations than is possible in conventional optical printing.

---- Image compression ----

As pointed out earlier, high resolution digital images require large amounts of memory. While computer memory capacity is now to the point where a single image can be held and processed internally, the large image size still plays havoc with transmission and archival of pictures.

At a communications channel bandwidth of one magabyte per second, a high resolution "A" size color image could take up to twenty seconds to transmit (to a printer or over a network to another computer). In reality, few computers or mass storage devices can support bandwidths larger than a few hundred kilobytes per second, resulting in even longer transmission times. Likewise, a 300 Mbyte hard disk might only hold fifteen to twenty images. Large image archives would require slow or expensive storage devices.

One approach to this problem is to compress the digital representation (file) of the image. Two broad categories of image compression are typically discussed: lossless and lossy. Lossless compression refers to schemes where the image is compressed and then reconstructed without any resulting loss of information. Typical compression ratios depend on the image content, but a simple image might compress by a factor of 2:1 or 3:1. Applications such as medical imaging, where an altered image could prove disastrous, typically require the use of lossless compression.

In the case of lossy compression, the reconstructed image can have suffered some (hopefully not significant) loss of information. Typical lossy schemes support compression rates of 20:1 to 40:1 with occasional claims of 100:1. The current driver behind lossy compression is color facsimile where, at the current bandwidths, it would take forever to transmit an uncompressed image. Through the work of the JPEG (Joint Photographic Experts Group) and the CCITT Group 4 FAX activities we are close to having a color compression standard. The JPEG proposed standard has already been implemented in silicon (C-CUBE) and there are several software versions available on a variety of platforms.

Compression is typically made possible because, with respect to some representation, the image is characterized by a degree of correlation. In the case of lossless compression, regions of uniform shading or tone will compress quite well and run-length encoding is effective.

For lossy compression, the representations tend to be more complex. In this case, typically a few of the coefficients of some transform (two dimensional cosine transform or some fractal representation) tend to dominate in a given geographical region of the image. In the respective region, the dominant coefficients can be used to represent the compressed form of the image. Upon reconstruction, the information associated with the less dominant coefficients is lost. In many cases (perhaps most for the better compression schemes) this lost information has no perceptual effect on the quality of the image, even at compression ratios up to 40:1.

Lossy compression schemes tend to be computationally intensive. Recent developments in special purpose digital signal processors (such as the C-CUBE implementation mentioned above) have reached the point where television frames can be compressed and decompressed in real time. The implementation of such hardware in mass storage devices and printers is just around the corner.

Other methods of image compression involve describing the image in some higher level representation. An example is Postscript, a page description language which is used to generate high quality combinations of text and graphics from an ASCII description of the page (bit mapped graphics can also be imbedded within a postscript description). Color postscript is now coming of age and is expected to become a defacto standard for desktop publishing, short run printing and many other applications. The color and bitmapped image capabilities of postscript are currently quite limited. Therefore, we expect extensions to the language in the near future.

Another type of high level description of graphical images can be represented by vector-based languages such as HPGL which is widely used by plotters. HPGL has no convenient provisions for handling bitmapped image data.

----- VI A Tale of Two Printers -----

The two color printers I will describe produce digital color images on photosensitive media. Both printers employ a common exposure device (called a writing engine) shown in fig. 6. Exposed paper is passed from the writing engine to a paper processor (developer) appropriate to the type of medium that has been exposed. Figure 7 shows a side view drawing of a thermal processor which is used with Dry Silver Color Paper. Figure 8 shows a roller processor for carrying conventional photographic paper through the chemical baths required to develop the color image.

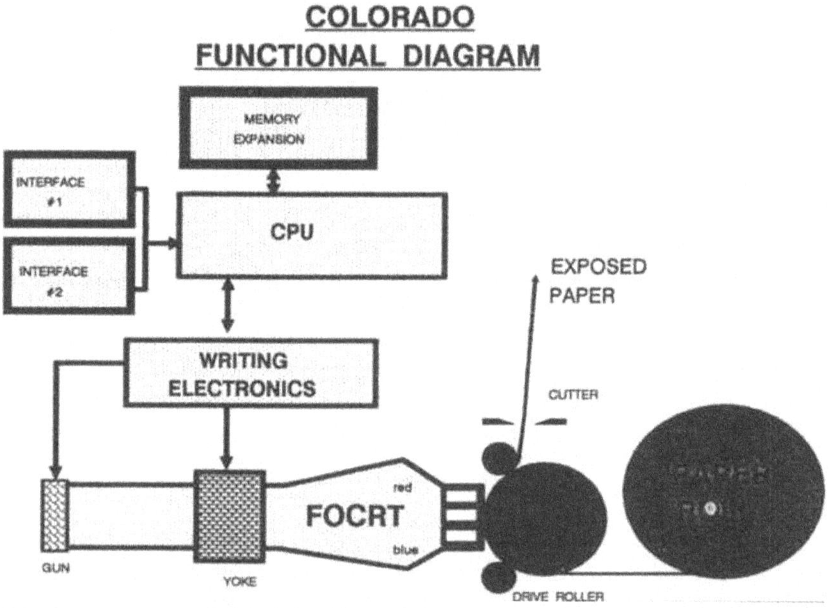

Fig. 6. A functional diagram of the COLORADO writing engine. This basic scheme is employed for exposing both Dry Silver COlor and conventional photographic paper.

The heart of the writing engine is a special 3-color CRT as is shown in fig. 9. The key element of the CRT is the fiber optic faceplate which was developed by Honeywell (now Metrum) nearly twenty years ago. The faceplate (8.75" X 0.75" X 0.5") is made up of a packed array of ten micron diameter fibers. The phosphor is deposited on the inner side of the faceplate where it can be excited by the electron beam of the CRT. The function of the fibers is to carry the light emitted from the phosphor to the outside if the CRT while preserving the focus (the electron beam has a spot diameter of 75 to 125 microns so the ten micron fibers are a good compromise). The photosensitive medium to be exposed is placed in contact with the outer surface of the faceplate where it receives a focused image of the excitation generated inside the CRT by the electron beam. If one were to attempt this type of contact exposure with a plain glass faceplate, the diverging optical rays would be well out of focus by the time it reached the paper.

The long rectangular aspect ratio of the faceplate permits exposure of only a narrow stripe on the paper at any given time. THerefore, the paper must be passed over the faceplate (the direction of travel being perpendicular to the long axis) in small increments with subsequent exposure at each increment.

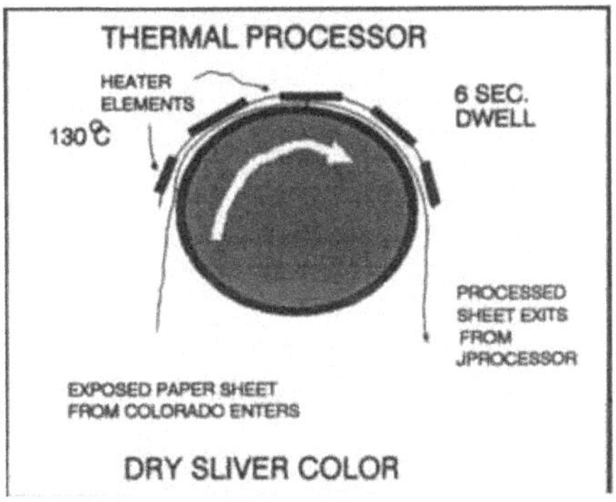

Fig. 7. A side view cutaway of the thermal processor used for processing (developing) images on Dry Silver Color paper.

The printing process proceeds as follows (refer to fig. 6):

1) A host computer is attached via a communications line to one of the two interface cards in the printer (currently an 8-bit parallel Centronics and an analog video interface are available).

2) A series of setup commands are sent to configure the printer for the size and type of image (color or B&W, data format etc.) to be printed.

3) The host computer begins sending the image in packed pixel format (i.e. the red, green, blue information for each pixel is sent in sequence)

4) The CPU (central processing unit) stores the image information in its memory in a line by line basis. The CPU typically does not have sufficient memory to store an entire image. Therefore, when the memory is filled, it will hold the host off until sufficient

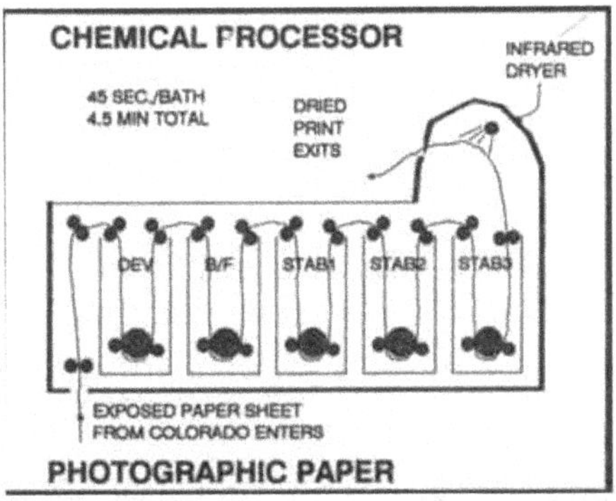

Fig. 8. A side view cutaway diagram of the chemical roller processor used for processing (developing) color images on conventional photographic paper. Rapid acces type paper and RA4 chemistry are used. The processor is mated to a COLORADO engine.

memory for another line of the image is freed.

5) As soon as the CPU has received a minimum number of lines (at least enough lines to cover the active region of the CRT face) it may begin the printing process which proceeds as follows:

a) The blue image information relating to line N is passed to the writing electronics. There the digital information is converted to analog form and coupled to the grid of the CRT synchronously with the motion of the electron beam along the long axis of the faceplate. The paper is stepped ahead by 0.00033" after the line is exposed.

b) The green information relating to line N - M is passed to the writing electronics where M is the number of lines separating the blue and green phosphors. The green line is then written to the paper as described above for the blue line. it commences the writing process

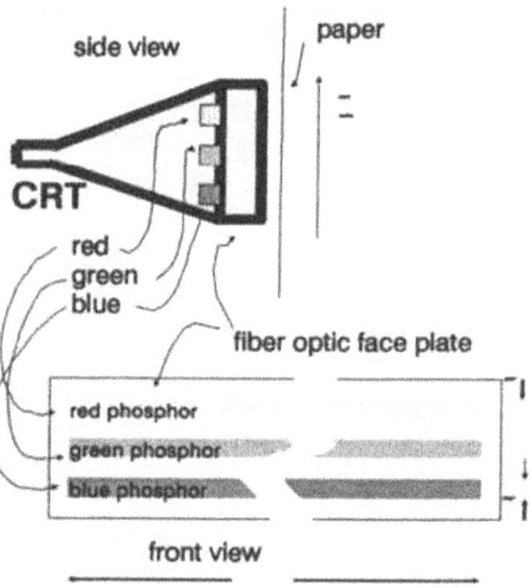

Fig. 9. Functional diagram of the fiberoptic CRT used for exposing color photosensitive media. Typical acceleration voltage is 12 KV with a beam current of from 1 to 100 micro amperes.

c) The red information relating to line N - 2M is passed to the writing electronics and then exposed as described above.

d) Information relating to line N - 2M is purged from the CPU memory since once the red part has been exposed it is no longer needed.

6) The above process continues until the entire image has been exposed. At that point, the paper is moved forward until the bottom edge of the current page is positioned at the cutter. The page is cut from the roll and the cut sheet is passed to the processing mechanism (thermal processor or wet chemical processor).

7) The writing engine is now ready to begin exposure of the next image.

The COLORADO exposure engine can print pictures at a rate of up to four pages per minute. The thermal processor used for the Dry Silver Color paper can likewise process images at the rate of four pages per minute. The chemical processor has been sized to process images at the rate of about 1.5 prints per minute.

The COLORADO writing electronics contains several interesting features:

1) The analog nature of the electron beam sweep and the small increment motion of the paper makes it quite easy to adjust the size of the pixel to obtain variable resolution, a capability not available from other color printing technologies which must perform some processing on the digital image to make it fit the resolution quantum of the appropriate technology.

2) The writing electronics also contains color control circuitry which adjusts the exposure of each phosphor color to achieve the desired color tones. The CPU, through the user interface (a LCD based front panel), permits the user to adjust the color control to his/her preference. Provision is also made for the densitometer measurement (internal for the Dry Silver Color implementation and external for the conventional photo paper) of a color test pattern, with the results being passed back to the CPU to affect a calibration of the the system to the paper lot to achieve consistent color quality.

3) The coupled data flow/paper motion provides an adaptive speed control mechanism which permits the printer to adjust the writing rate to match the incoming data rate.

4) The writing electronics has provisions for calibrating the current/voltage behavior of the CRT to a known state in order to obtain consistent images at low exposure (drift of the I/V characteristics of the electron gun due to thermal expansion has long been a problem for CRT based exposure systems).

----- VII Wrapup -----

This paper has provided a quick tour through some of the technology and thought underlying a new family of image quality color printers. Unfortunately, the current publication does not permit the use of color images to demonstrate the features of the printers. I guess we will have to find an excuse to publish another book that will be set in color using my machine to print it.

THE RESEARCH ON SUPERCONDUCTIVITY IN BRAZIL**

Oscar F. de Lima
Instituto de Fisica, Unicamp,13081 Campinas-SP, Brazil

1. INTRODUCTION

Superconductivity occurs below a certain critical temperature (T_c) and critical magnetic field, when the electrical resistance goes to zero and the field is spontaneously excluded from the sample volume. Since the first observation[1] of the zero resistance transition for mercury, some thousands of superconducting materials have been studied. Recently[2] the discovery of a new class of superconducting oxides opened the High-T_c era with T_c's as high as 130K, well above liquid nitrogen temperatures.

The huge amount of experimental and theoretical work done in the last 90 years on superconductivity has been reviewed[3] from time to time, and several textbooks[4] have also been published.

In Brazil, relatively modest research on superconductivity occurred mainly in the last 18 years. Most of the authors who worked in this field are still active, although some of them have changed their field of interest. Conversely, about 72% of the presently active authors started recently, in the High-T_c era.

A review of the published work done in Brazil, or by people somehow connected to Brazil, in the field of superconductivity, will be presented here. In spite of the author being an experimentalist, all effort was put into reviewing a wealthy variety of theoretical papers. So, I apologize in advance for any faults which are very likely to happen in such kind of work.

In section 2 a brief historical overview on superconductivity is presented. In section 3 the research related to Brazil is reviewed. For convenience the papers quoted were classified, more or less in chronological order, in one of the subsections: theoretical work, and experimental work. In section 4 some comments are presented.

2. HISTORICAL OVERVIEW ON SUPERCONDUCTIVITY

From the discovery of zero resistance transition[1], in 1911, until the discovery of the Meissner effect[5], in 1933, superconductivity was considered to be simply a case of infinite conductivity, and unsuccessful attempts were made to understand the lack of electronic scattering. The Meissner effect consists of a magnetic field exclusion from the superconductor except in a small penetration region near the sample surface,

---

**Dedicated to Professor Rogério C. C. Leite in celebration of his 60th birthday.

where currents flow so as to balance the applied field and make the total field vanish in the sample interior.

From 1933 until 1950 several phenomenological theories were developed and applied with success. Thermodynamics was used, for example, to calculate the condensation energy in the superconducting state, the jump in the specific heat at T_c, and the entropy behavior below T_c. The Gorter-Casimir two fluid model[6] was the first to introduce an order parameter to describe the relative fraction of superconducting (x) and normal (1-x) electrons. In 1935 the London brothers proposed their theory[7] for the electromagnetic properties, describing quite well the infinite conductivity and perfect diamagnetism (=Meissner effect). They postulated a local relationship ($J=c/4\pi\lambda^2 A$) between the current density (J) and vector potential (A), which together with the Maxwell equations led to a solution of exponentially decreasing field penetration at the normal-superconductor (N-S) interface.

In 1950 the Ginzburg-Landau (G-L) theory[8] was published, which introduced a complex order parameter (Ψ) with amplitude and phase, to describe the supercurrent flow. The two basic G-L equations are obtained by minimization of the free energy functional, following a general theory for second-order transitions. Here, the free energy is expressed as an expansion in powers of the complex order parameter. The G-L theory introduced the kappa factor: $\kappa = \lambda/\xi$, where λ is the field penetration depth and ξ is the coherence length, a kind of spacial measure of the long-range-order interaction responsible for the electrons condensation.

The concept of surface energy applied at the N-S interfaces led to the classification of two types of superconductors: a) Type I, when $\kappa< 1/\sqrt{2}$ (positive surface energy); b) Type II, when $\kappa > 1/\sqrt{2}$ (negative surface energy). Later, in 1957, A. A. Abrikosov presented a theory of type-II superconductivity[9], to account for the peculiar magnetization curves already measured by L. V. Shubnikov, earlier in 1937. These curves revealed that the magnetic flux starts to penetrate into the sample at a lower critical field, H_{c1}, but preserving superconductivity until an upper critical field, H_{c2}, is reached. Abrikosov's theory predicted that when $H_{c1}<H<H_{c2}$, flux penetrates in the form of an array of quantized vortex lines, establishing a triangular flux-line-lattice.

In 1950 Fritz London's book[10] was also published, where the idea of superconductivity as a quantum phenomena was introduced in his "program for the molecular theory of superconductivity". London was the first to suggest that superconductivity would be a quantum phenomena on a macroscopic scale that would require a kind of condensation of the average momentum distribution.

An important step towards a microscopic theory was the suggestion of H. Fröhlich[11], in 1950, that superconductivity would be somehow connected with the interaction between electrons and lattice vibrations. This electron-phonon mechanism would give an effective attractive interaction between the electrons. This idea was strengthened by the independent discovery of the influence on T_c caused by a variation in the ionic mass, the so-called isotopic effect, also published[12] in 1950.

Starting about 1953, a number of experiments were carried out that indicated an energy gap in the quasi-particle excitation spectrum of superconductors. In 1956, L.N. Cooper showed that in the presence of an attractive interaction, a pair of electrons close to the Fermi Surface will form a bound state, no matter how weak the interaction is. Culminating all this work, in 1957 a microscopic theory[13] by Bardeen, Cooper and Schrieffer (BCS) appeared, based on the electronic pairing idea. The BCS theory solved a many-body quantum mechanical problem whose ground state consists of condensed cooper pairs. This theory was successfully applied for all kinds of superconductors until 1986.

The original BCS theory was subsequently improved in several senses. In special, different formulations were proposed by N.N.Bogoliubov, who employed quantum field theory , and L.P. Gorkov, who introduced Green's function method to superconductivity.

A very striking effect was predicted by B.D.Josephson in 1962. He found[14] that a supercurrent can flow through a tunnel junction when no voltage is applied; this is the so-called dc Josephson effect. When a voltage is applied across the junction, the phase difference on both sides increases linearly with time, leading to an alternating current of frequency $f=eV/h$; this is the so-called ac Josephson effect. Both sides of a Josephson tunneling junction consist of superconductors, separated by a thin insulating layer.

By the end of 1986 the High-T_c era started, after the discovery[2] by Bednorz and Müller of superconductivity in a family of perovskite-like compounds. Now, dozens of different superconducting oxides have already been identified, although the basic systems are: 1) La-Ba-Cu-O[2] (T_c <30K), 2) Y-Ba-Cu-O[15] (T_c <93K), 3) Bi-Sr-Ca-Cu-O[16] (T_c <110K), 4) Tl-Ba-Ca-Cu-O[17] (T_c <130K), 5) Ba-K-Bi-O[18] (T_c <30K), 6) Nd-Ce-Cu-O[19] (T_c <25K). An enormous number of papers (>15000!) has been published on these new ceramics, dealing with all imaginable aspects of sample preparation, characterization, theory, futurology, and application. There is no room here even for a timid review of the international literature in this area.

So far, no theory has been successful in presenting a microscopic mechanism to account for superconductivity in the high-T_c oxides. On the other hand, they are type-II superconductors, displaying almost the same behavior as the old ones, except for the extremely small coherence length ($\xi \sim 5$ - 30Å) and very large anisotropy (e.g. H_{c2} > 50 H_{c2}). The G-L mean field theory has been widely used with reasonable success, helping to interpret many of the experimental results.

A very small coherence length implies that all sort of small defects or inhomogeneities in the material would work as weak-links, with depressed superconductivity in these regions. The large anisotropy is due to superconductivity being almost confined in some planes (CuO_2 layers). Consequently, a polycrystalline material exhibits relatively small values for the critical fields and critical current density. These intrinsic properties represent a real challenge in the area of technological applications.

3. SUPERCONDUCTIVITY IN BRAZIL

3.1. Theoretical Work

In 1957 the first work done by a Brazilian on superconductivity was published. That was a theory of the specific heat of superconductors based on an energy-gap model, proposed by N.BERNARDES[20] (IFUSP/SP) while he was a graduate student under J.Bardeen, at the University of Illinois-Urbana. In his phenomenological theory a two-fluid model was proposed, based on an approximation in which one-electron states within a range of the order kT_c from the Fermi level are forbidden. It was assumed that the energy gap decreases with temperature and vanishes at T_c. The results for the electronic specific heat showed the general exponential dependence on temperature in agreement with the experimental observations.

Most of the theoretical work done by 1987 was in the area of superconducting electronics. Here, the basic object is a tunneling junction, consisting of two superconducting regions separated by a thin insulator layer (<20Å), or a normal metal bridge. It is very common to appear in the literature the term weak-link, which refers to a more general weak-coupling phenomena like, for instance, the proximity effect in a N-S interface.

Between 1976-1985 S.FROTA-PESSOA (IFUSP/SP) et al. studied the following topics: a) Spin density of states of proximity effect between S-N metal sandwiches in high magnetic fields[21]. They included the effects of Pauli spin paramagnetism and did the calculation of the field dependence for the electronic density of states and order parameter. A little before, they had also shown[22] that the transition of a single thin film superconductor is an example of a tricritical point. b) Space and time variations of the order parameter, and other relevant quantities within a short superconducting weak link[23]. Here, instead of using the conventional time dependent G-L equations (TDGL) they applied the Golub model, which includes the contribution of relaxation mechanisms, allowing the description of the dynamical behavior of the weak links. c) Properties of a short weak-link with distinct chemical potentials at the boundary[24]. Here, the TDGL equations were numerically solved taking into account the differences between chemical potentials, when imposing the boundary conditions on the phase of Ψ, the G-L complex order parameter. d) Time delay and transients in the response of one-dimensional superconductors to supercritical current pulses[25,26,27,28]. Here, they used the TDGL equations to calculate the delay time between the time a current pulse is applied to a link and the time a voltage first appears. e) Model for small three-dimensional superconducting variable-thickness bridge (VTB) constrictions[29], whose dimensions are smaller than the coherence length, ξ. Here, they solved the linearized G-L equations for the VTB model, and found expressions for the order parameter and superconducting current as a function of the geometrical parameters and the phase difference across the bridge. f) Voltage in a charge-imbalance experiment, taking into account the approximate charge neutrality in the superconductor[30]. Here, a two-fluid model proposed by Pethick and Smith[31] was used in the case of charge neutrality.

Still in the field of superconducting electronics R.NICOLSKY (UFRJ/RJ) et al. published several works[32,33,34,35,36,37,38] on the subject of Andreev Reflections (AR) and Current-Voltage characteristics of S-N-S sandwiches. Considering that (1) AR are a phenomenon typical of a inhomogeneous BCS-like superconductor, and (2) AR are predicted to cause negative differential conductivity for sufficiently large mean free paths in the N-region and $T \ll T_c$, they set the conditions which would indicate the presence of cooper pairs in the new high-T_c oxide superconductors. Later, Nicolsky[34-36] pointed out some evidences for a BCS-like superconductivity in the oxides. He established a criterion for categorizing Josephson junctions as tunneling or metallic, and found a dominant metallic behavior for the intrinsic weak-links in the oxide-superconductors, stressing the important role played by AR in these materials. Further, he concluded that the intergrain junctions should not limit the current-carrying properties of single-crystals and grains. Instead, the intergrain junctions occurring in polycrystalline and granular samples, should be responsible for the strong current-carrying limitation, as was already well known in the literature. More recently Nicolsky et al.[39] have proposed new possibilities for non-Josephson superconducting electronics (e.g. radiofrequency generation) using metallic weak links.

In 1980 A.O.CALDEIRA (UNICAMP/SP) and A.J.Leggett published[40,41] their theory on the influence of dissipation on Quantum Tunneling in Macroscopic Systems. That was an important contribution of particular interest in the context of tunneling of a macroscopic entity, such as the trapped flux in a SQUID (Superconducting Quantum Interference Device). A more detailed discussion on this subject appeared in a review paper published[42] in 1983.

E.GRANATO (INPE/SP) et al. studied several aspects of Josephson-junction arrays, as follows: a) The resistance oscillations as well as the effects of disorder under a transverse magnetic field[43,44]. b) The effects of weak disorder on arrays of superconducting grains coupled by weak-links in which fluctuations play a dominant role[45,46]. c) Zero temperature analysis of the superconductor-insulator transition and its universal resistance behavior[47]. d) Zero temperature analysis of a model for one-dimensional periodic Josephson-junction ladders, in a perpendicular magnetic field[48]. More recently Granato et al. have worked[49,50] on subjects more closely related to the high-T_c superconductors. By using a microscopic theory of particle diffusion, they studied[49] the contribution to the linear resistivity arising from the pinning of flux lines by dilute macroscopic defects. The real picture of flux lines moving with respect to a fixed pin structure was mapped into the diffusion process of a particle interacting with the static flux line lattice.

A more recent contribution to the subject of Josephson tunneling is the work by J.ALIAGA, H.CERDEIRA (UNICAMP/SP) et al.[51], where they apply an information-theoretical approach and find a subalgebra that allows them to obtain naturally the dependence of the Josephson current on the phase, as well as a geometric interpretation of the phase. They used a simplified Hamiltonian model of pair tunneling between two weakly coupled superconductors to calculate the ac Josephson current.

In 1979 E.Z.DA SILVA (UNICAMP/SP) et al. presented a theory[52,53] of the fluctuation-induced magnetoconductivity for an anisotropic bulk in the dirty limit, and for fibrous superconductors. Based on the Maki-deGennes generalization of the TDGL theory they were able to calculate the magnetoconductivity as a function of the applied field, H, the angle, θ, between the electrical current and H, and temperature, T. The theory was applied to experimental data[54] on polysulphur nitride, which is a polymeric superconductor ($T_c \approx 0.3K$) with fibrous structure.

A study of the superconducting transition in many-valley semiconductors, like the chalcogenide PbTe-Tl (T_c<1.4K), was done by H.V.SILVEIRA et al.[55,56] (UNICAMP/SP). They calculated the order parameter, the ac-conductivity, and T_c as a function of the carrier concentration.

3.1.1. High-T_c era. After 1986 several theorists started giving their contributions in the field of superconductivity.

M.M.DORIA (IFUSP/SP, now at PUC/RJ) et al. have contributed in several topics: a) Using generalized G-L equations, they[57] discuss the possibility that the high-T_c superconductors do not have a scalar, but a tensor order parameter. b) Using a kind of London approach they[58] analyze the equilibrium vortex structure for an arbitrary field orientation with respect to an anisotropic uniaxial crystal ($YBa_2Cu_3O_{7-\delta}$). c) They propose[59] a microscopic model which produces hole attraction in the CuO_2 planes, based on competing magnetic interactions, involving the oxygen holes and the copper spins. d) In the framework of the G-L theory, they[60] derive and discuss the virial theorem: $H.B = 4\pi(F_k + 2F_f - F_{in}/2)$, where H is the external field, B the average of the magnetic induction, F_k the average kinetic energy, F_f the average field energy, and F_{in} is an energy term which is nonzero only for inhomogeneous systems. Later[61], these authors applied this theorem to calculate magnetization curves through a numerical technique labeled as Simulated Annealing Method. e) Using a generalized G-L free energy for two components (s+d) wave pairing they[62] obtain an upward curvature in the upper critical field, in qualitative agreement with experimental results. f) Based on a model of layered superconducting planes interacting through Josephson coupling without reflection symmetry perpendicular to the planes, they[63] identify an intrinsic modulation of the phase of the order parameter for Nd-Ce-Sr-Cu-O type compounds.

A.LATGE and E.V.ANDA (UFF/RJ) et al.[64,65] presented a detailed study of the effect that oxygen disorder has on the electronic structure of the highly correlated high-T_c superconductors. Their analysis is based on a Hamiltonian that includes the interatomic and intra-atomic electronic repulsion at the various Cu and O sites of $YBa_2Cu_3O_{7-\delta}$. Under the assumption that T_c is proportional to the number of holes in the CuO_2 planes, they were able to reproduce features of the experimentally observed result for the oxygen concentration dependence of T_c.

G.A.LARA and G.G.CABRERA (UNICAMP/SP) recently published[66] an extended one-band Hubbard model, including Coulomb repulsion and correlated hopping between nearest neighbors in a square lattice. The interplay between antiferromagnetism and superconductivity is discussed, and the possible application to the high-T_c oxides is commented.

3.2. Experimental Work

E.LERNER (UFRJ/RJ) is probably the first Brazilian who got a Ph.D. doing an experimental work[67] on superconductivity. He studied the thermal and electrical conductivities[68], as well as the magnetic properties[69] of superconducting Mo-Re alloys. The samples were formed as fine wires, commercially available, and suitable for winding high-field superconducting magnets (H<40kOe). In 1973 Lerner founded the Low Temperature Laboratory of Universidade Federal do Rio de Janeiro, where the experimental work on superconductivity was carried on by other people, whose work is quoted below.

R.F.PEREIRA, M.F. da SILVEIRA, and E.MEYER (UFRJ/RJ) studied[70,71,72] A-15 compounds (Nb_3Al and Nb_3Au) which were prepared through a technique of melting by condenser discharge, very suitable for synthesizing metastable phases. In this way they were able to produce small spheres (<0.5mm diameter) of the alloy in a fast and clean operation. The A-15 phases were obtained after a reaction heat treatment. Their T_c values, measured with an inductive bridge[73], indicated the stabilization of the metastable compounds Nb_3Al and Nb_3Au.

More recently, J.BARRETO et al. (UFRJ/RJ) observed[74] a degradation of the superconducting compound $YBa_2Cu_3O_{7-\delta}$ after irradiation by α-particles, of 28 MeV, at room temperature.

In 1969 D.G.PINATTI (UNICAMP/SP) took his Ph.D. doing experimental work[75] on the superconductivity of Molybdenum. Back to Brazil, he founded the Low Temperature Laboratory of UNICAMP, starting in 1970. Between 1973-1975 Prof. ROGERIO C.C.LEITE was the Physics Institute Director, and his precious support was instrumental for the Low Temp. Lab., as well as the Cryogenic Center of UNICAMP. Following, a review of the published work done on superconductivity, by the people from UNICAMP.

N.BOBOSHKO et al.[76,77,78,79] (UNICAMP/SP) studied the energy dissipation due to the movement of the magnetic flux in superconducting tin, by means of the torsion pendulum technique. A phenomenological analysis of the experimental data of flux-flow resistivity as a function of temperature was proposed[79], and successfully tested.

S.MOEHLECKE (UNICAMP/SP) et al. made several contributions: a) The effect of hydrogen doping on the superconductivity of Nb single crystals was studied[80]. A decrease of the electronic density of states at the Fermi level was suggested to be responsible for an observed T_c depression. b) The effects of composition and neutron irradiation on the superconducting properties of several A15 compounds were studied[81,82,83,84,85,86,87,88]. Large depressions (up to 80%) in T_c were observed for all samples (Nb_3Ge, Nb_3Al, Nb_3Pt) with the exception of Mo_3Os, where much smaller decreases in T_c occurred. The original T_c values were fully restored by an appropriate annealing. Their results were discussed in terms of site-exchange disorder, and a correlation between radiation and thermally induced disorder was noted.
c) The electrical resistivity of Nb_3Pt in a broad homogeneity range $Nb_{100-x}Pt_x$ ($20 \leq x \leq 28.9$) was studied[89] as a function of composition x, from room temperature down to T_c. A closer correlation with the resistivity slope ($d\rho/dT$) at high temperatures and with the electronic specific heat coefficient was noted. The results were interpreted in terms of an order-disorder model. d) They studied[90] the homogeneity range of the superconducting Chevrel phase $SnMo_6S_x$ ($6.9 \leq x \leq 8.0$) and found it to be very narrow, around the composition $SnMo_6S_{7.5}$, for samples doped with oxygen up to the solubility limit. e) More recently[91] they studied the effects of heat treatment on the superconducting Bi-Ca-Sr-Cu-O system.

G.ZERWECK (UNICAMP/SP) took his Ph.D. in Germany and moved to Unicamp in 1974. He gave an important contribution through his original writings[92,93,94] and also by introducing new students into superconductivity, until he went back to Germany in 1979. His research dealt mainly with the mixed state properties of type-II superconductors[93]. Within the framework of the elastic theory for the flux line lattice he derived[92] an expression for the shear modulus, C_{66}, for low-k type-II superconductors. He also developed a microscopic model to account for the pinning of flux lines by grain boundaries in a metal. This model describes the reduction of the mean free path of conduction electrons which leads to a decrease of the self-energy of flux lines, in a layer which is considerably thicker than the perturbed zone of the boundary itself.

O.F.de LIMA (UNICAMP/SP) et al. made the following contributions: a) Using an inductive quasi-dc technique the magnetic flux density profile and the magnetization curves were measured in Nb singlecrystals[95]. A Field History Effect[96], and the influence of plastic deformation on the flux density gradient, were analyzed and discussed in view of the Critical State Model. The inductive technique was also used by A.S.BRITO et al.[97] (UNICAMP/SP) to study the enhanced flux density gradients near the surface of superconducting Niobium, caused by surface pinning. Later, J.L. de CASTRO (UNICAMP/SP) studied[98] the flux distribution in cylindrical-model-samples made out of coaxial layers of Cu-Nb-...-Cu-Nb. b) Using a liquid phase sintering approach and starting from a mixture of Nb-H and Cu powders, a new method for preparing Cu-Nb_3Sn superconductor wires was introduced[99,100,101,102,103]. This required a systematic study about wetting of niobium by liquid copper[104]. A.T. de REZENDE[103] (UNICAMP/SP) improved the method and was able to get critical current density values around 10^5 A/cm^2, at H=7 Tesla and T=4.2K. c) Thin films (3000-4000 Å) of the high-T_c oxide $La_{2-x}Sr_xCuO_4$ were prepared by a combined evaporation reaction method showing T_c onsets around 25K. This work[105] was done at the very beginning of the High-T_c era, through Jan.-Feb./1987. d) The effects of oxygen on the superconductivity of $YBa_2Cu_3O_x$ ($6.3<x<7.0$) were studied[106] through a method of oxygen extraction under vacuum. T_c was shown to decrease with x, displaying a plateau around 70K for $6.6<x<6.8$, though a sharp transition was observed at T_c=93K for x=7.0. Later[107], samples with different compositions in the Y-Ba-Cu-O system were aged and the transition measured under an oxygen atmosphere.

Between 1975 and 1985 there was a continued effort at UNICAMP towards the preparation of Nb-Ti alloys and the A15 compound Cu-Nb_3Sn, aimed at the production and characterization of superconducting wires. From 1980 to 1984 some people from Pirelli S/A (S^{to} Andre,SP) were introduced into this area through a technical cooperation which involved the Fundação de Tecnologia Industrial (FTI) of Lorena/SP.

Recently, the people from FTI/SP (D.RODRIGUES, C.A.BALDAN, C.BORMIO, C.Y.SHIGUE, M.J.RAMOS, and D.G.PINATTI) reported[108,109,110] on their program of superconducting wires (NbTi and Cu-Nb_3Sn) and magnets fabrication.

At the Physics Institute of Universidade de São Paulo (IFUSP), the experimental research on superconductivity also started about 18 years ago. There, F.P.MISSELL and N.F.OLIVEIRA Jr.(IFUSP/SP) et al. made several contributions as follows: a) They determined H_{c2} through resistive transition measurements, and found no influence of Pauli spin paramagnetism for a Ti-11at%Nb alloy[111]. Studies on the temperature dependence of H_{c2}, as well as on other superconducting properties for Ta-Zr alloys were carried out[112] by W.A.ORTIZ (now at UFSCAR/SP) et al. b) Studies of the ultrasonic attenuation in the mixed state of Nb-26at%Hf[113,114,115] and V-42at%Ti[116,117,118] were performed, using shear and longitudinal waves with frequencies between 5-140 MHz, by means of conventional pulse-echo techniques. Their experimental data were incompatible with some current microscopic theories. However, they agreed very well with a phenomenological model[119] in which magneto-hydrodynamic effects, like flux-flow resistivity, were included. Then, an effective mixed-state resistivity was used instead of the usual normal-state resistivity, to calculate the magnetic-field-dependent change in the ultrasonic velocity, given by the Alpher-Rubin theory[120].

Some research on amorphous superconductors was also carried out by the people of IFUSP. F.P.MISSELL et al. studied the pressure dependence of T_c for amorphous[121] $(Mo_{0.6}Ru_{0.4})_{86}B_{14}$ and La-based alloys[122,123], under hydrostatic pressures up to $p \sim 10$ kbar. Their results suggested that variations of T_c could be accounted for by volume effects, through the electron-phonon coupling constant. Later, they also studied the composition influence on the electronic density of states and its relation with T_c, in amorphous alloys of Zr-Rh, Zr-Ru, Zr-Re[124,125], and of Zr-Pd, Zr-Ni[126].

More recently, at IFUSP they started to work on the production and characterization of Nb-Ti superconducting wires[127,128,129], in cooperation with the company Pirelli S/A.

J.OTUBO (Pirelli S/A) et al[130,131,132]. developed a Cu-Nb_3Sn wire fabrication method, based on cold powder-metallurgy processing of large particle powders, leading into a submicron multifilamentary structure. Typically they started with a Cu tube filled with a mixture of Cu+Nb powders, whose particle size ranged between 0.25 to 0.50 mm in diameter, and reached a final wire (0.25 mm diameter) after an areal reduction factor around 10^6. During the cold work deformation the soft Nb particles are elongated into tiny ribbon-like filaments embedded in the Cu matrix. Both, external tin deposition and internal tin core processing were employed, followed by a final reaction heat treatment ($\sim 650^\circ$C). Typical current densities as high as 3×10^4 A/cm^2 at H=14T, T=4.2K, were obtained.

So far, no experimental research has been done in Brazil on the subject of Josephson junctions. However, the use of SQUID gradiometers in detecting biomagnetic signals has been explored by P.C.RIBEIRO (PUC/RJ) et al. for many years[133,134,135,136,137,138].

3.2.1. High-T_c era. After 1986 several experimentalists started giving their contributions in the study of superconducting materials.

H.A.BORGES (PUC/RJ) et al. studied[139,140] the effect of pressure on the resistance and T_c of $RBa_2Cu_3O_x$ (R=Gd,Er,Yb,Y), and also of $EuBa_2(Cu_{1-y}Zn_y)O_x$, $y \leq 0.05$. For the first system, T_c, which is 90K or more at ambient pressure, is enhanced by pressure at rates between 0.09 and 0.19K/kbar, giving a positive dT_c/dp. For the second system, doping with Zn produced a strong depression of T_c ($\sim$ 19K/at%Zn). However, the application of pressure tends to compensate the effect of doping by raising T_c towards the undoped $EuBa_2Cu_3O_x$ behavior. Borges et al.[141,142] also did studies about the pressure effects on superconducting heavy-electron Uranium compounds, like $U_{1-x}Th_xBe_{13}$.

P.PUREUR, J.SCHAF et al.[143,144,145,146,147,148] (UFRGS) have done neat work in discussing the magnetic glass behavior and the H-T phase diagram, based on resistivity and dc-susceptibility data for $RBa_2Cu_3O_7$ (R= Y,Eu,Gd) samples. They have found evidences that support the description of these systems in terms of a granular superconducting network, where disorder and frustration play an important role. Recently they proposed a low-field-high-temperature H-T phase diagram for the polycrystalline oxide superconductors; three regions below the $T_c(H)$ transition line were identified: 1) A reversible region between $T_c(H)$ and $T_g(H)$, where $T_g(H)$ defines the reversibility line. Here, the system would be composed of disconnected superconducting grains. 2) A region corresponding to the pure superconducting glass regime, delimited above by $T_g(H)$ and below by $T_M(H)$, the latter being an upper limit line for the saturated Meissner state. 3) A region corresponding to a conventional Abrikosov state, where flux creep and flux pinning become increasingly important as the temperature is lowered.

R.F.JARDIM et al.[149,150,151,152,153] (UNICAMP/SP) did a systematic study of the effects produced by a partial substitution of Mn for Cu in $YBa_2(Cu_{1-x}Mn_x)_3O_{7-y}$, $x \leq 0.20$, using the solid diffusion method. Apparently the Mn solubility in the $YBa_2Cu_3O_{7-\delta}$ matrix is very small ($x \sim 0.05$). Then, most of the Mn is incorporated into other secondary phases which precipitates in the matrix. This leads to a transition broadening due to an increased inhomogeneity of the sample. By increasing the reaction temperature (T=970°C) they also did a kind of Liquid Phase Sintering, which favored an enhanced grain growth of the $YBa_2Cu_3O_{7-\delta}$ phase, dependent on the Mn concentration, and peaking at x=0.025. More recently[154] a phonon mean free path in the high-T_c oxides was evaluated at room temperature, through measurements of the sample density and thermal diffusivity and conductivity. They found a high value of 5μm for $YBa_2Cu_3O_{7-\delta}$, which led them to identify large defects (e.g. grain boundaries, precipitates, voids) as the main phonon scattering centers. R.F.JARDIM et al.[155] (now at IFUSP/SP) are also carrying out studies on magnetic properties of $(Nd_{1-x}Gd_x)_2CuO_4$ solid solutions.

E.BAGGIO-SAITOVITCH et al. (CBPF/RJ) have performed[156,157,158,159] extensive work on ^{57}Fe Mössbauer spectroscopy of $YBa_2(Fe_xCu_{1-x})_3O_{7-\delta}$ and La-Ba-Cu-O, at low (T<295K) and high (T>295K) temperatures. They found that ^{57}Fe atoms substitute for Cu preferentially in the Cu(1) positions of $YBa_2Cu_3O_{7-\delta}$, allowing for different oxygen configurations. A depression on T_c, probably associated with changes of the local structure in the Cu-O chains, was also observed. In situ Mössbauer experiments performed in vacuum at high temperatures[160], gave interesting information on the kinetics of oxygen loss in ^{57}Fe doped samples of $YBa_2Cu_3O_{7-\delta}$. Furthermore, an ordered magnetic phase of the Fe atoms was established through Mössbauer studies at room temperature[161,162], for the oxygen deficient phase $YBa_2(Fe_xCu_{1-x})_3O_{6+\delta}$. A magnetic ordering was also observed[163] at T=4.2K for samples of $La_{2-x}Ba_xCuO_4$ ($x \leq 0.35$) with 0.5at%Fe, even in the superconducting phase $La_{1.85}Ba_{0.15}CuO_4$.

E.MATTIEVICH et al. (UFRJ/RJ) also did Mössbauer studies[164] on $YBa_2Cu_3O_{7-\delta}$ doped with 1% ^{57}Fe. They arrived at the generic conclusion that Fe ions were occupying Cu sites.

C.RETTORI (UNICAMP/SP) et al. did Electron Spin Resonance (ESR) studies[165,166,167] on several phases of the Y-Ba-Cu-O system. Initially they found a reverse relation between magnetism (ESR intensity) and superconductivity, interpreted as a competition between (antiferromagnetic) pairing and magnetic fluctuations. The ESR observed in these copper compounds were assigned to the $Cu^{2+}(3d^9)$ ions located in an anisotropic environment. Later , by fitting theoretical spectra simulations to ESR data measured in several samples of the superconductor $YBa_2Cu_3O_{7-\delta}$, and in the semiconductors Y_2BaCuO_5 and $BaCuO_{2+x}$, as well as in their corresponding oxygen-deficient stoichiometry, they found that no ESR signal was coming from any of them. Instead, the resonance observed should be coming from another residual copper compound. Rettori et al.[168] also did an interesting experiment where they measured the magnitude and phase of the microwave transmission through thin films of Y-Ba-Cu-O in the presence of a dc magnetic field and an oscillating magnetic field. They showed that the observed non linear behavior could be explained by a model which assumes an array of Josephson junctions and vortex motion in the presence of a magnetic field.

R.S. de BIASI et al.(IME/RJ) also did ESR measurements[169,170] in the orthorhombic (superconducting) and tetragonal (non-superconducting) forms of $YBa_2Cu_3O_{7-\delta}$. Further, they studied[171,172] the low-field microwave (9.1 GHz) absorption in the Y-Ba-Cu-O and Bi-Sr-Ca-Cu-O systems. Using the microwave absorption technique to register the Normal-Superconductor transition, they recently did a study[173] on T_c variation with Gd content in the solid solution $Y_xGd_{1-x}Ba_2Cu_3O_7$.

J.A.O. de AGUIAR (UFPE/PE) et al. also did ESR studies on $YBa_2Cu_3O_{7-\delta}$ crystals[174,175] as well as on $YBa_2(Cu_{1-x}Fe_x)_3O_{7-\delta}$ polycrystalline[176] samples. In all cases no Cu- or Fe-EPR signal was observed.

S.M.REZENDE et al.(UFPE/PE) showed[177,178] that the microwave absorption of polycrystalline $YBa_2Cu_3O_{7-\delta}$ in the superconducting state is highly nonlinear in the microwave power. They observed a non-negligible loss, even at moderate power levels ($\geq$1 mW), in the absence of a dc field. An interpretation based on a model for the interaction between the microwave currents and the flux lines was presented.

J.A.SANJURJO (UNICAMP/SP) et al. did Raman measurements and lattice dynamics studies[179] on $YBa_2Cu_3O_7$ as a function of temperature (10-295K). A group theoretical analysis of the basis vectors allowed them to present a tentative assignment of the normal modes of vibration. A critical review[180] of the literature on Raman measurements was also presented by Sanjurjo in 1988.

J.M.FERREIRA (UFPE/PE) et al. made several contributions: a) In the very beginning of the high-T_c race they did a thorough study on the $RBa_2Cu_3O_{7-\delta}$ (R=rare earth) magnetic superconductors[181,182,183]. Except for R=Ce,Pr and Tb all the other compositions showed a superconducting transition around T_c=90K. For R=Tm they found a very high slope $(dH_{c2}/dT)_{T_c} \sim -2.8$ Tesla/K. Using the standard WHHM theory[184] this would indicate the possibility for extremely high H_{c2}, around 35 Tesla at liquid Nitrogen (77K) and 100 Tesla at liquid Helium (4.2K). b) They performed very careful low temperature specific heat measurements for $La_{1.8}Sr_{0.2}CuO_4$ and $RBa_2Cu_3O_{7-\delta}$ (R=Y,Eu,Ho,Tm,Yb)[185], (R=Nd,Sm,Gd,Dy, Er)[186]. For the latter set of rare earth substitutions they found a magnetic ordering of the trivalent R ions. The temperature of the peak of the specific heat anomaly due to magnetic ordering, were T_N=0.52, 0.61,2.25, 0.90,0.60K for the compounds with R=Nd,Sm,Gd,Dy, and Er, respectively. Antiferromagnetic ordering was determined[187] in the case of superconducting (δ = 0.1) and oxygen-deficient non-superconducting (δ = 0.5) $YBa_2Cu_3O_{7-\delta}$ compounds, where R=Nd and Sm. c) They studied[188] partially aligned samples of $RBa_2Cu_3O_{7-\delta}$ (R=Y,Nd,Sm,Eu,Dy,Ho,Er,Tm,Yb) obtained from a mixture of finely ground powders with epoxy, which were then allowed to harden in applied magnetic fields H_A=18kOe. For samples R=Y,Nd,Sm,Dy and Ho the single-crystal-grain c-axes were partially aligned parallel to H_A, and perpendicular to H_A for R=Eu,Er,Tm,Yb.

J.P.ANDREETA et al.[189] (IFQSC/SP) prepared the superconducting compound $TmBa_2Cu_3O_{6+x}$ (T_c~92K) and did a careful X-ray study to characterize the crystallographic phases occurring in their impure sample.

E.E.CASTELHANO (IFQSC/SP) et al.[190] studied the room temperature infrared reflectance spectrum and the derived ϵ'' dieletric function of $Bi_2Sr_2CaCu_2O_{8+x}$ together with a factor group analysis and tentative assignment of its crystal optic modes.

N.J.H.GALLO et al.[191] (IFQSC/SP) showed that it is possible to directly observe the structural defects (e.g. dislocations) in Bi-Sr-Ca-Cu-O single crystal compounds, using the traditional etching technique.

J.R.DELBEN and H.C.BASSO[192] (IFQSC/SP) did a substitution of Zn for Cu in $YBa_2Cu_3O_{7-\delta}$ and found segregation of impurity phases at the grain boundaries, dependent on the Zn concentration and method of preparation.

D.I.dos SANTOS (IFQSC/SP) et al.[193] have worked on the preparation and characterization of samples in the (Bi,Pb)-Sr-Ca-Cu-O system. Both the solid state reaction technique and the solution[194] technique have been used to synthesize the samples.

L.GOMES (IPEN/SP) and A.J.P.BRAGA (UNICAMP/SP) et al. reported[195] on thermal diffusivity (α) studies for $YBa_2Cu_3O_{7-\delta}$. α describes a bulk behavior by relating in a simple way the thermal conductivity, K, to the specific heat, C_p, through the relation: $\alpha = K/\rho C_p$, where ρ is the mass density. Two laser excitation techniques were used: The photoacoustic phase-lag and the flash method. They observed an anomalous jump in α at T_c, which is related to a jump in the specific heat expected at the superconducting transition.

Thermal diffusivity studies were also done by E.C.SILVA et al.[196] (UNICAMP/SP) on samples of the La-Ba-Cu-O and Y-Ba-Cu-O systems, using the photoacoustic thermal characterization technique. They found that α is very sensitive to the composition and porosity of the samples.

M.T.A.ORLANDO et al.(IPEN/SP) recently did[197] ac magnetic susceptibility studies on the Bi-Sr-Ca-Cu-O system, where they found indications of three superconducting transitions.

4. SOME COMMENTS

A considerable portion (39%) of the works cited here were done while the Brazil-linked-author was visiting some research centers abroad. Although this counts positively, there is also no doubt that more effort is still needed to establish regular research groups on superconductivity in Brazil. Compared with other third world countries (e.g. India, China, Mexico) Brazil still ranks in a very modest position.

It is surprising the relatively high quantity and quality of theoretical studies on Josephson tunneling phenomena, opposed to almost no experimental work in this area. On the other hand, there has been a reasonable amount of experimental work in the area of metallurgical effects on the superconducting properties (e.g. T_c, J_c, H_{c1}, H_{c2}), with almost no theoretical support. This latter is traditionally provided by a more phenomenological approach, in order to treat a great variety of mixed state properties involving the flux line lattice, like pinning, creep, magnetic relaxation, and so on. Apparently, from the theoretical point of view, this sounds like a kind of dirty environment. However, there is no other approach, so far, which allows the interpretation of magnetization curves and critical current measurements, of paramount importance for materials characterization.

To finish, I would like to apologize for those works that unintentionally were not quoted here. Although my intention was to include everybody and everything, I am aware of the difficulty to accomplish this goal when doing this kind of review paper.

The main purpose of this review is to give an idea of the effective effort in the area of superconductivity in Brazil, and provide a kind of up-to-dated Who's Who list. I hope this could become of some value for the newcomers in the field, as well as for the veterans.

5. ACKNOWLEDGEMENTS

The author is grateful to the following people, who kindly gave him much valuable information: N. Bernardes, E. Lerner, F. P. Missell, S. F. Pessôa, N. F. de Oliveira Jr., A. Caldeira, S. Moehlecke, C. Rettori, R. Nicolsky, S. M. Rezende, J. M. Ferreira, and J. A. O. de Aguiar. A fellowship from Conselho Nacional de Desenvolvimento Cientifico e Tecnologico (CNPq) is also acknowledged.

6. REFERENCES

1.Kamerlingh Onnes,H., Akad. van Wetenschappen (Amsterdam) 14, 113, 818 (1911)
2.Bednorz,J.G. and Müller,K.A., Z.Phys. B64, 189 (1986)
3.a) Bardeen,J., Encyclopedia of Physics (Springer-Verlag,Berlin, 1956),vol.15,p.274. b) Parks,R.D.,ed.,Superconductivity,M.Dekker, 1969. c) NATO vols. d) Physical Properties of High Temperature Superconductors, vol.I & II, ed.D.M.Ginsberg (World Sci.Publ.Co., Singapore, 1989)
4.a) E.A.Lynton, Superconductivity, Methuen,1969,3rd ed. b) M.Tinkham, Introduction to Superconductivity, McGraw-Hill,Kogakusha,1975. c) P.G. de Gennes, Superconductivity of metals and alloys, Benjamin, 1966.
5.Meissner,W. and Ochsenfeld,R., Naturwiss. 21, 787 (1933)
6.Gorter,C.J. and Casimir,H.B.G., Physica 1, 305 (1934)
7.London,H., London,F., Proc.Roy.Soc.(London) A149, 71 (1935)
8.Ginzburg,V.L. and Landau,L.D., Zh.Eksp.Teor.Fiz. 20, 1064 (1950)
9.Abrikosov, A.A., Zh.Eksp.Teor.Fiz. 32, 1442 (1957)
10.London,F., Superfluids, vol.1, John Wiley & Sons, N.York (1950)
11.Frölich,H., Phys.Rev. 79, 845 (1950)

12.a) Maxwell,E., Phys.Rev. 78, 477 (1950). b) Reinolds, Serin, Wright and Nesbitt, Phys.Rev. 78, 487 (1950)
13.Bardeen,J., Cooper,L.N. and Schrieffer,J.R.,Phys.Rev. 108,1175(1957)
14.Josephson, B.D., Phys.Rev.Lett. 1, 251 (1962)
15.Wu,M.K. et. al, Phys.Rev.Lett. 58, 908 (1987)
16.Maeda,H. et al. Jpn.J.Appl.Phys.Lett. 27, L209 (1988)
17.Sheng,Z.Z. and Hermann,A.M., Nature 332, 55 (1988)
18.Cava,R.J., Batlogg,B., Krajewski,J.J., Farrow,R., Rupp Jr.,L.W., White,A.E., Short,K., Peck,W.F. and Kometani,T., Nature 332, 814 (1988)
19.Tokura,Y., Takagi,H. and Uchida,S., Nature 337, 345 (1989)
20.Bernardes,N.,"Theory of the specific heat of superconductors based on an energy-gap model", Phys.Rev. 107(2), 354 (1957)
21.Gallagher,W.J., Frota-Pessôa,S. and Schwartz,B.B.,"Spin Density of States of Proximity Effects Superconducting Sandwiches in High Fields", Solid State Commun. 23, 837 (1977)
22.Frota-Pessôa,S. and Schwartz,B.B.,"Explicit Calculation of a Tri-critical Point in a Pauli Paramagnetically Limited Superconductor", Solid State Commun. 20, 505 (1976)
23.Frota-Pessôa,S.,Blackburn,J.A. and Schwartz,B.B.,"Space and Time Variations of the Order Parameter and Other Relevant Quantities Within Superconducting Link", Solid State Commun. 27, 629 (1978)
24.Frota-Pessôa,S., Blackburn,A.A. and Schwartz,B.B.,"Short weak link with distinct chemical potentials at the boundary", Phys.Rev. B20, 993 (1979)
25.Octavio,M., Frota-Pessôa,S., Oppenheim,I.F. and Blackburn,J.A., "Transients in One-Dimensional Superconductors", Phys.Lett. 82A(7), 365 (1981)
26.Oppenheim,I.F., Frota-Pessôa,S. and Octavio,M.,"Time Delay Response of Superconducting Filaments to Supercritical Current Pulses", Phys.Rev. B25, 4495 (1982)
27.Oppenheim,I.F.,"Estudo de Conexões Enfraquecidas em Supercondutores", Master Thesis, IFUSP,SP (1981)
28.Nicolsky,R.,"Estudo sobre conexões enfraquecidas e não equilíbrio em superconductores", Master Thesis, IFUSP,SP (1981)
29.Nicolsky,R., Octavio,M. and Frota-Pessôa,S.,"Model for Small Superconducting Constrictions", J.Low Temp.Phys. 58, 11 (1985)
30.Nicolsky,R. and Frota-Pessôa,S.,"Voltage in a charge-imbalance experiment, taking into account the approximate charge neutrality in the superconductor", J.Low Temp.Phys. 46, 107 (1982)
31.Pethick,C.J. and Smith,H., Ann.Phys.(NY) 119, 133(1979)
32.Kümmel,R., Huckestein,B. and Nicolsky,R.,"Andreev reflections and the current-voltage characteristic of the superconductor-normal metal-superconductor junctions", Jpn.J.Appl.Phys. 26,Suppl.26-3, 1471 (1987)
33.Kümmel,R.,Huckestein and Nicolsky,R.,"Negative differential conductivity in SNS-junctions: a test of pairing configurations in high-T_c superconductors", Solid State Commun. 65, 1567 (1988)
34.Nicolsky,R.,"Superconductor-Normal metal interfaces in high-T_c superconductors",in: Progress in High Temp.Supercon. 5, ed. (World Sci. Publ.,Singapore, 1988),p.219

35.Nicolsky,R.,"Metallic interfaces in high-T_c superconductors", Physica C153-155, 992 (1988)
36.Nicolsky,R.,"Andreev reflections and critical currents in high-T_c superconductors", Cryogenics 29, 388 (1989)
37.Kümmel,R., Gunsenheimer,U.,Plehn,H. and Nicolsky,R.,"Density of States and Charge Transport in SNS junctions", in: Progr. in High Temp. Supercon. 25, ed.Nicolsky,R.(World Sci.Publ.Co., Singapore, 1990) p.344
38.Kümmel,R., Gunsenheimer,U. and Nicolsky,R.,"Andreev scattering of quasiparticle wave packets and current voltage characteristics of superconducting metallic weak links", Phys.Rev. B42, 3992 (1990)
39.Luiz,A.M. and Nicolsky,R.,"Radiofrequency generation using a SNS microbridge", IEEE Trans.Magn. (1990)
40.Caldeira,A.O. and Leggett,A.J.,"Influence of dissipation on quantum tunneling in macroscopic systems", Phys.Rev.Lett. 46, 211 (1981)
41.Caldeira,A.O, Ph.D. Thesis, University of Sussex, Brighton, England (1980)
42.Caldeira,A.O. and Leggett,A.J.,"Quantum Tunneling in Macroscopic Systems", Annals of Phys. 149, 374 (1983)
43.Granato,E. and Kosterlitz,J.M.,"Quenched disorder in Josephson-junction arrays in a transverse magnetic field", Phys.Rev. B33, 6533 (1986)
44.Kosterlitz,J.M. and Granato,E.,"Resistance oscillations in a Josephson-junction array in a magnetic field", Phys.Rev. B34, 2026 (1986)
45.Kosterlitz,J.M. and Granato,E.,"Disorder in Josephson junction arrays", Physica B152, 62 (1988)
46.Granato,E.,Kosterlitz,J.M.,"Disorder in Josephson junction arrays in a magnetic field", Phys.Rev.Lett. 62(7), 823 (1989)
47.Granato,E. and Kosterlitz,J.M.,"Superconductor-Insulator transition and universal resistance in Josephson-junction arrays in a magnetic field", Phys.Rev.Lett. 65(10), 1267 (1990)
48.Granato,E.,"Phase transitions in Josephson-junction ladders in a magnetic field", Phys.Rev. B42, 4797 (1990)
49.Ala-Nissila,T., Granato,E. and Ying,S.C.,"Anomalous diffusion near the flux lattice melting transition in high-T_c superconductors", J.Phys.-Condens.Matter 2, 8537 (1990)
50.Kishore,R. and Granato,E.,"Hubbard Model for metal non-metal transition in high-T_c superconductors", J.Phys.-Condens.Matter 2, 5633 (1990)
51.Aliaga,J.,Cerdeira,H.,Proto,A.N. and Otero,D.,"Information-Theoretical Approach to Josephson Tunneling", Phys.Rev. B40, 4375 (1989)
52.da Silva,E.Z. and Gyorffy,B.L.,"On the fluctuation magnetoconductivity of an anisotropic dirty superconductor", J.de Physique C6, Supl.au#8,tome 39, 494 (1978)
53.da Silva,E.Z. and Gyorffy,B.L."Theory of magnetoconductivity induced by superconducting fluctuations in anisotropic bulk and fibrous samples", Phys.Rev. B20, 147 (1979)
54.Barret,P., Herrenden-Harker,W.G., Priestley,M.G. and da Silva,E.Z., "Superconducting fluctuation conductivity in $(SN)_x$: temperature and field dependence", J. of Physics F11, 873 (1981)

55.da Silveira,H.V. and Cerdeira,H.,"Conductivity of Superconducting Chalcogenides",Solid State Commun. 49, 1053 (1984)
56.da Silveira,H.V.,"Estudo de Propriedades Supercondutoras em Alguns Semicondutores Multivales", Ph.D. Thesis,UNICAMP,Campinas,SP (1986)
57.Brand,H.R and Doria,M.M.,"Theory of how to distinguish a scalar from a tensor order parameter in the high-T_c superconductors", Phys.Rev. B37, 9722 (1988)
58.Campbell,L.J., Doria,M.M. and Kogan,V.G.,"Vortex lattice structures in uniaxial superconductors", Phys.Rev. B38, 2439 (1988)
59.Doria,M.M. and Borges,H.A.,"Spin-singlet pairing mechanism in magnon-pairing theories", Phys.Rev. B38, 4999 (1988)
60.Doria,M.M., Gubernatis,J.E. and Rainer,D.,"Virial theorem for Ginzburg -Landau theories with potential applications to numerical studies of type-II superconductors", Phys.Rev. B39,9573 (1989)
61.Doria,M.M., Gubernatis,J.E. and Rainer,D.,"A virial theorem and the method of simulated annealing in the Ginzburg-Landau theory", Physica C162-164, 1521 (1989)
62.Doria,M.M., Brand,H.R. and Pleiner,H.,"Possible explanation for the curvature in the upper critical field for ceramic superconductors", Physica C159, 46 (1989)
63.Doria,M.M. and Brand,H.R.," Spatial modulation of the order parameter in the T* phase of ceramic superconductors", Physica C165, 457 (1990)
64.Latgé,A., Anda,E. and Morán-Lopez,J.L.,"Effects of correlation and disorder on the phase diagram of $YBa_2Cu_3O_{6+x}$", Phys.Rev. B42, 2543 (1990)
65.Latgé,A., Anda,E. and Morán-Lopez,J.L.,"Oxygen-disorder and correlation effects on the hole concentration and metal-nonmetal transition in $YBa_2Cu_3O_{6+x}$", Phys.Rev. B42, 4288 (1990)
66.Lara,G.A. and Cabrera,G.G.,"Covalent bond pairing in an extended Hubbard model with correlated hopping: Interplay of antiferromagnetism and superconductivity",Solid State Commun. 76, 1121 (1990)
67.Lerner,E.,Ph.D. Thesis, Ohio State University,USA (1966)
68.Lerner,E. and Daunt,J.G.,"Thermal and Electrical Conductivities of Mo-Re Alloys in the Superconducting and Normal States", Phys.Rev. 142, 251 (1966)
69.Lerner,E., Daunt,J.G. and Maxwell,E.,"Magnetic Properties of Superconducting Mo-Re Alloys", Phys.Rev. 153, 487 (1967)
70.Pereira,R.F.R., Master Thesis, UFRJ, RJ (1978)
71.da Silveira,M.F., Pereira,R.F.R. and Meyer,E.,"Produção de Nb_3Au por descarga de condensadores", Rev.Bras. de Fis. 12, 867 (1982)
72.da Silveira,M.F., Master Thesis, UFRJ, RJ (1982)
73.Pereira,R.F., Meyer,E. and da Silveira,M.F.,"A simple technique for measuring the superconducting critical temperature for small (>10 g) samples", Rev.Sci.Instrum. 54, 899 (1983)
74.Barreto,J., Mattievich,E., Pereira,R.F.R., da Silveira,M.F., Gonçalves,Z.C. and dos Santos,G.R.,"Charged particle damage studies of $YBa_2Cu_3O_{7-\delta}$", J.Mat.Sci.Lett. 8, 689 (1989)
75.Pinatti,D.G.,"The superconductivity of Molybdenum", Ph.D. Thesis, Rice University, Houston-TX, USA (1969)
76.Boboshko,N.,"Movimento de Fluxo em Supercondutores", Master Thesis, UNICAMP,Campinas,SP (1973)

77.Boboshko,N.,"Dissipação de Energia Devida ao Movimento de Fluxo Magnético em Supercondutores",Ph.D.Thesis,UNICAMP,Campinas,SP (1976)
78.Boboshko,N. and Pinatti,D.G.,"Temperature dependence of energy dissipation in type I superconductors", Solid State Commun. 26, 575 (1978)
79.Boboshko,N.,"Analysis of flux-flow curves in superconductors as a function of temperature", Solid State Commun. 27, 25 (1978)
80.Moehlecke,S.,"Hidrogenio no Nióbio e sua Supercondutividade", Master Thesis, UNICAMP,Campinas,SP (1974)
81.Sweedler,A.R., Cox,D.E., Moehlecke,S., Newkirk,L.R. and Valencia, F.A.,"Superconductivity and Phase Stability of Nb^3Ge", J.Low Temp.Phys. 24, 645 (1976)
82.Sweedler,A.R., Moehlecke,S., Jones,R.H., Viswanathan,R. and Johnston,D.C., "Superconductivity of Neutron Irradiated Mo_3Os", Solid State Commun. 21, 1007 (1977)
83.Moehlecke,S., Cox,D.E. and Sweedler,A.R.,"The effect of composition and order on the superconducting properties of the A15 phase on Nb-Pt", Solid State Commun. 23, 703 (1977)
84.Dew-Hughes,D., Moehlecke,S. and Welch,D.O.,"Recovery of T_c by annealing of irradiated A15 compounds", J. of Nuclear Materials 72, 50 (1978)
85.Sweedler,A.R., Cox,D.E. and Moehlecke,S.,"Neutron irradiation of superconducting compounds", J. of Nuclear Materials 72, 50 (1978)
86.Moehlecke,S., Cox,D.E. and Sweedler,A.R.,"The effects of neutron irradiation of the superconducting properties of Nb_3Pt", J.Less-Common Met. 62, 111 (1978)
87.Moehlecke,S., Sweedler,A.R. and Cox,D.E.,"Effects of commposition and neutron irradiation on the superconducting porperties of Nb_3Al", Phys.Rev. B21, 2712 (1980)
88.Foner,S., Mcniff Jr.,E.J., Moehlecke,S. and Sweedler,A.R.,"Effect of composition on the upper critical field on V-Ga", Solid State Commun. 39, 773 (1981)
89.Moehlecke,S, Borges,H.A., Santos,C. and Sweedler,A.R.,"Resistivity as a function of composition in the superconducting Nb-Pt A15 phase", J.Low Temp.Phys. 70, 449 (1988)
90.Scarminio,J. and Moehlecke,S.,"Homogeneity range of the superconducting Chevrel phase $SnMo_6S_8$", Phys.Lett. 134A, 75 (1988)
91.Moehlecke,S. and Andrade,M.C.,"Effects of heat treatment on the superconducting Bi-Ca-Sr-Cu-O compounds", in: Progr.in High Temp. Supercon. 9 , ed. Nicolsky et al.(World Sci. Publ.Co.,Singapore, 1988) p.458
92.Zerweck,G.,"The shear modulus of the flux line lattice in low-k type-II superconductors", Phys.Stat.Sol.(b) 78, K15 (1976)
93.Zerweck,G.,"Magnetic Hysteresis and Flux Profiles in Type-II Superconductors", Rev.Bras.Fis. 8, 225 (1978)
94.Zerweck,G.,"On pinning of superconducting flux lines by grain boundaries", J.Low Temp.Phys. 42(1/2), 1 (1981)
95.de Lima,O.F.,"Perfis da Densidade de Fluxóides em Monocristais de Nióbio no Estado Supercondutor", Master Thesis, UNICAMP,Campinas,SP (1977)

96.de Lima,O.F., Zerweck,G. and Brito,A.S., "Field History Effect of the Flux Gradient in Superconducting Niobium", J. Low Temp.Phys., 32(5/6), 717 (1978)
97.Brito,A.S., Zerweck,G. and de Lima,O.F.,"High Critical Flux Density Gradients Near the Surface of Superconducting Niobium", J.Low Temp. Phys. 36(1/2), 33 (1979)
98.de Castro,J.L.,"Distribuição do campo magnètico no interior de supercondutores do tipo-II", Master Thesis,UNICAMP,Campinas,SP (1986)
99.de Lima,O.F.,"Fios Supercondutores Multifilamentares de Nb_3Sn",Ph.D. Thesis, UNICAMP,Campinas,SP (1981)
100.de Lima,O.F.,"Fios Supercondutores Multifilamentares de Cu-Nb-Al e Cu-Nb_3Sn Obtidos por um Novo Mètodo", Rev. Fís. Apl. e Instr., 1, 1 (1985)
101.de Lima,O.F. and de Rezende,A.T.,"Superconducting wires of Cu-Nb_3Sn: The Liquid Phase Sintering Method", J.Mat.Sci., 21, 2855 (1986)
102.de Lima,O.F. and de Rezende A.T.,"Liquid Phase Sintering and the use of Nb-H powder for preparing Cu-Nb_3Sn wires", IEEE Trans.on Magn. 23(2), 968 (1987)
103.de Rezende,A.T.,"Fios supercondutores de Cu-Nb_3Sn preparados pelo mètodo da Sinterização em Fase Liquida de Cobre com o uso de Nb-H", Master Thesis, UNICAMP,Campinas,SP (1984)
104.de Lima,O.F., Krehl,M. and Schulze,K.,"Wetting characteristics of copper on niobium", J.Mat.Sci., 20, 2464 (1985)
105.de Lima,O.F., Mattson,J., Sowers,C.H. and Brodsky,M.B.,"Superconducting Thin Films Based on $La_{2-x}Sr_xCuO_4$", Appl.Phys.Lett. 51, 355 (1987)
106.de Lima,O.F., Jardim,R.F., Gama,S. and Torriani,I.,"Oxygen kinetics and superconductivity in the high-T_c oxide $YBa_2Cu_3O_x$", IEEE Trans.on Magn. MAG-25(2), 2171 (1989)
107.de Lima,O.F., Martins,A.C.G., Torriani,I and de Campos,S.P., "Superconducting Transition for Y-Ba-Cu-O samples aged and measured under oxygen", in: Progress in High Temp. Supercon. 25, ed. R. Nicolsky (World Sci.Publ.,Singapore, 1990) p.507
108.Rodrigues Jr.,D. and Pinatti,D.G.,"Multifilamentary Cu-Nb_3Sn Superconductor wires: Fabrication, Characterization and Theoretical Analysis Using Collective Pinning Theories",in: Progress in High Temp.Supercon. 25 ed.R.Nicolsky (World Sci.,Singapore, 1990)p.757
109.Baldan,C.A. and Shigue,C.Y.,"Fabrication program of NbTi and Nb_3Sn superconducting magnets for mineral separation and high magnetic field generation",in: Progr.in High Temp.Supercon.25,ib.,765 (1990)
110.Bormio,C., Ramos,M.J. and Pinatti,D.G.,"Nb-46.5at%Ti EB-Melting for AC and DC Superconducting Applications",in: Progr.in High Temp. Supercon. 25,ib.,772 (1990)
111.Missell,F.P., Oliveira Jr.,N.F. and Shapira,Y.,"Absence of the influence of Puli spin paramagnetism on H_{c2} for a Ti-11at%Nb alloy", Phys.Status Solid (b) 67, K11 (1975)
112.Ortiz,W.A. and Missell,F.P.,"Superconducting properties of Ta-19at%Zr and Ta-29at%Zr", Phys.Status Solid (a) 35, K135 (1976)
113.Missell,F.P., Oliveira Jr.,N.F., and Shapira,Y.,"Ultrasonic attenuation in the mixed state of Nb-26at%Hf", Solid State Commun. 18, 1553 (1976)

114.Missell,F.P., Oliveira Jr. N.F., and Shapira,Y.,"Magnetoacoustic attenuation in the mixed-state of Nd-26at%Hf", Phys.Rev. B14, 2255 (1976)
115.Missell,F.P.,"Ultrasonic velocity in the mixed state of Nb-26at%Hf", Phys.Rev. B19, 1322 (1979)
116.Salém-Sugui Jr.,S.,"Atenuação do som na liga supercondutora V-42at%Ti", Master Thesis, IFUSP,S.Paulo,SP (1979)
117.Salém-Sugui Jr.,S. and Missell,F.P.,"Ultrasonic attenuation in the mixed state of V-42at%Ti", J.Low Temp.Phys. 37, 257 (1979)
118.Missell,F.P., and Salém-Sugui Jr.,S.,"Ultrasonic velocity in the mixed state of V-42at%Ti", J.Phys.F 10, 1213 (1980)
119.Shapira,Y. and Neuringer,L.J., Phys.Rev. 154, 375 (1967)
120.Alpher,R.A. and Rubin,R.J., J.Acoust.Soc.Am. 26, 452 (1954)
121.Missell,F.P. and Frota-Pessôa,S.,"Pressure dependence of the superconducting transition temperature of a $(Mo_{0.6}Ru_{0.4})_{86}B_{14}$ metallic glass", Solid State Commun. 40, 379 (1981)
122.Bindilatti,V.,"Efeito da pressão hidrostática sobre a supercondutividade em ligas amorfas de Lantanio", Master Thesis, IFUSP, S.Paulo,SP (1981)
123.Bindilatti,V., Missell,F.P. and Tenhover,M.,"Pressure dependence of the superconducting transition temperatures of amorphous La-based alloys", Physica 107B-C, 391 (1981)
124.Missell,F.P., Frota-Pessôa,S., Wood,J., Tyler,J. and Keem,J.E., "Electronic density of states in amorphous zirconium alloys", Phys.Rev. B27, 1596 (1983)
125.Missell,F.P., Bergeron,R., Keem,J.E. and Ovshinsky,S.R.,"Comparison of amorphous melt-spun and vapor-quenched Zr-Rh alloys", Solid State Commun. 47, 177 (1983)
126.Missell,F.P. and Keem,J.E.,"Electronic density of states in amorphous Zr-Pd and Zr-Ni alloys", Phys.Rev. B29, 5207 (1984)
127.Otubo,J.,"Fabricação de fios supercondutores multifilamentares de Nb-Ti", Master Thesis, IFUSP,S.Paulo,SP (1986)
128.Vellego,G.,Master Thesis, IFUSP,S.Paulo,SP (1988)
129.Vellego,G., Bindilatti,V. and Oliveira Jr.,N.F.,"System for critical current measurements on Nb-Ti superconducting wires", Rev.Fis.Apl. e Instr. 4(2), 172 (1989)
130.Otubo,J., Pourrahimi,S., Zhang,H., Thieme,C.L.H. and Foner,S., "Submicron multifilamentary high performance Nb_3Sn produced by powder metallurgy processing of large powders", Appl.Phys.Lett. 42(5), 469 (1983)
131.Otubo,J., Pourrahimi,S., Thieme,C.L.H., Zhang,H., Schartz,B.B. and Foner,S.,"Submicron filament multistrand powder metallurgy processed Cu-Nb-Sn wire", IEEE Tran.Magn. MAG-19, 764 (1983)
132.Zhang,H., Pourrahimi,S. Otubo,J., Thieme,C.L.H., Schwartz,B.B. and Foner,S.,"Tin core processing of powder metallurgy Cu-Nb-Sn", ib., MAG-19, 769 (1983)
133.Bruno,A.C. and Costa Ribeiro,P., Cryogenics 23, 346 (1983)
134.Bruno,A.C., Costa Ribeiro,P., von der Weid,J.P. and Symko,O.G., "Discrete spatial filtering with SQUID gradiometers in biomagnetism", J.Appl.Phys. 59, 2584 (1986)
135.Bruno,A.C. and Costa Ribeiro,P.C.,"Digital filter design approach for SQUID gradiometers", J.Appl.Phys. 63, 2820 (1988)

136.Costa Ribeiro,P.C., Paulsen,C.C. and Symko,O.G.,"Spatial Fourier-transform method for evaluating SQUID gradiometers", Reev.Sci.Instr. 58(8), 1510 (1987)
137.Costa Ribeiro,P., Williams,S.J.W. and Kaufman,L.,"SQUID arrays for simultaneous magnetic measurements: Calibration and source localization performance", IEEE Biomed. 35(7), 551 (1988)
138.Bruno,A.C. and Costa Ribeiro,P.,"Spatial deconvolution algorithm for superconducting planar gradiometer arrays", IEEE Trans.on Magn. MAG-25(2), 1219 (1989)
139.Borges,H.A., Kwok,R., Thompson,J.D., Wells,G.L., Smith,J.L.,Fisk, Z. and Peterson,D.E.,"Comparison of the pressure dependencies of T_c in 90K superconductors $ReBa_2Cu_3O_x$ (Re=Gd,Er, and Yb) and$YBa_2Cu_3O_x$", Phys.Rev. B36, 2404 (1987)
140.Borges,H.A., Wells,G.L., Cheong,S-W., Kwok,R.S., Thompson,J.D., Fisk,Z., Smith,J.L. and Oseroff,S.,"Resistivity of $EuBa_2(Cu_{1-x}Zn_x)_3O_y$ as a function of temperature, magnetic field, pressure and Zn concentration", Physica A&B, 411 (1987)
141.Fisk,Z., Borges,H.A., McElfresh,M., Smith,J.L., Thompson,J.D., Ott,H.R., Aeppli,G., Bucher,E., Lambert,S.E., Maple,M.B., Broholm,C. and Kjems,J.K.,"Superconductivity of heavy-electron uranium compounds", Physica C153-155, 1728 (1988)
142.Borges,H.A., Thompson,J.D., Aronson,M.C., Smith,J.L. and Fisk,Z., "Electrical resistivity of $U_{1-x}M_xBe_{13}$ under pressure", J.Magn.Magn.Mat. 76&77, 235 (1988)
143.Pureur,P. and Schaf,J.,"Electrical and magnetic properties of the high-T_c oxide superconductor $YBa_2Cu_3O_7$", J.Magn.Magn.Mat. 69, L215 (1987)
144.Pureur,P., Schaf,J., Kunzler,J.V. and Fraga,E.R.,"Breaking of ergodicity in high-T_c superconductor $YBa_2Cu_3O_7$", Solid State Commun. 66, 931 (1988)
145.Pureur,P., Schaf.J. and Kunzler,J.V.,"Electric and Magnetic measurements on the high-T_c oxide superconductors $YBa_2Cu_3O_7$, $EuBa_2Cu_3O_7$ and $GdBa_2Cu_3O_7$", in: Progress in High Temp. Supercon. 9, ed. R.Nicolsky et al.(World Sci.,Singapore, 1988) p.137
146.Schaf,J., Pureur,P. and Kunzler,J.V.,"Glass behavior and the H-T phase diagram of the high-T_c ceramic superconductors $YBa_2Cu_3O_7$, $EuBa_2Cu_3O_7$, and $GdBa_2Cu_3O_7$", Phys.Rev. B40, 6948 (1989)
147.Pureur,P., Schaf,J., Gusmão,M.A. and Kunzler,J.V.,"Inhomogeneous superconductivity in polycrystalline $YBa_2Cu_3O_x$: Resistive transition and fluctuation conductivity", in: Progress in High Temp. Supercon. 25, ed. R.Nicolsky (World Sci.Publ.Co., Singapore, 1990) p.236
148.Schaf,J., Pureur,P. and Kunzler,J.V.,"Field induced magnetic crossover in the BiSrCaCuO system", in: Progress in High Temp. Supercon. 25, ed.R.Nicolsky (World Sci.,1990) p.263
149.Jardim,R.F., Gama,S., de Lima,O.F. and Torriani,I.,"Effect of Mn on the Superconductivity of $YBa_2Cu_3O_7$", Phys.Rev. B38, 4580 (1988)
150.Jardim,R.F., Gama,S., de Lima,O.F. and Zacarelli,S.,"Microstructure and Superconducting Fraction in the $YBa_2(Cu_{1-x}Mn_x)_3O_{7-\delta}$", Solid State Comm., 68(9), 835 (1988)
151.Jardim,R.F., Gama,S., de Lima,O.F. and Torriani,I.,"Substitution of Mn for Cu in the High-Tc Superconductor $YBa_2Cu_3O_7$", IEEE Trans.on Magn.,MAG-25(2), 2307 (1989)

152.Jardim,R.F.and Gama,S.,"Enhanced Grain Growth in $YBa_2(Cu_{1-x}Mn_x)_3O_{7-\delta}$ Compounds", Physica C 159, 306 (1989)
153.Jardim,R.F.,"Efeitos da substituição do Cu por Mn no composto $YBa_2Cu_3O_{7-\delta}$", Ph.D. Thesis, UNICAMP,Campinas,SP (1989)
154.Jardim,R.F., Gama,S., Bento,A.C. and Machado,F.L.A.,"Phonon mean free path evaluation in high-T_c oxides", in: Progress in High Temp. Supercon. 25, ed. R.Nicolsky (World Sci.,Singapore, 1990) p.590
155.Jardim,R.F., Westphal,C.H., Becerra,C.C., Paduan-Filho,A. and Oliveira Jr.,N.F.,"Magnetic properties of $(Nd_{1-x}Gd_x)_2CuO_4$ solid solution", in:Progress in High Temp. Supercon. 25, ed.R.Nicolsky (World Sci.,Singapore, 1990) p.183
156.Baggio-Saitovitch,E., Azevedo,I.S., Socorzelli,R.B., Saitovitch,H., da Cunha,S.F., Guimarães,A.P., Silva,P.R. and Takeuchi,A.Y.,"^{57}Fe Mössbauer study of the superconductor $YBa_2(Fe_xCu_{1-x})_3O_7$", Phys.Rev. B37, 7967 (1988)
157.Baggio-Saitovitch,E.,"Mössbauer Studies of ^{57}Fe Substitution of Cu Ions in Superconducting Oxides",in: Progress in High Temp.Supercon. 9, ed.R.Nicolsky et al.(World Sci.,Singapore, 1988) p.246
158.Baggio-Saitovitch,E., Azevedo,I.S. and Scorzelli,R.B.,"Mössbauer studies of ^{57}Fe: $YBa_2Cu_3O_{7-\delta}$ at high temperatures", Trans.Jpn.Inst.Met. 29, 569 (1988)
159.Baggio-Saitovitch,E., Scorzelli,R.B., Azevedo,I.S. and Micklitz,H., "Dynamical information for Fe-doped $YBa_2Cu_3O_{7-\delta}$ superconducting oxides obtained from high-temperature Mössbauer studies", Phys.Rev. B41, 2103 (1990)
160.Baggio-Saitovitch,E., Scorzelli,R.B., Azevedo,I.S. and dos Santos,C.A.,"On the mechanism of oxygen desorption in Fe:Y-Ba-Cu-O superconducting oxides", Solid State Commun. 74, 27 (1990)
161.Baggio-Saitovitch,E. Litterst,F.J., Azevedo,I.S. and Scorzelli, R.B.,"Magnetic ordering of $YBa_2(Cu_{1-x}Fe_x)_3O_7$", Hyp.Int. 50, 529 (1989)
162.Baggio-Sitovitch,E., Azevedo,I.S. and Scorzelli,R.B.,"Magnetic phase present in $YBa_2(Cu_{1-x}Fe_x)_3O_{6+\delta}$ oxides", Hyp.Int. 50, 521 (1989)
163.Baggio-Saitovitch,E., Scorzelli,R.B., Azevedo,I.S., dos Santos,C.A. and Litterst,F.J.,"^{57}Fe Mössbauer studies on semiconducting and superconducting La-Ba-Cu-O", Physica C166, 518 (1990)
164.Mattievich,E., Moreira,L.F., da Silveira,M.F., Pereira,R.F.R, de Amorim,H.S., Amaral Jr.,M.R. and Meyer,E.,"Mössbauer results of the $YBa_2Cu_3O_{7-\delta}$ superconductor doped with 1% ^{57}Fe", Hyp.Int. 42, 1254 (1988)
165.Rettori,C., Davidov,D., Belaish,I. and Felner,I.,"Magnetism and critical fields in the high-T_c superconductors $YBa_2Cu_3O_{7-x}S_x$ (x=0,1): An ESR study", Phys.Rev. B36, 4028 (1987)
166.Castilho,J.H., Venegas,P.A., Barberis,G.E., Rettori,C., Jardim,R.F. and Gama,S.,"Evidence for dynamic effects in the ESR spectra of the high-T_c superconductor $YBa_2Cu_3O_7$", Solid State Commun. 64(7), 1043 (1987)
167.de Mesquita,R.N., Castilho,J.H., Barberis,G.E., Rettori,C., Torriani,I., de Lima,O.F., Gama,S., Jardim,R.F., Terrile,M.C., Basso,H.C. and Nascimento,O.R.,"ESR experiments and spectra simulations in $YBa_2Cu_3O_{7-\delta}$", Phys.Rev. B39, 6694 (1989)

168.Golosovsky,M., Davidov,D., Rettori,C. and Stern,A.,"Magnetic field modulation effects on the microwave transmission through superconducting thin films of Y-Ba-Cu-O", Phys.Rev. B40, 9299 (1989)
169.de Biasi,R.S., Fernandes,A.A.R., Ferreira,C.L., Pascual,R. and Saavedra,A.,"Electron Spin Resonance of $YBa_2Cu_3O_{9-\delta}$ in the normal and superconducting states", J.Mater.Sci.Lett., 7(3), 263 (1988)
170.de Biasi,R.S., Fernandes,A.A.R., Ferreira,C.L., Pascual,R. and Saavedra,A.A.,"Electron Spin Resonance in the tetragonal phase of $YBa_2Cu_3O_{9-\delta}$", J.Mater.Sci.Lett. 7(8), 819 (1988)
171.de Biasi,R.S. and Fernandes,A.A.R.,"Low-field microwave absorption in the Bi-Sr-Ca-Cu-O system", J.Mater.Sci. 25, 54 (1990)
172.de Biasi,R.S., Fernandes,A.A.R. and Pereira,R.F.R.,"Temperature dependence of the microwave absorption in the Y-Ba-Cu-O and Bi-Sr-Ca-Cu-O system", J.Mater.Sci.Lett. 9(8), 919 (1990)
173.de Biasi,R.S., D'Almeida,F.M.R., Brisola,R.S. and Fernandes,A.A.R., "Critical temperature in $Y_xGd_{1-x}Ba_2Cu_3O_y$ solid solutions", in: Progress in High Temp.Supercon. 25, ed.R.Nicolsky (World Sci., Singapore, 1990) p.645
174.de Aguiar,J.A.O., Monovsky,A.A., van den Berg,J. and Brom,H.B.,"ESR in $YBa_2Cu_3O_{7-\delta}$ single crystals", J.Phys.C: Solid State Phys. 21, L237 (1988)
175.de Aguiar,J.A.O., Menovsky,A.A., van den Berg,J. and Brom,H.B.,"ESR in $YBa_2Cu_3O_{7-\delta}$ crystals", Physica C153-155, 743 (1988)
176.de Aguiar,J.A.O, van den Berg,J., Brom,H.B., Nieuwwenhuys,G.J., Mydosh,J.A., van Berkel,F.P.F. and Zndbergen,H.W.,"EPR on $YBa_2(Cu_{1-x}Fe_x)_3O_{7-\delta}$ ceramics", Physica C156, 571 (1988)
177.Rezende,S.M. and de Aguiar,F.M.,"Nonlinear microwave absorption in ceramic superconducting Y-Ba-Cu-O", Phys.Rev. B39, 9715 (1989)
178.Macêdo,M.A., Machado,F.L.A. and Rezende,S.M.,"Nonlinear microwave absorption in ceramic superconducting Y-Ba-Cu-O at low power levels", in: Progress in High Temp.Supercon. 25, ed.R.Nicolsky (World Sci., Singapore, 1990) p.339
179.Sanjurjo,J.A., López-Cruz,E., Katyar,R.S., Torriani,I. and Rettori,C.,"Raman measurements and lattice dynamics in the high-T_c superconductor $YBa_2Cu_3O_7$", Solid State Commun. 64(4), 505 (1987)
180.Sanjurjo,J.S.,"Raman measurements in $YBa_2Cu_3O_{7-\delta}$:A critical review",in: Progress in High Temp.Supercon. 9, ed.R.Nicolsky et al.(World Sci., Singapore, 1988) p.386
181.Yang,K.N., Dalichaouch,Y., Ferreira,J.M., Lee,B.W., Neumeier,J.J., Torikachvili,M.S., Zhou,H. and Maple,M.B.,"High temperature superconductivity in rare earth (R) barium copper oxides $RBa_2Cu_3O_{9-\delta}$", Solid State Commun. 63, 515 (1987)
182.Maple,M.B., Dalichouch,Y., Ferreira,J.M., Hake,R.R., Lee,B.W., Neumeier,J.J., Torikachvili,M.S., Yang,K.N., Guertin,R.P. and Kuric,M.V.,"$RBa_2Cu_3O_{7-\delta}$ (R=Rare Earth) high T_c magnetic superconductors", Physica 148B, 155 (1987)
183.Neumeier,J.J., Dalichaouch,Y., Ferreira,J.M., Hake,R.R., Lee,B.W., Maple,M.B., Torikachvili,M.S., Yang,K.N. and Zhou,H.,"Thulium barium copper oxide: A 90K superconductor with a potential 1-MG upper critical field", Appl.Phys.Lett. 51, 371 (1987)
184.a) Werthamer,N.R., Helfand,E. and Hohenberg,P.C., Phys.Rev. 147, 362 (1966). b) Maki,K., Phys.Rev. 148, 362 (1966).

185.Ferreira,J.M., Lee,B.W., Dalichaouch,Y., Torikachvili,M.S., Yang,K.N. and Maple,M.B.,"Low temperature specific heat of the high T_c superconductors $La_{1.8}Sr_{0.2}CuO_4$ and $RBa_2Cu_3O_{7-\delta}$ (R=Y,Eu,Ho,Tm, and Yb)", Phys.Rev. B37, 1580 (1988)
186.Lee,B.W., Ferreira,J.M., Dalichaouch,Y., Torikachvili,M.S., Yang,K.N. and Maple,M.B.,"Long-range magnetic ordering in the high T_c superconductors $RBa_2Cu_3O_{7-\delta}$ (R=Nd,Sm,Gd,Dy, and Er)", Phys.Rev. B37, 2368 (1988)
187.Yang,K.N. Ferreira,J.M., Lee,B.W., Maple,M.B., Li,W.H., Lynn,J.W. and Erwin,R.W.,"Antiferromagnetic ordering in superconducting and oxygen deficient nonsuperconducting $RBa_2Cu_3O_{7-\delta}$ compounds (R=Nd and Sm)", Phys.Rev. B40, 10963 (1989)
188.Ferreira,J.M., Maple,M.B., Zhou,H., Hake,R.R., Lee,B.W., Seaman,C.L., Kuric,M.V. and Guertin,R.P.,"Magnetic field alignment of high T_c superconductors $RBa_2Cu_3O_{7-\delta}$ (R=Rare Earth)", Appl.Phys. A47, 105 (1988)
189.Andreeta,J.P., Basso,H.C., Castelhano,E.E., Gallo,J.N.H., Martin, A.A. and Piro,O.E.,"Crystallographic phases and transport properties in the superconducting Tm-Ba-Cu-O compound system", Phys.Rev. B36, 5588 (1987)
190.Piro,O.E., Guida,J.A., Massa,N.E., Aymonino,P.J., Castelhano,E.E., Basso,H.C., Gallo,J.N.H. and Martin,A.A.,"Infrared reflectivity and vibrational structure of superconducting $Bi_2Sr_2CaCu_2O_{8+x}$", Phys.Rev. B39, 7255 (1989)
191.Gallo,N.J., Martin,A.A., Andreeta,J.P., Botura,V.A. and Fernandes, A.C.,"Direct observations of the structural defects in oxide superconductors of high T_c single crystal", J.Mater.Sci.Lett. 8(11), 1262 (1989)
192.Delben,J.R. and Basso,H.C.,"Zinc doping effect on Y-Ba-Cu-O", in: Progress in High Temp.Supercon. 25, ed.R.Nicolsky (World Sci., Singapore, 1990) p.636
193.Balachandran,U., dos Santos,D.I., van Stumberg,A.W., Graham,S.W., Singh,J.P., Youngdahl,C.A., Goretta,K.C., Shi,D. and Poeppel,R.B., "Processing Y- and Bi-based superconductors", X^{th} winter meeting on Low Temp.Phys.-High-T_c Supercon. (World Sci.,Singapore,1989) p.129
194.dos Santos,D.I., Balachandran,U., Guttschow,R.A. and Poeppel,R.B., "Bi-based superconductor prepared by solution technique", J.non-Crystalline Solids, 121(1/3), 448 (1990)
195.Gomes,L., Vieira,M.M.F., Baldochi,S.L., Lima,N.B., Novak,M.A., Vieira Jr.,N.D., Morato,S.P., Braga,A.J.P., Cesar,C.L., Penna,A.F.S. and Filho,J.M.,"Thermal diffusivity of superconducting $YBa_2Cu_3O_{7-\delta}$", J.Appl.Phys. 63(10), 5044 (1988)
196.Silva,E.C., Vargas,H., Bento,A.C., Jardim,R.F., Gama,S., Pinheiro, E.A. and Galembeck,F.,"Photoacoustic characterization of superconducting oxides: Thermal diffusivity",in: Progress in High Temp. Supercon. 9, ed.R.Nicolsky et al.(World Sci.,Singapore, 1988) p.331
197.Orlando,M.T.A., Gomes,L., Morato,S.P., Muccillo,E.N.S. and Muccillo,R.,"AC magnetic susceptibility behavior of BiSrCaCuO high T_c superconductor", in: Progress in High Temp.Supercon. 25,ed.R.Nicolsky (World Sci.,Singapore, 1990) p.188

The Laboratório Nacional de Luz Síncrotron: a Brazilian Synchrotron Light Source

C.E.T. Gonçalves da Silva
Instituto de Física, Universidade Estadual de Campinas
and
Laboratório Nacional de Luz Síncrotron, CNPq.

INTRODUCTION.

Rogério Leite's career as a physicist was centered on the study of the interaction of light with matter. In the early sixties, when he started research work, the laser was a brand new source of electromagnetic radiation, which led to the opening up of whole new fields of investigation. Raman spectroscopy, for instance, was transformed by the laser from a virtuoso experiment on selected materials to one of the most powerful tools for the investigation of elementary excitations in solids.

Synchrotron light sources promise to be as revolutionary as the laser. Although in use for more than twenty years, it was only in the last decade that their full impact on spectroscopic work began to be appreciated. Initially considered an undesirable by-product of the acceleration of electrons (or positrons) in circular accelerators dedicated to high energy physics, the demand for synchrotron radiation grew to the point where large accelerators (1 GeV and more electron energies) were and are specifically designed and built for the production of photons. Their tunability from far infrared to hard X-rays, high flux and brilliance, polarization and time structure make these the ideal sources for all those regions of the spectrum where no lasers are available. It is interesting to note that in the Third International Conference on Light Scattering in Solids, which took place in Campinas in 1975, presided by Sérgio Porto and organized by Rogério, P. Eisenberger[1] was already alerting his colleagues to the immense potential of this new light source, calling it "the "laser" of the 1970's and 1980's".

The large facilities required for synchrotron light sources have brought an additional benefit to science: the cross-fertilization from interactions among researchers from many different disciplines working in a common experimental hall. The somewhat artificial barriers between disciplines, enforced by the environment of the "small science" university laboratories, tend to vanish when the brick walls and the distance separating physicists, chemists, biologists, materials scientists, disappear around the beam lines of a synchrotron light source storage ring. Unlike high energy physics installations, these are, in fact, not Big Science, but Small Science institutions. Although no systematic research has been done on the econometrics of synchrotron laboratories, the indications are that they are more cost efficient than the traditional small departmental research groups[2].

In the early 80's, a group of Brazilian physicists, led by Roberto Lobo, then Director of the Centro Brasileiro de Pesquisas Físicas (CBPF) in Rio, raised the issue of the need of a synchrotron light source in Brazil, as a means to stimulate experimental investigation. This idea was widely discussed with the local scientific community and faced strong opposition on the grounds that Brazil was not ready for "Big Science". When the author first contacted Rogério, to obtain his support, his reaction was also far from enthusiastic, until one crucial argument was given: the source was to be designed and built in Brazil, by a local team[3].

The project, as a matter of fact, had much to please Rogério: a national engineering effort to produce photons of high quality. It hit a resonance. Since 1987, Rogério has served as Chairman of the Board of the Laboratório Nacional de Luz Síncrotron.

The aim of this paper, dedicated to Rogério on his sixtieth birthday, is to make a brief presentation of the project for the Brazilian synchrotron light source.

BRIEF HISTORY OF THE PROJECT.

The first goal of the Laboratory was to build an electron linear accelerator to be used as an injector into the storage ring, the synchrotron light source. In fact, the LINAC was intended to be a "grand graduate school project" for training of the technical staff, since none of its members had any previous experience with the design and construction of particle accelerators. Although some external help was generously available from several laboratories around the world, most of the training was in-house and the initial emphasis on a substantial degree of autonomous development was maintained throughout this period.

The evolution of the LINAC project is also the history of the evolution of LNLS since the end of 1986. The first funds started coming in in November 1986. For the next six months, the main task was that of selecting and hiring the first members of the technical staff as well as setting up a minimal infra-structure. In July 1987, LNLS was able to move into its first, temporary, building. During the next two years, work concentrated on the LINAC. Some components, in particular, the gun, the accelerating sections, and the klystrons, were purchased abroad. The gun and the sections were manufactured by the Institute of High Energy Physics in Beijing, under a collaborative agreement. The other components were purchased from commercial suppliers. Finally, by November 10th, 1989, all the components and the biological protection tunnel were ready. Assembling the LINAC was completed by December 15th. On December 22nd, a memorable evening for those present, the electron beam was observed at the end of the second structure. The first part of the mission had been accomplished. LNLS had demonstrated that it was possible, on a time frame compatible with international standards, starting essentially from scratch, to build a particle accelerator in Brazil.

As mentioned above, for training purposes, LNLS chose to build many parts locally. The following (major) components were built in house:

- Modulators (medium and high power supplies for the klystrons).
- Power supplies for klystron solenoids.
- All magnets and their power supplies.
- Current monitors.
- Control system, including custom hardware.
- RF couplers, phase shifters, attenuators.
- RF detectors, VSWR detectors.
- Temperature stabilization system (± 0.05 K).

The LINAC design characteristics and the results of the commissioning are summarized in the Table below:

Technical specifications:	
SLAC-type TW, disk-loaded, constant gradient, accelerating structures.	
Section length	3.01 m.
Injection energy	80 keV.
Frequency -	2856 MHz.
Mode -	$2\pi/3$
Repetition rate -	0.5 to 33 Hz.
Output energy (2 sections)-	50 MeV.
Energy dispersion (FWHM)-	< 10%.
Macropulse current -	200 mA.
Pulse length -	100 ns.
Results of the commissioning:	
RF Power (total):	25 MW.
Energy:	55 MeV.
Gun current:	880 mA, 105 ns.
End-of LINAC current:	225 mA.
Emittance:	1.4π mm·mrad.
Energy dispersion:	3,5 % @ 51 MeV, 200 mA.

The official switching on of the LINAC took place on February 19th, 1990. Professor Carlo Rubbia, the Director General of CERN, one of the laboratories which most collaborated with LNLS, was present at the ceremony.

On May 14th, 1990, the State of São Paulo officialized the donation of the 380.000 m^2 (93 acre) plot of land for the permanent campus of LNLS. This is located not far from the University of Campinas (UNICAMP). The completion of the first building on campus, initially dedicated to the production of magnets and vacuum chambers, is schedule for June 1991. The Laboratory is expected to move from its present, temporary, installations to its permanent buildings by the end of 1992.

THE UVX ELECTRON STORAGE RING.

The parameters for the storage ring are described in detail in the following paper by Lin *et al.*[3] . The scientific and technological requirements which determined their choice can be summarized by the following:

- Critical photon wavelength of 10 Å.
- Lattice with long straight sections for insertion devices.
- Long electron beam lifetime (>10 hrs.).
- Low energy injection from the LINAC (100 to 200 MeV).
- Conservative magnet design. Small number of magnetic elements.
- Reliable operation of the ring.

The choice of critical photon wavelength was dictated by the wish that the UVX ring be a research tool for materials science. Figure 1 shows the photon flux from the bending magnets, normalized to 100 mA current in the ring. The straight sections allow for the installation of insertion devices, wigglers and undulators, which will add flexibility to the photon spectrum available to the users. This will give the UVX ring a fairly long useful life, well into the next decade. For cost reasons, it was decided to have low energy injection. However, the ring is planned to be operated with full energy injection, once a booster synchrotron is available. In order to minimize the time of construction, a reasonably conservative and "simple" magnet lattice was chosen. Since this is discussed in detail elsewhere [3], we do not dwell on the subject here.

The main reason for the interest in synchrotron light sources can be seen in Figure 1: the broad tunability of the source, from the far infrared to the X-ray region of the electromagnetic spectrum. Other important characteristics are:

- vertical colimation of the beam; its vertical aperture is inversely proportional to the electron energy, being on the order of 0,5 mrad for a 1 GeV electron beam.

- high degree of polarization; the radiation emitted in the plane of the electron orbit is plane polarized. The off-plane radiation has a small vertical polarization.

- time structure of the light pulses; due to the RF acceleration process, the electrons are bunched into packets inside the ring. This characteristic is very much ring dependent. For the UVX ring the light pulse frequency can be varied between 3.88 MHz (single bunch) to about 500 MHz (all bunches filled). The pulse length depends on a variety of factors, including the total current circulating in each bunch. For the UVX ring it is estimated to vary between 30 and 130 ps.

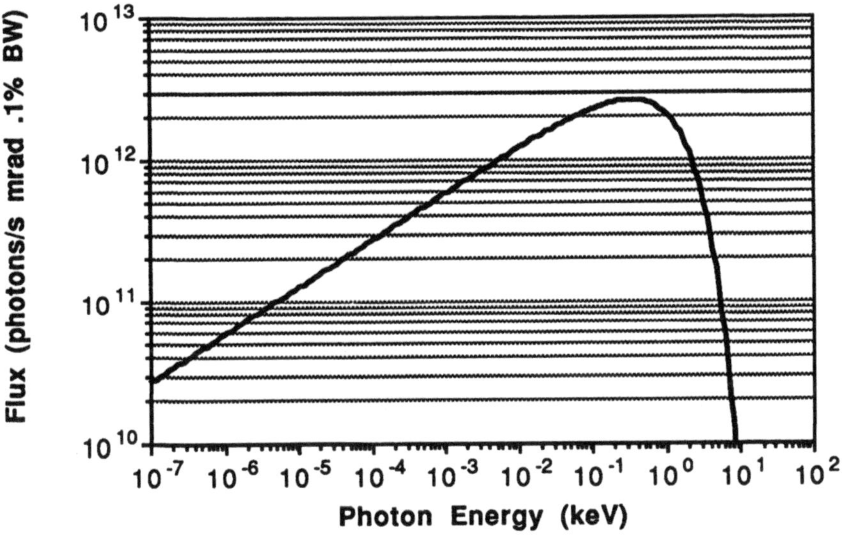

Figure 1 - The spectral photon flux from the UVX ring bending magnets, at 1.15 GeV and 100 mA electron current, integrated over the whole vertical angle, per horizontal milliradian and per 0.1 % bandwidth.

Work is now under way on the prototypes of major components for the storage ring. The magnets will be built from laminated (1.5 mm thickness) low carbon steel cores, in order to minimize eddy currents during ramping, and excited with water cooled copper coils. LNLS opted for the somewhat unconventional laser cutting of the laminations and has purchased a machine for its own use, to be delivered early in 1991. Although this decision was taken because of the costs of alternative stamping processes, it is befitting that light should be used to produce light. Both the steel for the laminations and the copper for the coils are already on order. These materials were developed by local suppliers according to the specifications provided by LNLS, thus further accomplishing one of the aims of the project: to contribute to the improvement of local industrial standards and products. A first mechanical prototype of the dipole will be ready by March '91. Its laminations were specially laser cut for LNLS by a local machine tool manufacturer. Combined function quadrupole/sextupole magnets have been chosen for the storage ring and are currently being designed. The power supplies of the magnets, in particular that of the dipoles, have very stringent technical requirements in terms of stability and ripple. Prototypes are being developed at LNLS for these supplies. The dipole power supply is a 240 kW (800 V @ 300 A or 400V @ 600 A), with long term (24 hr.) current stability less than 10^{-4} full scale and ripple on the same order. A satisfactory prototype is already in operation. The low power RF system is operational, but the 60 kW CW klystron and its power supply have not yet been ordered. The 500 MHz RF cavities have been donated to LNLS by DESY. There are two vacuum chamber prototypes being built: one in aluminum and the other in stainless steel. Their performance, as well as the availability of the components in Brazil, will determine the final choice of material.

THE SCIENTIFIC CASE.

One important characteristic of the radiation from storage rings is its brilliance (or brightness, as some prefer), the photon flux per unit solid angle per unit source area. The brilliance is determined by the quality of the electron optics of the ring and is the figure of merit for those experiments which require a small angular divergence and small irradiated area, such as the novel spectromicroscopies, with high lateral resolution (aiming at sub-micron resolutions). Figure 2 shows the brilliance for one of the modes of operation of the UVX ring.

Although the LNLS ring will not be competitive with comparable energy rings being built at Berkeley and Trieste, which aim at brilliances one order of magnitude higher, it will be on a par with the Louisiana State

University and Taiwan machines, due to come into operation roughly at the same time as the UVX ring. Most importantly, though, it will provide the training grounds for a generation of accelerator builders and synchrotron radiation users in Brazil and Latin America. The next generation will have a very solid foundation upon which to move forward.

LNLS is being designed to be a national laboratory. This means that most of its users will come from outside, from Brazilian and foreign institutions. The relatively long time delay between the beginning and the completion of the project has not been a drawback. There is a need to train not only the machine builders, but also the future users. With the parameters of the storage ring known well in advance, the users' community can direct its effort in the right direction. This requires

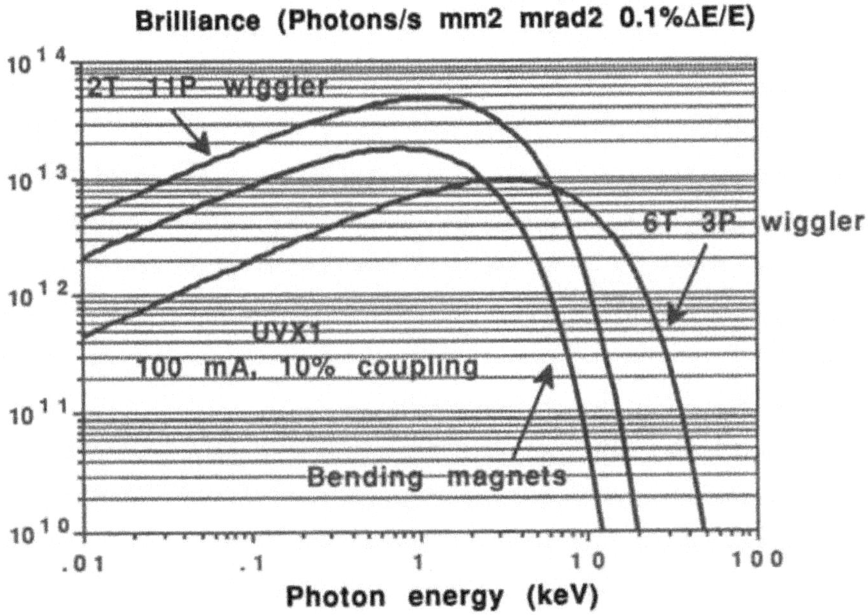

Figure 2 - Brilliance of the "normal" mode of operation of the UVX ring, calculated for radiation from the bending magnets; from a 2 Tesla, conventional magnets, 11 pole wiggler; and for a 6 Tesla superconducting wavelength shifter. The UVX ring has a second mode of operation (not shown), for which the brilliance can be increased by a factor of 2.

a substantial amount of planning from all concerned. To successfully achieve the goal of planning. building and operating a large research facility, in a developing country unused to projects of such magnitude for science, is one of the least visible, but more fundamental strategic objectives of LNLS.

The Brazilian user community met with experienced users from abroad during two Workshops organized in Campinas in July 1988 [4] and September 1989 [5]. These events were useful for defining the first stations to be installed in the UVX ring.

At this moment, the planned beam lines are shown on Table 1.

Beam line	Monochromator	Spectral range	Workstations
BM1	TGM (3 gratings)	10-300 eV	Photoelectron Spectroscopy, Molecular Spectr., Surface Physics
BM2	SEM, high resolution	200-1000 eV	Molecular Spectroscopy, Surface Physics
BM3	Two-crystal UHV	600 eV - 10 keV	X-Ray Spectr., EXAFS, X-Ray fluorescence
BM4	Two-crystal RV	1.5-10 keV	EXAFS, Perfect crystal crystallography, Powder diffraction, X-Ray fluorescence
BM5	One crystal, curved,high flux	5 -9 keV	Small Angle X-Ray Scattering (SAXS), Diffuse scattering
BM6	None		Microlithography

Table 1 - Planned beam lines for the UVX ring initial operation.

The construction of the instrumentation for use of synchrotron radiation has already begun both by in-house staff and by external users. LNLS assumes responsibility for the construction of the front-ends, i.e., for the transfer of photons from the vacuum chamber to the experimental stations. Obviously this implies a fair amount of coordination between the users' needs and the engineering to be provided by LNLS. which is done through the recently organized LNLS Users' Group. The construction of the first prototype front-end is well advanced and should be ready by the end of the first semester of 1991.

CONCLUSIONS.

Within a few years, in time for Rogério's 65th birthday, Brazilian and Latin American physicists, chemists, biologists, and materials scientists will have available high quality photons from the UVX storage ring. This will open new research opportunities in the old problem of radiation-matter interaction and create a powerful stimulus to the development of experimental science in this area of the world. After all, for the moment being, LNLS is the only planned synchrotron light source South of the Equator. The technical staff of LNLS will be ready to tackle greater challenges in the design and construction of ever more efficient sources of electromagnetic radiation.

The schedule for LNLS, over the next few years is shown below:

1991

Finalize prototype work for UVX ring.
Start construction on final site.
Install production facilities.
Start production of main components.
Standard front-ends in production.
Three monochromators ready.
Development of EXAFS and SAXS stations.

1992

End production of main components.
Characterization of magnets.
Begin assembling 100 MeV LINAC.
Begin assembling transport line.
Continuing work on experimental stations.

1993

100 MeV LINAC commissioning.
Transport line commissioning.

Begin assembling UVX ring.
First external users' experimental stations.

1994

Commissioning of UVX ring.
Begin development of insertion devices.
First users come in.

1995

Routine operation of UVX ring.
Begin development of booster synchrotron.
Install first insertion device.

Acknowledgements - The author would like to thank the staff of LNLS for the continuing discussions which helped to introduce a theoretical condensed matter physicist to the mysteries of design and operation of accelerators and synchrotron light sources. He would also like to express his thanks to Roberto Lobo, who, with infinite patience, through what seemed to be endless discussions with the Brazilian scientific community, laid the foundations for LNLS in the early 80's.

---

[1] P. Eisenberger, in Proc. of the Third International Conference on Light Scattering in Solids, M. Balkanski, R.C.C. Leite, and S.P.S. Porto (eds.), Flammarion ,1976, p. 959.

[2] Y. Petroff, private communication.

[3] To this day, the author has the strong feeling that Rogério did not quite believe the source could be built as the original proponents foresaw it. However, it is to his credit that he strongly supported the project, because he believed that, even if they failed to achieve their aims in the proposed time frame, the net gain to the country would be positive.

[4] Synchrotron Light: Applications and Related Instrumentation, A. Craievich (ed.), World Scientific, 1989.

[5] Synchrotron Light: Applications and Related Instrumentation II, A. Craievich (ed.), World Scientific, 1990.

A SOLID STATE PARALLEL DETECTION SYSTEM FOR IR, VISIBLE AND X-RAY PHOTONS

A. Rubens B. de Castro (*)
Paulo de Tarso Fonseca (*)
J. Eduardo Verdugo

Laboratório Nacional de Luz Síncrotron
(*) also at
IFGW/UNICAMP
13085 - Campinas, SP - Brasil

ABSTRACT

We describe a modular, solid-state, parallel detection system and its interface to IBM PC's. The detector can be fitted with a cooled image intensifier. As applications, we show Raman spectra obtained with excitation from a small HeNe laser, and laser beam position measurements with .003 mm resolution.

1. INTRODUCTION

One-dimensional position sensitive photon detector systems are useful for spectroscopy in the IR[1], visible, UV[2] and X-ray[3] spectral regions, in addition to applications other than light spectroscopy, such as quantitative microtomography[4]. One of the authors has reviewed the subject recently[5].

We describe the design and characteristics of the system built at LNLS in collaboration with IFGW-UNICAMP. It has been used both with and without an image intensifier, for Raman spectroscopy and as a high precision beam position detector.

It comprises (1) cooled detector head, (2) interface and control electronics, (3) control software, (4) cooled chamber for the image intensifier, (5) shutter assembly and its control electronics, (6) closed-loop water cooling.

In what regards the electronics and computer interfacing, our design is related to that of Parkinson[6], differing in so far as we wish to use the diode array detector over a large dynamic range, even when preceded by an intensifier.

The detector itself is a EGG Reticon RL1024S (or RL512SF)[7]. The intensifier was bought from ITL[8]. Most other components are locally available.

Our design philosophy was to make the system modular, so that the image intensifier or the shutter assembly could be left out or included, depending on convenience, and each part of the system could be modified at a later time, without having to redesign everything.

We now proceed to describe its parts.

2. DESCRIPTION

2.1 Detector head

The physics of focal plane Si diode arrays have been discussed in several places[5]. The RL1024S is a collection of 1024 p-i-n diodes with .025 mm center-to-center spacing and 2.5 mm high. All diodes are initially reverse biased to, say, 5 volts. They behave then as small charged capacitors. The array is then exposed to light, and the photo-induced current discharges the sites in proportion to the total number of photons absorbed in each Si diode. The interface electronics scans the array in order to detect the charge that was lost in each site, storing the value in computer memory.

The detector head is shown in figure 1. The diode array is supported by a printed circuit board providing rigid shielded electric connections from the RL1024 chip to the video amplifiers, from the computer to the RL1024 control lines ("START", "CLOCKS", "RESETS", etc), and from the detector thermoelectric cooler to the cooler control and power supply. Cooling is needed to reduce the thermally-activated electron- hole pair production ("dark current") in the diodes, if long exposure times are desired. We used a 3-stage stack manufactured by Marlow Ind.[9], which lowers the detector array underside temperature to about - 40°C. The heat pumped out of the chamber is dissipated in the water-cooled bottom plate of the chamber.

The detector chamber is not evacuated, but care was taken to make it vacuum tight. In order to prevent moisture condensation under cooled operation, all joints are "O" ring sealed. The internal dead volume is sufficiently small that the water held in the air trapped in the chamber after assembly under normal lab conditions is negligible. Contrarily to some of our previous designs, we have had no moisture problems with these chambers.

2.2 Interface Electronics

The interface electronics includes (i) digital buffers for the scan control lines, (ii) video pre-amplifiers, (iii) intermediate analog and digital processing, (iv) interface board for IBM

PC compatibles, (v) power supply, (vi) control electronics for the detector thermo-electric cooler.

(i) The digital buffers translate the TTL levels available at the PC interface board to CMOS levels (-10, +5 v) required to drive the RL chip. Care was taken to minimize cross-talk between lines, which produces faulty scans and latch-up of the device in undetermined logical states. This in practice requires very short connections between the translation circuits and the RL detector. The buffers are, in consequence, housed in a separate compartment of the same metal box that houses the video pre-amplifiers. The box is bolted to the detector assembly. PCB edge connectors make direct electrical connections to the RL chip printed circuit board in the detector chamber.

(ii) There are separate, identical, video pre-amplifiers for the odd and the even diodes. They use FET differential pairs to take advantage of the fact that the RL series of array detectors provides, in addition to the "video" outputs, also "dummy" outputs not connected to light sensors but subject to the same switching noise present in the video outputs. This noise is cancelled out using a differential circuit. The total voltage gain is 100x. Rise time is less than 4 microsec, but the maximum scan rate is limited by other system components to 45 microsec/sensor. In this more limited bandpass, the (indirectly) measured equivalent input noise is about 1000 electronic charges. This is to be compared to a theoretical lower bound of 700 electronic charges resulting from statistical fluctuation in reseting the sensor sites to the nominal starting charge. Since the noise performance of the RL device is critically dependent on the quality of the bias supply used to precharge all the diodes to some nominal voltage, the pre-amp board includes an active low-pass filter with a roll-off of 40 db/decade, which also insures smooth turn-on and turn-off.

(iii) The intermediate electronics recombines the video signals from even and odd channels after DC restoration. The signal from each channel is then held in a sample-hold circuit during the time needed for analog-digital conversion.

DC restoration is a technique whereby the signal present in the video line immediately before sampling a light sensor is subtracted analogically from the signal present in the video line during sampling[10]. This technique cancels out all of the noise caused by statistical fluctuations in the charge stored in the video lines after nominal reset to zero (which is always done prior to sampling a sensor), and also the low frequency noise in the front-end electronics. The DC restore circuit also allows one to shift the baseline of the "even" and "odd" pixel sequences, so that they overlap each other in the final graphic display of the images stored in memory. The video signal can then be further amplified (10x) if desired by removing a jumper in the board.

The analog-digital conversion is performed with an Analog Devices AD5240KD (12 bits, 5 microsec total conversion time).

All connections to the detector head, to the power supply and to the computer originate in this board.

(iv) The interface board for IBM PC's uses two 8255 PIO LSI buffers and is of conventional design. All peripheral signal lines have open collector buffers for improved noise immunity.

(v) The power supply must provide plus minus 15 volts for the ADC, plus 5 volts, plus minus 10 volts for the RL chip logic drive and for the analogic processing of the video signals. This supply has conventional requirements in what concerns voltage precision and low amplitude noise. It has most stringent requirements regarding absence of spikes on turn-on and turn-off.

Low cost commercial supplies (and even some high priced ones as well) are unsuitable. The supplies we built use transformers with electrostatic shielding between windings; noise suppression RC networks were installed on the power switch; the transformer and the switch were separated from the rest of the supply by enclosing the former in an additional metal shield.

In this way, turn-on and turn-off spikes on any output have been limited to less than one spike with amplitude exceeding 1 volt per 10 trials of turn-on or turn-off. Given our choice of supply voltages and the max ratings for the RL device, the max tolerable spike amplitude would, in our case, be 6 volts, which is easily exceeded in unprotected supplies.

(vi) The detector thermo-electric cooler is a 3-stage Peltier stack Marlow SP-1056[9] which dissipates circa 30 W. The heat is rejected to the water-cooled bottom plate of the detector head. The temperature differential achievable in our detector head is about 60°C.

Two thermistors were installed close to the detector bottom surface. One is used in the closed-loop proportional temperature controller and the second is available for independent temperature measurements.

The proportional temperature controller compares the thermistor resistance with the reference set-point and changes the current in the cooler in proportion to the error. Both voltage and current are independently limited to prevent exceeding the device maximum ratings. The measured temperature drift was .05°C/hour.

The DC output is heavily filtered, and usual practices of noise reduction were followed, as it was found that the TE cooler connections were introducing electrical noise into the video lines, due to their close physical proximity.

2.3 Control Software

A package was prepared covering the following functions: actual chip read-out, video display of spectra, spectra storage (in memory or disk), spectral data processing, data printing, control and status monitoring of accessories.

The routine for actual chip read-out is written in Assembly Language and was optimized for best spectral signal to noise ratio and shortest scan time at a given clock rate. Spectral quality is dependent on the delay between the time the charge is dumped on the video line and the time it is sampled and "DC restored". Read-out time is 45 microsec/sensor, with clock running at 4.77 MHz.

The user is involved only with a Pascal program that is menu driven. It can be easily altered to suit specific applications, if it is deemed that the resources already available are inadequate. These resources allow signal averaging, image background automatic subtraction, exposure time control, storage of images on disk or on provisory image buffers A B C D, retrieval of images previously stored, point product, division, addition or subtraction of images, etc. When the shutter and its electronic board are installed, the program takes complete control of this accessory, monitoring and displaying its status continuously. When the image intensifier is installed, the program checks for excessive incident light intensity and inhibits both the application of DC voltage to the intensifier and shutter opening.

A "Help" library is always available and can be accessed without destroying the currently displayed data.

2.4 Cooled chamber for the intensifier

The whole body of the intensifier is cooled, using a double-walled chamber. The inner wall is a copper tube inside which the intensifier fits snugly. This tube is insulated from the outer wall with poliurethane foam, and is thermo- electrically cooled with two Marlow MI1063T single stage units capable of pumping more than 8 Watts each, at a temperature differential of 40°C. They reject heat to water-cooled plates which are "O" ring sealed to the sides of the chamber.

The TE cooler power supply is a simple full-wave rectifier with capacitive filtering, designed to work close to the maximum heat pumping rate of the MI1063T.

The intensifier itself is gently pressed against the fiber optic face of the RL512SF array detector by means of a bent steel washer.

The detector chamber is separated from the intensifier chamber by an aluminum foil diaphragm; a 3 x 13 mm^2 rectangular hole exposes the detector active area to the output phosphor screen of the intensifier. This diaphragm is actually sandwiched between detector and intensifier, after it is carefully centered. Its purpose is to eliminate the RF noise caused by the miniature switching power supply enclosed in the intensifier body, which generates the HV necessary for its micro channel plates. Such noise is picked up by the focal plane detector array and overloads the video pre-amplifiers.

2.5 Shutter assembly

This accessory is useful for spectral manipulations and as a protection device for the image intensifier.

The shutter is a blade 5 x 35 mm^2 moving behind a screen with a rectangular hole 3 x 30 mm^2. Both screen and blade are made of thin aluminum sheet, black anodized. The blade is driven by a small solenoid over a course of 5 mm. The whole assembly is only 1 inch wide and replaces the front mounting flange of the standard detector head.

Two photo-diodes were installed close to the rectangular aperture in order to monitor the incident light level. On exposure to excessive light levels, the first diode inhibits opening of the shutter, while the second inhibits powering the image intensifier. It is hoped that the redundancy will afford a higher degree of protection for the intensifier, which is expensive. The threshold was set to a luminance of 10 lux, the light level outdoors in early evening.

The position of the shutter blade is monitored by a third photo-diode and a LED.

The power supply for the solenoid, its relay switch, the sense electronics for the light intensity and blade position were all enclosed in the same box, that connects to the computer via an extra 8255 PIO installed in the interface board.

2.6 Water chiller

The water chiller is an inexpensive drinking fountain Elege PR040[11] modified to work in closed loop. We replaced the top (with its water dispensers and water sink) with a plain board, and added a standard water pump (1/8 HP, 3500 RPM) and suitable fittings for 1/4" polyflo tubing. We also added two instrumentation fans right in front of the freon compressor in order to allow it to work at high duty cycle. The water temperature is 12°C plus or minus 4°C, and the cooling capability is about 100 Watts, which is adequate for our needs. This was found to cost ten times less than the cheapest thermostat commercially available.

3. APPLICATIONS

3.1 Raman spectroscopy

In a Raman experiment, an incident beam of monochromatic light is scattered by a sample. The spectral distribution of the scattered light shows "sidebands" characteristic of the quantum energy level structure of the sample. Raman spectra are therefore potentially useful for clinical assays of aqueous solutions[12] and chemical analysis of transparent fluids, a development that has been made possible by the pioneering work of Leite et al[13,14] on laser-excited Raman scattering.

In practice, photo-assisted sample decomposition and interfering sample luminescence become a hindrance as one tries to circumvent the usually slow spectral data rate by increasing the exciting laser power or photon energy.

Improving the data rate to the point that small, inexpensive laser sources could be used to gather high quality Raman spectra in a short time was one of the reasons for building the system we just described.

Figure 2 shows a block diagram of the spectrograph we built. The light source for excitation is a HeNe alignment laser (5 mW, unpolarized). The parallel detection system was mounted tangent to the focal surface of a Jobin Yvon 522 002 20 aberration corrected concave holographic grating on a substrate 45 x 60 mm^2, with linear spectral dispersion 40 angs/mm. The spectral range covered is 520 angs wide (more than 1000 inverse cm in the red region). Its center can be selected with precision of 1 channel, by rotating the grating about its pole. Spectral calibration was done using a Hg lamp.

Figure 3 shows spectra of various fluids recorded in times ranging from 20 to 200 sec. The signal to noise ratio could be further improved by background subtration, at the expense of longer acquisition time (needed to record the background) and by various smoothing procedures, at the expense of resolution.

3.2 High precision beam position detector

In order to test the accuracy of angular motion in a mechanism to be used in a grazing incidence monochromator for soft X-rays at LNLS, a weakly focussed HeNe laser beam was deflected by a mirror attached to the mechanism and made to hit a RL1024S detector. The Gaussian spot at the detector surface was .5 mm wide. The approximate center of the Gaussian profile was determined by finding the maximum in the peak. Better precision was achieved by means of an interpolation procedure using 9 points close to the center, performed in real time by the same computer which controlled the RL1024S. The estimated resolution with the 9 point fitting is .003 mm.

Figure 4 shows a record of the tracking error for our prototype mechanism. In order to get these data, the computer drove both the mechanism and the detector, compared after each angular step the measured deflection and the expected one, and stored all data in disk files.

A RMS fluctuation of 2 arcsec in the mirror rotation angle can be inferred from the data. This is quite compatible with the mechanism design specifications, but seems to include a large contribution of optical artifacts, mainly distorsion of the beam Gaussian profile due to surface figure errors in the flat mirror and non-uniform optical density of the attenuators placed in front of the array detector. These artifacts are being further studied, since we intend to use the same technique to test mechanisms rated for higher precision.

CONCLUSION

In conclusion, we have described a modular, solid-state parallel detection system built at LNLS for various applications. It has been successfully used as part of a small Raman spectrograph, and as a high precision beam position detector.

Financial support from the São Paulo State Secretary of Science and Technology is acknowledged.

It is the authors' wish that no Agency should derive military benefit from this paper.

REFERENCES

[1] A. R. B. de Castro and P. R. B. Pedreira, "IR Raman spectroscopy with semiconductor devices for excitation and detection", Opt. Commun., 62, 348 (1987).

[2] A. R. B. de Castro, "Grazing incidence spectrograph with classically ruled toroidal grating and focal plane detector", Phys. Scripta 41, 499 (1990).

[3] H. Tolentino, E. Dartyge, A. Fontaine and G. Tourillon, "X-ray absorption spectroscopy in the dispersive mode with Synchrotron Radiation: Optical considerations", J. Appl. Cryst., 21, 15 (1988).

[4] W. R. Dix, K. Engelke, W. Graeff, M. Lohmann and R. Reumann, "Quantitative mikrotomographie mit Synchrotronstrahlung", Hasylab Jahresbericht 1988, 471, publ. DESY Hamburg FRG (1989).

[5] A. R. B. de Castro, "Solid State Parallel Detectors", in "Synchrotron Light: Applications and related instrumentation", Proc. of I Workshop, Campinas, SP, Brasil, 25-28 July 1988, ed. A. Craievich, publ. World Scientific Publ. Ltd., Singapore (1989).

[6] M. Parkinson, "Photoelectron-counting electronics for a microchannel-plate-intensified integrated diode array", Rev. Sci. Instrum., 60, 608 (1989).

[7] EGG Reticon, 345 Potrero Ave., Sunnyvale, Cal. 94086 USA

[8] Instrument Technology Ltd., 29 Castleham Road, St. Leonards-on-sea, East Sussex TN38 9NS, England.

[9] Marlow Industries, Inc., 10451 Vista Park Road, Dallas Texas 75238-1645, USA.

[10] G. R. Hopkinson and D. H. Lumb, "Noise reduction techniques for CCD image sensors", J. Phys. E: Sci. Instrum., 15, 1214 (1982).

[11] Geltec Comércio e Indústria S/A, C.P. 3265, 21530, Rio de Janeiro, RJ, Brasil.

[12] A. R. B. de Castro and J. G. R. S. Franco, "Use of correlation technique in carbohydrate Raman assays", 170-171, Proc. Conf. on Lasers and ElectroOptics, Anaheim Cal USA, June 1984.

[13] R. C. C. Leite and S. P. S. Porto, "Continuous photoelectric recording of the Raman effect in liquids excited by the He-Ne red laser", J. Opt. Soc. Am., 54, 981 (1964).

[14] T. C. Damen, R. C. C. Leite and S. P. S. Porto, "Angular dependence of the Raman scattering from benzene excited by the He-Ne cw laser", Phys. Rev. Lett., 14, 9 (1965).

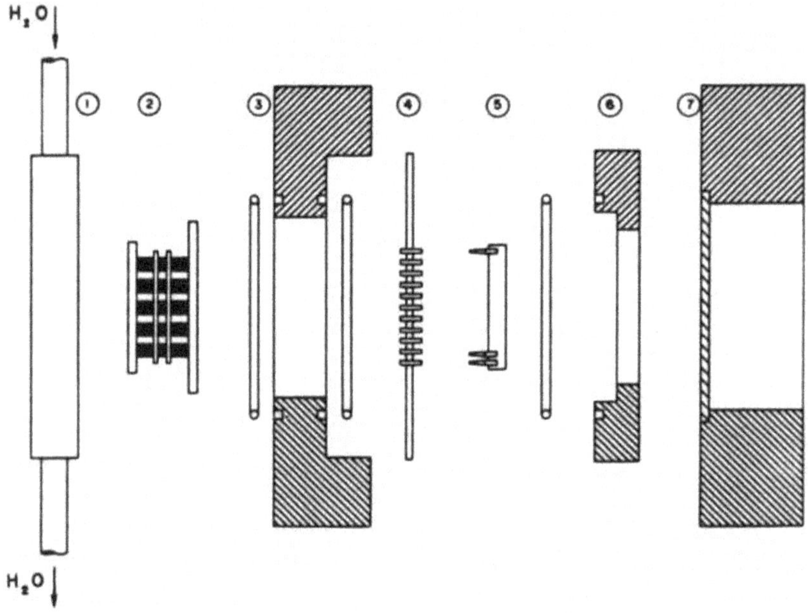

Figure 1: Exploded view of the detector head:
1 - water-cooled bottom plate,
2 - TE cooler,
3 - body,
4 - printed circuit board,
5 - diode array,
6 - retainer and
7 - flange with air-tight glass window.

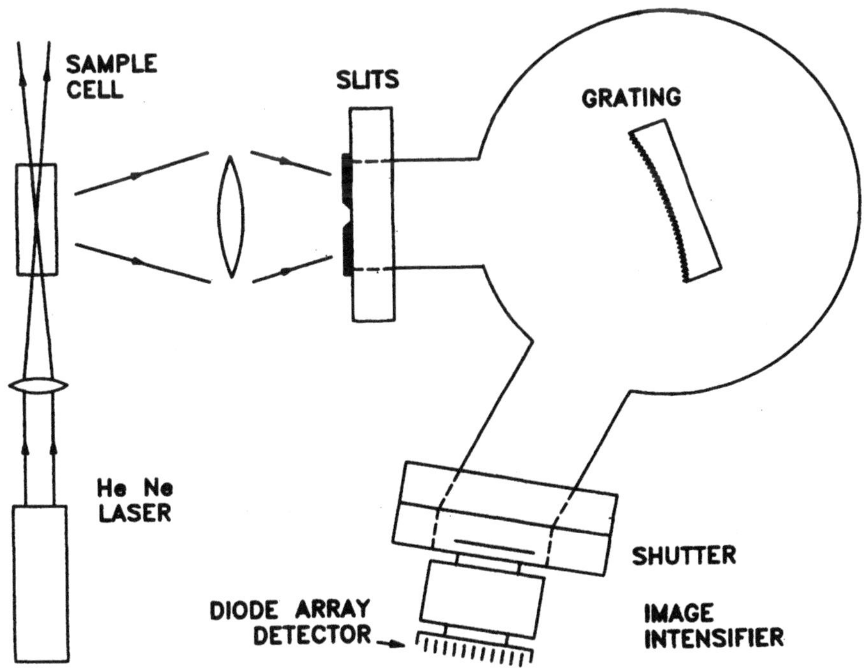

Figure 2: Block diagram of the Raman Spectrograph.

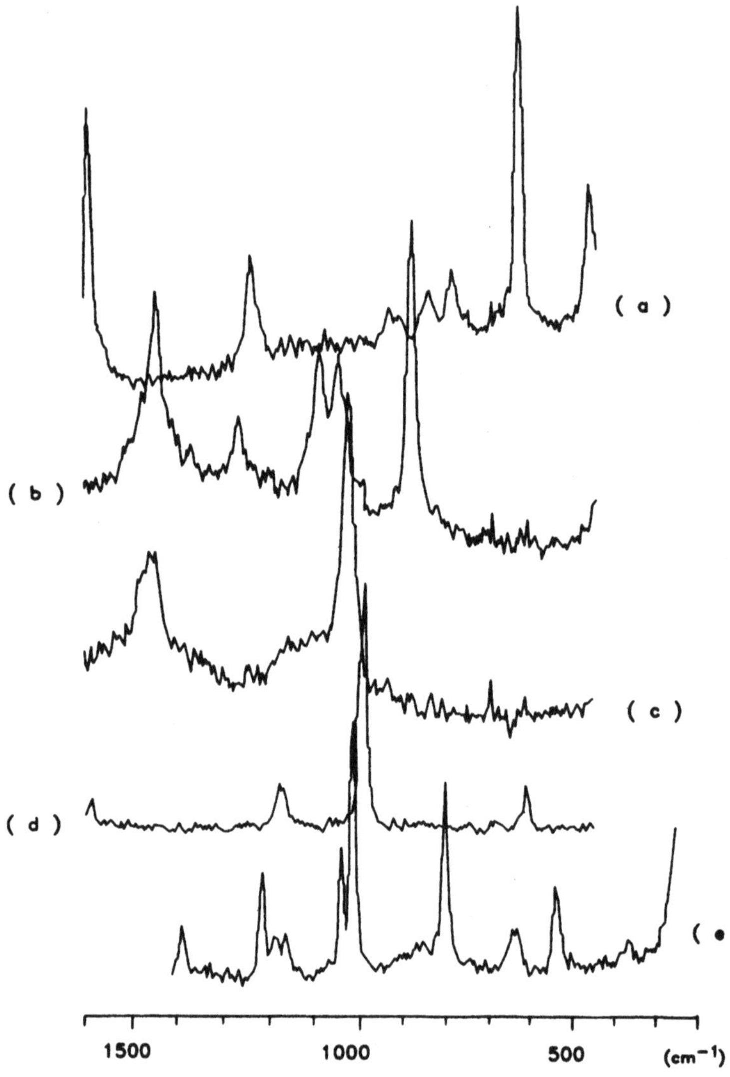

Figure 3: Raman spectra of organic fluids:

a - Trichloroethylene, 200 sec;

b - Ethanol, 200 sec;

c - Methanol, 200 sec;

d - Benzene, 20 sec;

e - Toluene, 50 sec.

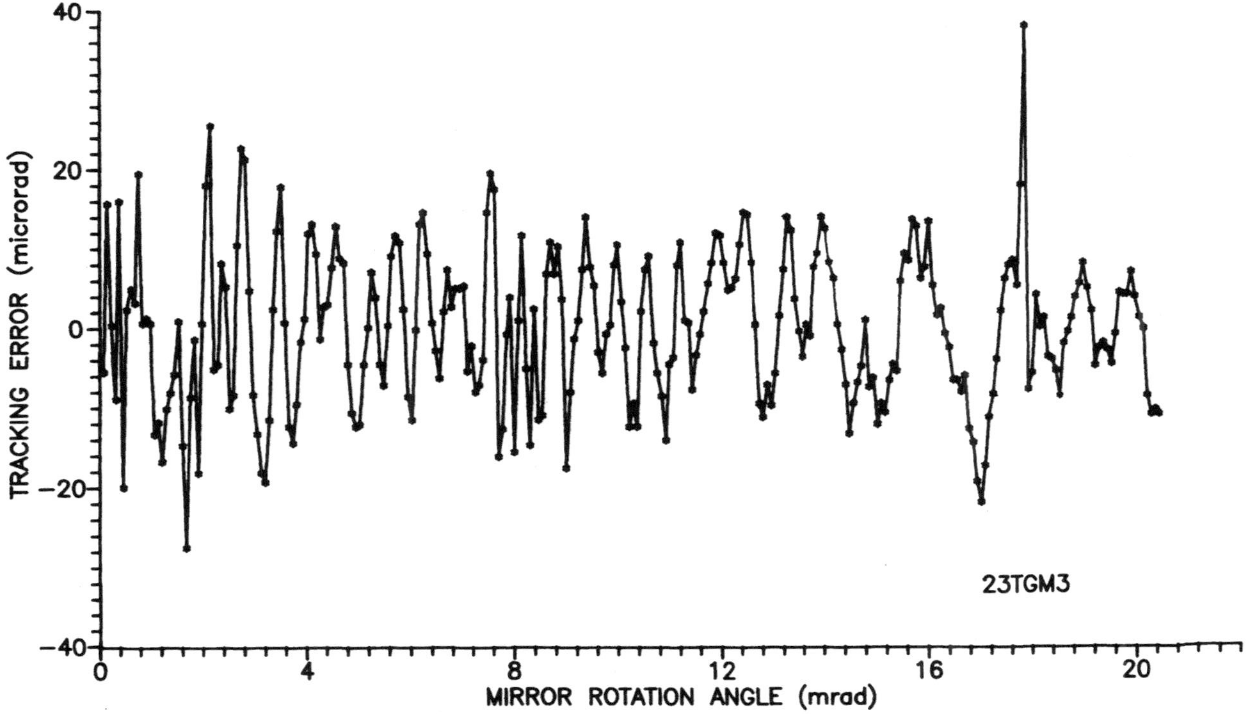

Figure 4: Tracking error of mechanism for grating rotation in Toroidal Grating Monochromator for soft X-rays, under construction at LNLS. The angular course shown is one third of the total monochromator course.

A MAGNET LATTICE FOR THE LNLS SOFT X-RAY SOURCE

Liu Lin, L.Jahnel and P.Tavares
Laboratório Nacional de Luz Síncrotron/CNPq
Caixa Postal 6192 - Campinas - SP - Brazil

1. INTRODUCTION

The design of a storage ring dedicated to the production of synchrotron radiation has to meet several requirements including performance specifications (determined by the users) and technological and cost constraints. The many options that are available to the designer, such as the choice of the lattice and the periodicity of the ring, strongly influence both requirements.Therefore, several candidate designs have to be analyzed.The proposal presented here UVX is the latest in the series of alternative designs being considered by LNLS.The main requirements for the VUV-Soft X-Ray electron storage ring are:

- the critical photon wavelength from the bending magnets, which should be 10 Å;
- the lattice which should include long straight sections to accommodate insertion devices, including high-field wigglers. If possible, the lattice should also envisage the use of micro-undulators or other devices with very small gaps;
- a long electron beam lifetime on the order of 10 hours, to be achieved;
- the design of the bending magnets, which should be conservative;
- reliable operation of the ring, which is of paramount importance. This includes low sensitivity of the lattice to errors and flexibility enough to allow for other operation modes (especially high emittance modes, interesting during commissioning), correction of the effects of insertion devices, etc;
- feasible low energy (100-200 MeV) injection. The injector adopted is a SLAC type electron Linac,
- small number of magnetic elements.

A new feature of UVX, as compared to the previous designs, is the combined quadrupole and sextupole fields in the quadrupoles.This feature has the main advantage of fewer different types of magnets to fabricate. It also has other benefits, such as the weaker chromatic sextupole strengths and the natural availability of sextupoles in the non-dispersive region to minimize the amplitude-dependent tune shift.

2. MAGNET LATTICE

The proposed solution UVX is a six fold symmetric double-bend Chasman-Green achromat lattice. See figure 1. It has twelve 1.4 Tesla rectangular bending magnets[1]. The energy of the ring is 1.15 GeV, providing λ_c =10 Å. The achromats connect six 3-meter long dispersion-free straight sections, four of which allow for insertion devices (one is used for injection and another for the RF cavity). The photon flux from the bending magnets at 1.15 GeV and 100 mA is shown in figure 2.

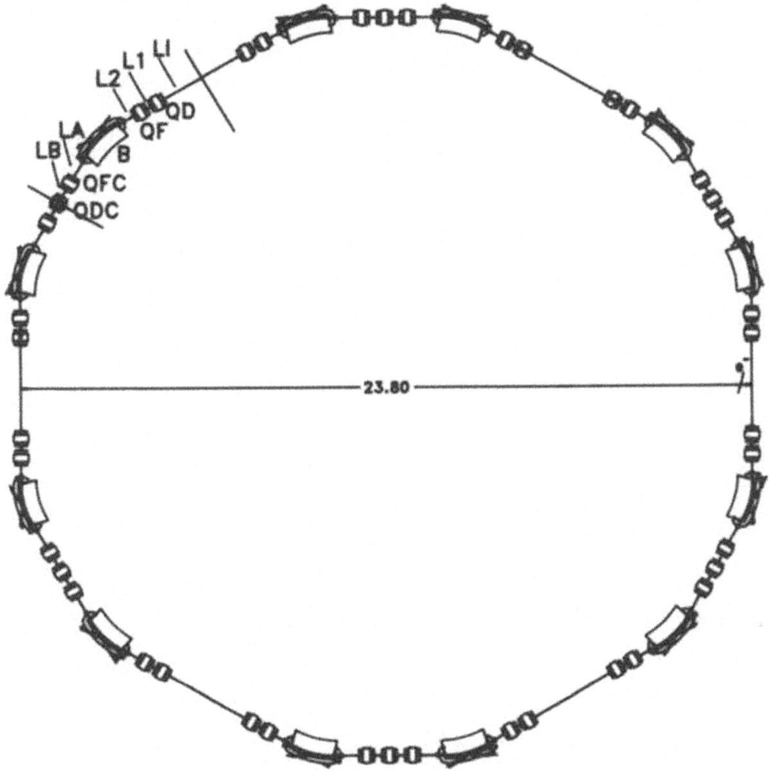

Figure 1: Lay-out of the 1.15 GeV UVX storage ring.

Table I: Main Parameters of the UVX storage ring for three different modes of operation.

	UVX	UVX-H	UVX-L	
Energy	1.15	1.15	1.15	GeV
Current	100	100	100	mA
Circumference	77.34	77.34	77.34	m
Magnetic Structure	CG-6-fold	CG-6-fold	CG-6-fold	
Revolution Frequency	3876.05	3876.05	3876.05	kHz
Harmonic Number	129	129	129	
RF-Frequency	500	500	500	MHz
Natural Emittance	63.0	115.0	33.0	nm.rad
Horizontal betatron tune	5.23	4.75	5.72	
Vertical betatron tune	2.12	1.82	2.81	
Synchrotron tune	4.054	4.054	4.054	1/1000 (@108 kV)
Momentum Compaction	0.010	0.010	0.010	
Natural energy spread	0.059	0.059	0.059	%
Nat. hor. chromaticity	-8.3	-5.4	-21.2	
Nat. vert. chromaticity	-6.2	-5.5	-5.7	
Hor. betatron damping time	11.0	11.0	11.0	ms
Ver. betatron damping time	10.5	10.5	10.5	ms
Synchrotron damping time	5.1	5.1	5.1	ms
Dipoles				
Bending radius	2.735	2.735	2.735	m
Bending field	1.4	1.4	1.4	Tesla
Number	12	12	12	
Quadrupoles				
Maximum gradient	3.6	4.0	4.1	m^{-2}
Number of families	4	4	4	
Number	42	42	42	
Sextupoles*				
Max. int. strength	6.7	3.6	5.1	m^{-2}
Number of families	4	4	4	
Number of sextupoles	42	42	42	
Injection energy	100	100	100	MeV

* Values for zero chromaticity

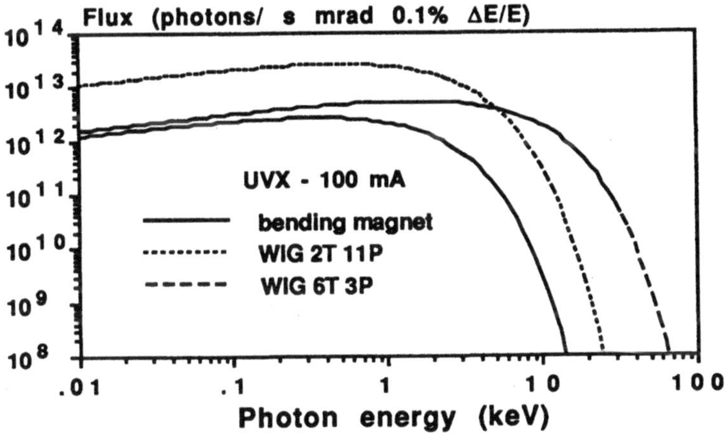

Figure 2: Photon flux from the bending magnets and wigglers for the UVX ring at 1.15 GeV and 100 mA.

Table II: Parameters of the lattice elements for normal operation mode.

Element	Length (m)	Strength (ρ:m, K:m^{-2}, s:m^{-2})*
Drift LI	1.548	
Quadrupole QF	0.250	K=2.80, s=0.3
Drift L1	0.390	
Quadrupole QD	0.250	K=-2.60, s=-1.0
Drift L2	0.750	
Rect. Bending B	1.432	ρ=2.735
Drift LA	0.950	
Quadrupole QFC	0.250	K=3.64, s=7.26
Drift LB	0.500	
Quadrupole QDC	0.125	K=-3.09, s=-8.66
Reflection Symmetry ------		

* Positive strength implies horizontal focussing. s is the sextupole integrated strength for ξ=+2.0.

The double-bend achromatic arc consists of two 30 degree bending magnets separated by one vertically focusing and two horizontally focusing quadrupoles. These three quadrupoles are needed to produce the desired separation of beta functions at the location where chromatic sextupoles are placed, that is inside the quadrupoles. The sextupolar fields are achieved with additional coils, as is done for Super-Aco[2] .

Two additional operating modes, a higher emittance (UVX-H) and a lower emittance (UVX-L) mode, are being investigated and are at different stages of study. The main parameters of UVX and the two other modes are listed in table I. In Section 2.1 a mode with low vertical betatron function in one of the insertion straights is presented, and the remainder of the report describes only the results for UVX normal operation mode. Figure 3 shows the optical functions for one superperiod. Table II gives the parameters of the magnetic elements.

Besides the two sextupole families for chromaticity correction, two other families in the non-dispersive region are used to compensate for the geometric aberration produced by the chromatic sextupoles. This can be done at almost no extra cost since the quadrupoles are already designed to allow for sextupolar coils.

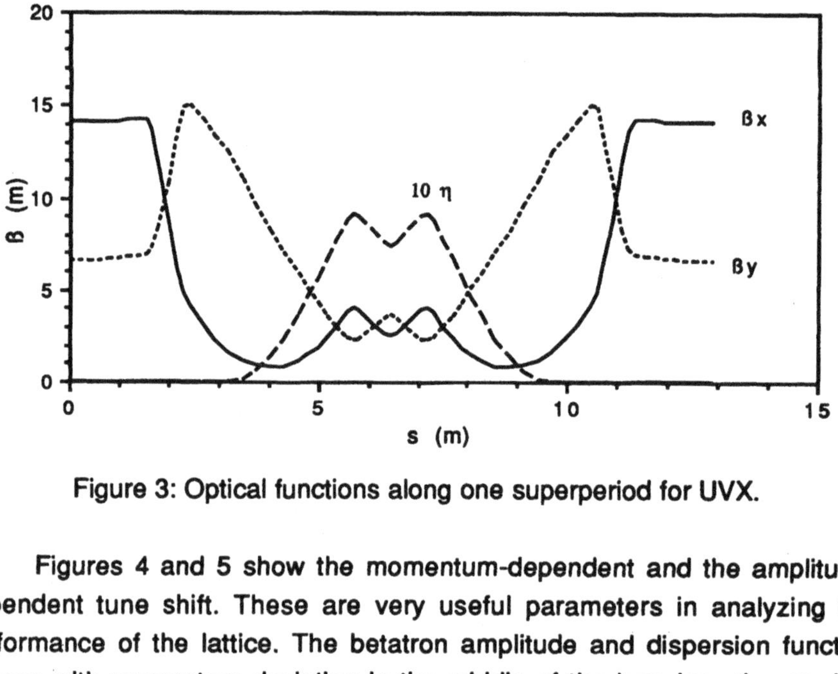

Figure 3: Optical functions along one superperiod for UVX.

Figures 4 and 5 show the momentum-dependent and the amplitude-dependent tune shift. These are very useful parameters in analyzing the performance of the lattice. The betatron amplitude and dispersion function change with momentum deviation in the middle of the long insertion straight are shown in figures 6 and 7.

All calculations described here are done with the following computer codes:

- MAD[3] - for linear lattice optimization.

• PATPET[4] - for chromaticity correction, tracking with multipoles, magnet imperfections, orbit distortions, tracking with energy oscillations and orbit correction.

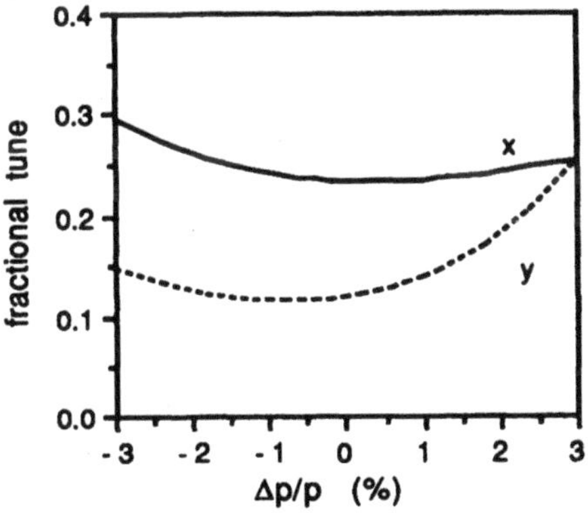

Figure 4: Tune variation with particle momentum.

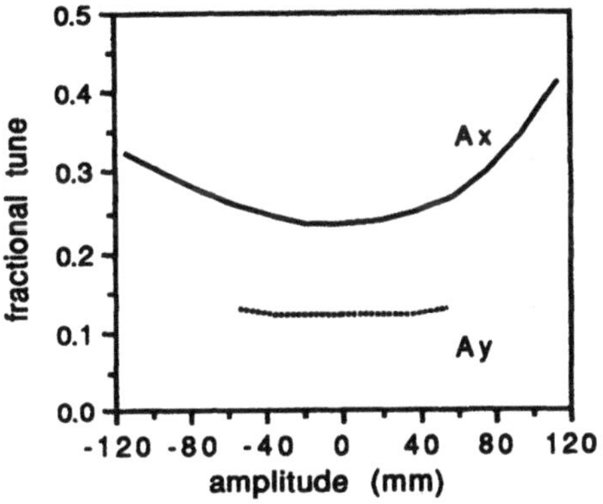

Figure 5: Tune variation with particle amplitude.

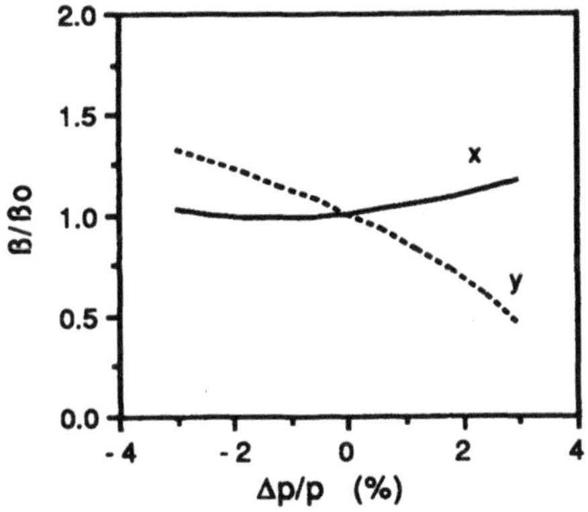

Figure 6: Betatron amplitude as a function of momentum deviation in the middle of the long insertion straight.

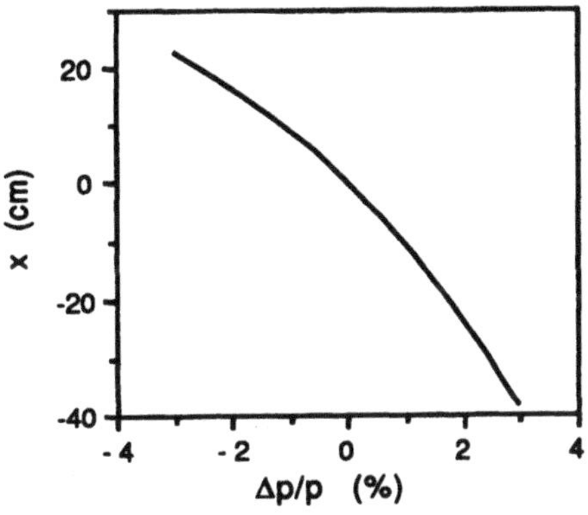

Figure 7: Dispersion as a function of momentum deviation in the middle of the insertion straight.

2.1 LOW VERTICAL β OPTICS

In this section a mode with low vertical betatron function in one of the dispersion free straight sections is investigated for applications with small gap insertions. This mode can be accomplished in the UVX lattice by independently powering the quadrupole doublets (QDW and QFW) adjacent to the low β straight. The values for the resulting quadrupole strengths in this mode are given in table III. The lattice functions for 3 superperiods of the ring are shown in figure 8. The vertical beta is matched to 0.69 m in the small gap insertion straight, a factor of 10 smaller than the value for the normal operation mode. Both the vertical and the radial tunes, as well as the emittance are kept the same as in the normal operation mode. The reduction in the vertical beta in the insertion straight causes the vertical phase advance to increase across this region. To keep the same tune, the vertical phase advance over the achromat must be reduced. This is not the case in the horizontal plane: the change in horizontal beta is very small, not significantly affecting the horizontal phase advance in the achromat, and this is the reason why the emittance of the lattice in this mode (ε_o=6.4 x 10^{-8} rad.m.) does not change.

Table III: Quadrupole strengths for the low vertical β mode.

Quadrupole	Strength K (m^{-2})
QFC	3.636
QDC	-3.092
QF	2.765
QD	-2.495
QFW	3.056
QDW	-3.372

We look now at the various dimensions at the point where β=0.69 m. The residual vertical emittance after closed orbit correction in ten simulations for the normal operation mode (see next section) is ε_y=(0.10±0.11) x 10^{-9} rad.m. and the maximum residual vertical orbit distortion y_{max}=0.19±0.06 mm. We assume the mean values plus 3 standard deviations, i.e, ε_y=0.43 nm.rad and y_{max}=0.37 mm. With these values the vertical beam size is σ_y=0.017 mm giving for a very small gap insertion, say, 2 mm, a distance of 95 σ from the center of the displaced closed orbit to the insertion wall. At 69 cm from this point of minimum, the vertical betatron function doubles its value and the beam size increases by a factor of 1.4. It is also interesting to note that the β scaled physical aperture at the small β symmetry point is 1.3 mm; thus any gap greater than this value will not limit the vertical acceptance of the ring.

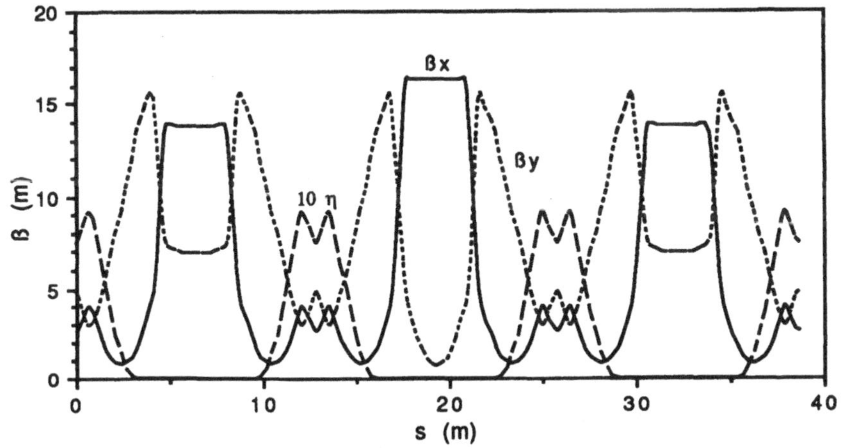

Figure 8: Lattice functions for three superperiods of UVX in the low vertical β mode.

2.2 FLEXIBILITY OF THE LATTICE

The flexibility of the UVX lattice is analyzed regarding two aspects: the range of variation of the vertical betatron tune and the ability to set the betatron functions to high/low values in the insertion straights.

The inclusion of insertion devices in the ring affects the beam in various aspects. In the case of wigglers and undulators with plane poles, the edge-focussing of these poles produces a vertical betatron tune shift which must be compensated to ensure operation far from resonances. A plane pole insertion device with length L and bending radius ρ produces a vertical tune shift

$$\Delta\nu_y = \frac{1}{4\pi}\beta_y\frac{L}{\rho^2}$$

where β_y is the betatron function at the wiggler position, assumed to be constant over the length of the magnet. A 1 m, 2 Tesla wiggler increases the vertical tune by 0.14; and a 25 cm, 5 Tesla one by 0.22. Let us take a worst case estimate of $\Delta\nu_y$=0.4. The UVX lattice can be tuned to compensate this vertical tune-shift without changing the horizontal tune. Larger tune shifts were not investigated up to now.

It is always desirable to set the betatron values in the insertion straights to values suitable for particular applications. Compared to the previous design (VUV-III) , UVX has an extra quadrupole family in the dispersive region. This allows for the changing of the betatron function in the insertions while keeping the same tune. An example is the low vertical beta mode described above. Lowering the horizontal beta while keeping the tune has the price of increasing the emittance. Further flexibility to tailor the beta values can be achieved using quadrupole triplets to match the betatron functions in the insertion regions.

3. SENSITIVITY TO ERRORS AND DYNAMIC APERTURE

Dynamic aperture studies are performed with the code PATPET. Particles are tracked for 500 turns for various situations:

- without errors,
- with systematic multipole errors,
- with random multipole errors,
- with random excitation (strength) errors,
- with random alignment errors, and
- with momentum deviation.

The systematic multipole errors are those caused by the finite dimensions of the poles and are equal for every magnet of the same kind, whereas random multipole errors arise because of constructional tolerances and are different for each magnet. The values of multipole errors assumed are given in table IV. These data are worst case values taken from similar calculations used in other projects [5],[6].

Table V shows the assumed random strength and alignment errors. The alignment errors are horizontal x and vertical y displacements and rotation α about the longitudinal axis. The random errors in tables IV and V are r.m.s. values for gaussian distributions, which are truncated at two sigmas.

For the cases where excitation and alignment errors are included, tracking is done after closed orbit corrections. Some of the results are shown in figures 9 to 11. The hatched regions in these figures correspond to the uncertainty region of dynamic aperture for five different sets of random errors used.

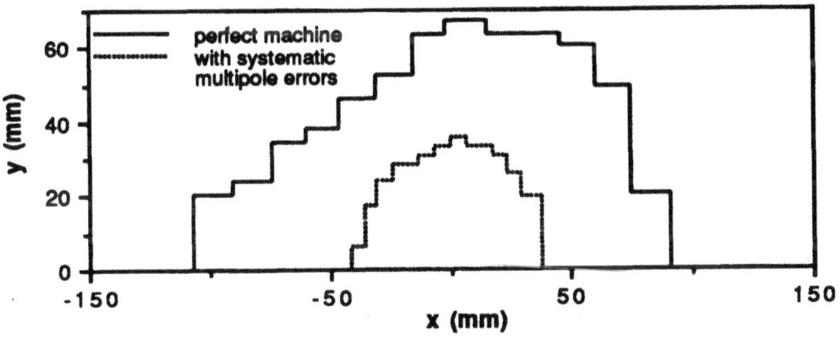

Figure 9: Dynamic aperture for the perfect machine and with systematic multipole errors only.

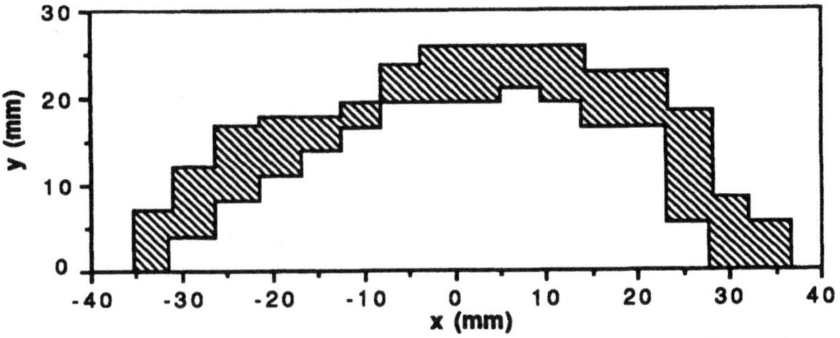

Figure 10: Dynamic aperture for the machine with systematic and random multipole errors, strength and alignment errors for on-momentum particles.

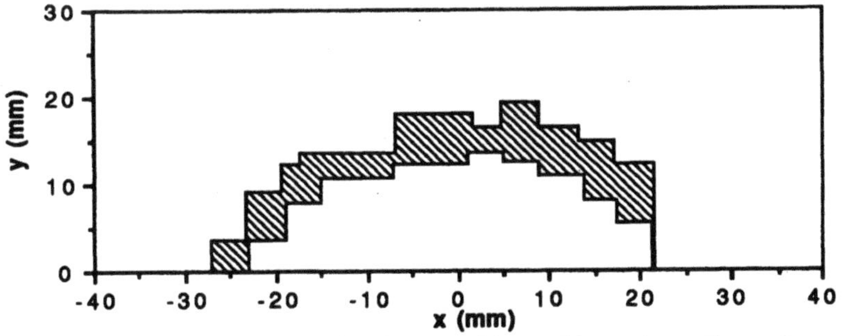

Figure 11: Dynamic aperture for the machine with systematic and random multipole errors, strength and alignment errors and 2% momentum deviation.

The simulations show that in the presence of multipole and alignment errors the dynamic aperture shrinks to dimensions less than the vacuum chamber aperture. However, it is still sufficiently large for injection and the gas Coulomb scattering lifetime is long enough (on the same order as the Touschek lifetime) to give an overall lifetime greater than 10 hours.

Table IV: Magnetic multipole errors.

$$B(x) = B_o\rho \sum_n \frac{k_n x^n}{n!} = \sum_n B_n$$

n	Dipole B_n/B_o*		Quadrupole B_n/B_1*	
	Systematic	Random	Systematic	Random
1	-	5×10^{-5}	-	-
2	2×10^{-5}	1×10^{-4}	-	5×10^{-4}
3	-	2×10^{-5}	-	2×10^{-4}
4	2×10^{-6}	1×10^{-5}	-	6×10^{-5}
5	-	-	1×10^{-4}	1×10^{-5}
9	-	-	1×10^{-9}	1×10^{-8}

*Values at x=1 cm.

Table V : Alignment and strength random errors (one standard deviation) in magnetic elements.

$<\Delta x>$	0.2 mm
$<\Delta y>$	0.2 mm
$<\Delta\alpha>$	0.02 °
$<\Delta S/S>$	0.01 %

For the closed orbit correction we use 18 monitors for both horizontal and vertical readings, 18 horizontal and 18 vertical correctors. The correctors are placed inside the quadrupoles. Figure 12 and table VI give the position of these elements in one superperiod.

The algorithm used for the orbit correction minimizes the r.m.s. displacements of the monitor readings. The correction of the orbit is iterated until convergence in the closed orbit displacement is reached for both transverse planes.

Ten random machines are studied for the nominal errors given in table V. In table VII we give the average and standard deviation of various parameters before and after correction.

It should be emphasized that errors on the position of correctors and monitors were not included and can increase the residual closed orbit distortion.

Table VI : Monitors and correctors distribution.

Element	z(m)	β_x(m)	β_y(m)	$\mu_x(2\pi)$	$\mu_y(2\pi)$
Middle of ins. straight	0.0	14.1	6.6	0.0	0.0
CX	1.673	13.7	7.3	0.019	0.040
MON	1.948	9.5	10.2	0.023	0.045
CY	2.313	4.9	14.9	0.031	0.049
CX	5.696	4.1	2.3	0.399	0.135
MON	6.171	3.0	3.1	0.420	0.164
CY	6.446	2.5	3.7	0.436	0.177
CY	10.578	4.9	14.9	0.841	0.304
MON	10.943	9.5	10.2	0.850	0.308
CX	11.218	13.7	7.3	0.853	0.314
Middle of ins. straight	12.891	14.1	6.6	0.872	0.353

Table VII : Closed orbit correction . Comparison between unperturbed, perturbed and corrected machines.

Parameter	Unperturbed machine	Perturbed machine	Corrected machine
ν_x	5.233	5.237±0.006	5.232±0.002
ν_y	2.119	2.124±0.005	2.123±0.003
$<x^2>^{1/2}$ (mm)	0.0	1.9±0.9	0.15±0.05
x_{max} (mm)	0.0	4.5±2.0	0.5±0.2
$<y^2>^{1/2}$ (mm)	0.0	3.4±1.1	0.06±0.01
y_{max} (mm)	0.0	7.6±2.4	0.17±0.04
$<\eta_x^2>^{1/2}$(cm)	40.4	40.6±0.2	40.4±0.2
$<\eta_y^2>^{1/2}$(cm)	0.0	5.5±3.2	2.0±1.1
ε_x (10^{-9} rad.m)	63.3	69.1±10.9	65.2±3.4
ε_y (10^{-9} rad.m)	0.0	0.6±0.5	0.10±0.11
$\varepsilon_y/\varepsilon_x$ (%)	0.0	0.8±0.5	0.15±0.18
$<\|\theta_x\|>$(mrad)	-	-	0.15±0.04
$<\|\theta_x\|_{,max}>$(mrad)	-	-	0.38±0.11
$<\|\theta_y\|>$(mrad)	-	-	0.10±0.02
$<\|\theta_y\|_{,max}>$(mrad)	-	-	0.25±0.05

* The average values for parameter z (over 10 simulations) are given in the form $<z>\pm\sigma_z$.

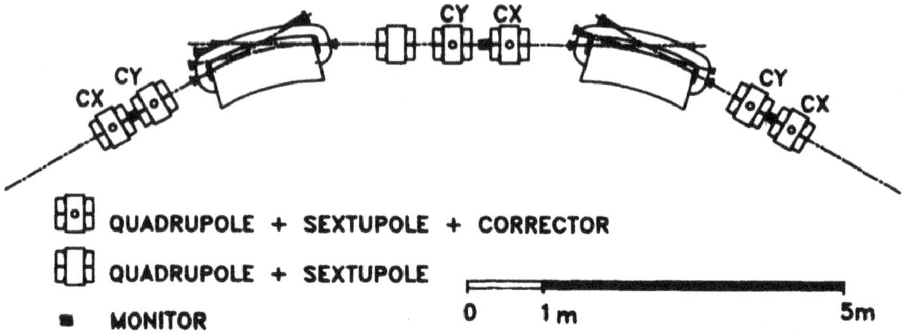

Figure 12: Distribution of correctors and monitors in one superperiod.

To check the misalignment tolerances, machines with up to four times the nominal errors are simulated. The percentage of stable machines is given in table VIII.

Table VIII : Stable machines over 10 simulations.

Error*	ε	2ε	3ε	4ε
Stable machines	10	9	6	2

* ε = nominal errors.

4. COLLECTIVE EFFECTS AND BEAM LIFETIME

Coherent collective effects can be divided into single and multi-bunch instabilities and can be related to the longitudinal and transverse coordinates. Longitudinal instabilities are not necessarily destructive - single-bunch longitudinal effects simply lead to lengthening and widening of the electron bunches (which can be beneficial in terms of Touschek lifetime), whereas coupled bunch longitudinal instabilities can be suppressed by means of feedback systems. On the other hand, transverse instabilities are often destructive and determine the maximum current that can be put in a single bunch. Incoherent collective effects are caused by the interaction of the beam with the molecules of the residual gas (gas scattering) or by multiple, small-angle (Intrabeam Scattering) and single large-angle (Touschek scattering) scattering of electrons by other electrons in the bunch. These effects influence the equilibrium beam dimensions and lifetime.

In this report, we shall be concerned with microwave instability (turbulent bunch lengthening) and potential-well distortion, emittance growth due to intra-beam scattering (IBS) and beam lifetime (only normal operation mode is considered). It is important to notice that these effects are correlated. Touschek lifetime, for example depends on bunch density, which in turn depends on the equilibrium emittance and bunch length determined by IBS and microwave instability.

Our aim is to perform a calculation of beam lifetime consistent with the equilibrium beam dimensions determined by these effects. This is done with the computer code ZAP[7]. All calculations are done for 100 MeV (injection energy) and 1.15 GeV assuming a 500 MHz RF system. Instability thresholds depend on the coupling impedance of the ring, which describes the electromagnetic interaction of the beam with the vacuum chamber metallic walls.

For the ring (broadband) impedance we use 13Ω (value measured at Alladin[8]) and 2Ω (measured at Super-Aco[9]). Although the value measured at Alladin is considerably higher than those reported for other rings, we feel it is advisable to have a worst case estimate for single bunch thresholds. However, a high impedance does not necessarily imply a poor Touschek lifetime, hence the calculations with the 2Ω impedance. For the HOM of the accelerating cavities we use the data of the KEK 500 MHz cavity[7]. Note that since we are only concerned with single bunch effects, these modes are only used to calculate an effective broadband impedance contribution to the overall ring impedance. The impedance of real machines decreases considerably at high frequencies. As a result, the broadband model gives unrealistic high impedances for short bunches. Although the exact frequency dependence is still subject to controversy an empirical formula known as *Spear scaling Law* has been suggested[7] as a means of calculating the effective impedance for bunches shorter than the average beam pipe radius. We have made calculations both with and without Spear scaling (ON/OFF in the figures).

Beam lifetime is determined by 3 processes: quantum emission of photons by the stored electrons (quantum lifetime), single large-angle collision events (Touschek lifetime) and single scattering of the stored electrons from residual gas molecules and ions (gas-scattering lifetime - inelastic or elastic). Quantum lifetime is on the order of 50 hours for 100 kV RF voltage at full energy and will be disregarded in what follows. In order to calculate the Touschek contribution to the overall lifetime, it is necessary to know the equilibrium values of bunch length , energy spread and transverse emittances. These will be different from the natural equilibrium values (i.e., those determined by the competing effects of quantum excitation and radiation damping) due to IBS and bunch lengthening (or shortening) and widening caused by microwave instability and potential well distortion. Figure 13 shows the Touschek lifetime at 100 MeV calculated using these equilibrium values.

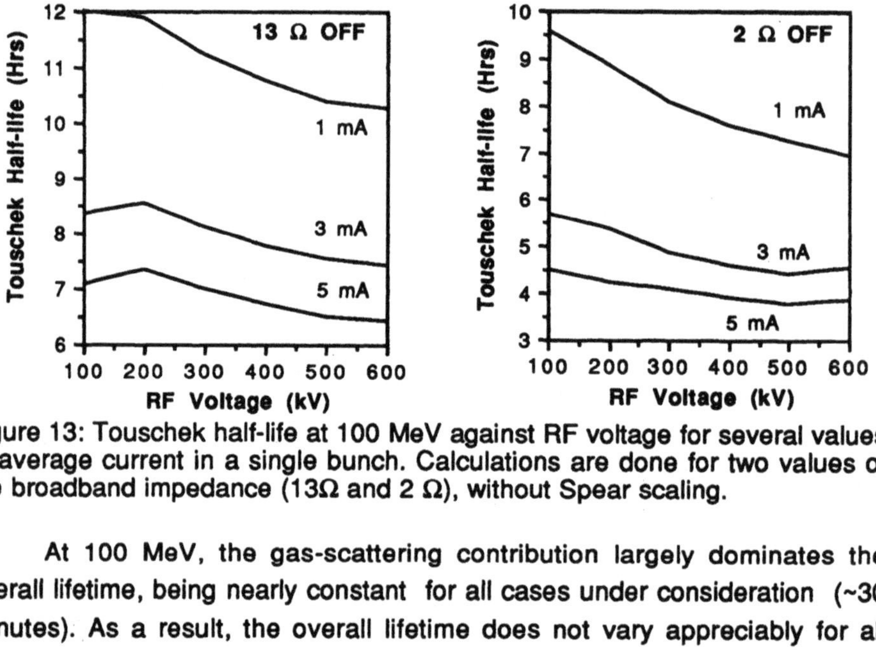

Figure 13: Touschek half-life at 100 MeV against RF voltage for several values of average current in a single bunch. Calculations are done for two values of the broadband impedance (13Ω and 2 Ω), without Spear scaling.

At 100 MeV, the gas-scattering contribution largely dominates the overall lifetime, being nearly constant for all cases under consideration (~30 minutes). As a result, the overall lifetime does not vary appreciably for all impedances considered, ranging from 25 to 28 minutes. Figure 14 shows the overall lifetime (which includes Touschek and gas-scattering lifetime contributions) at full energy.

We conclude that the overall lifetime is adequate, both at low and high energy. At 100 MeV, IBS guarantees that the lifetime is not reduced below the radiation damping times (few seconds) . At full energy, increasing the RF voltage up to 1 MV (but not much more than that) would be beneficial, although 500 kV seems quite enough to make for a half-life above 10 hours as originally intended.

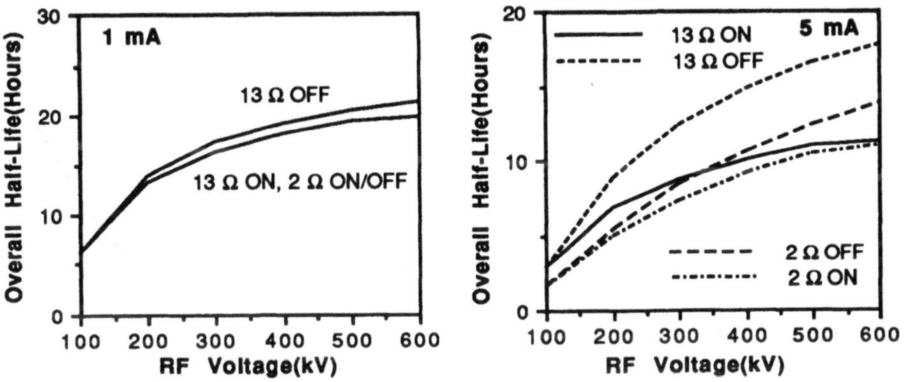

Figure 14: Overall half-life at full energy against RF voltage for various average currents in a single bunch. Calculations are shown for two values of the coupling impedance, without SPEAR scaling.

REFERENCES

[1] F.W.B.Talarico and R.T.Neuenschwnader, 'Dipole Magnet for the LNLS Synchrotron Light Source', IEEE PAC 1990.

[2] A. Dael et al, 'Magnet System for the Super-Aco - The New Orsay Radiation Source', presented at the 9th International Conference on Magnet Technology (1985).

[3] F.C.Iselin and J.Niederer, 'The Mad Program', CERN/LEP-TH/88-38.

[4] L.Emery, H.Wiedemann and J.Safranek, 'User's Guide for PatPet Version 88.2', SSRL ACD-NOTE 36.

[5] '1-2 GeV Synchrotron Radiation Source-Conceptual Design Report', PUB-5172 Rev.

[6] S.Tazzari, Editor, 'Design Study for the Synchrotron Light Source', LNF-87/6(R).

[7] M.S.Zisman, S. Chattopadhyay, J.J.Bisognano, 'ZAP User's Manual', LBL Report - 21270 (1986)

[8] S.Chattopadhyay, M.Cornacchia, A.Jackson and M.S.Zisman, 'Accelerator Physics Experiments at Alladin', LBL Report.

[9] P.Marin, P.Souchet, 'Vacuum Chamber and Related Problems' - Super ACO internal report ANNEUX/RT/88-03.

www.ingramcontent.com/pod-product-compliance
Lightning Source LLC
LaVergne TN
LVHW082255241026
850861LV00008B/47
* 9 7 8 9 8 1 0 2 0 5 7 2 0 *